Quantum
Monte Carlo Methods
in
Condensed Matter Physics

Quantum Monte Carlo Methods
in
Condensed Matter Physics

Editor: M. Suzuki

Dept. of Physics, University of Tokyo
Japan

World Scientific
Singapore • New Jersey • London • Hong Kong

Published by

World Scientific Publishing Co. Pte. Ltd.
5 Toh Tuck Link, Singapore 596224
USA office: 27 Warren Street, Suite 401-402, Hackensack, NJ 07601
UK office: 57 Shelton Street, Covent Garden, London WC2H 9HE

British Library Cataloguing-in-Publication Data
A catalogue record for this book is available from the British Library.

ISBN-13 978-981-02-1659-7
ISBN-10 981-02-1659-9
ISBN-13 978-981-02-3683-0 (pbk)
ISBN-10 981-02-3683-2 (pbk)

PREFACE

This book reviews recent developments of quantum Monte Carlo methods and some remarkable applications to quantum spin, fermion and boson systems.

The quantum Monte Carlo method is a very powerful and non-perturbational tool for studying thermal and quantal fluctuations in interacting quantum spin systems and strongly correlated electron systems. In the standard quantum Monte Carlo method based on the generalized Trotter formula, quantum fluctuations are expressed in terms of configurations in the additional Trotter dimension of the transformed system. This corresponds to the Feynman path integral. The path-summation scheme in quantum Monte Carlo simulations is particularly useful when a fairly large amount of machine time of high-speed computers is available.

This book contains twenty-two papers by thirty authors. The first paper gives the foundations of the standard quantum Monte Carlo method, including some recent results on higher-order decompositions of exponential operators and ordered exponentials. The second paper presents a general review of quantum Monte Carlo methods used in the present book. Some related methods such as the transfer-matrix method and the decoupled-cell method are also briefly explained. The negative-sign problem is also discussed. This is one of the most challenging problems in the field of quantum Monte Carlo techniques, as is also discussed in some other papers of the present book. Some new methods are proposed to overcome partially this difficult negative-sign problem.

Low-dimensional quantum spin systems are studied by several authors using quantum Monte Carlo methods or the quantum transfer-matrix method based on the equivalence theorem that a d-dimensional quantum system is equivalent to the corresponding $(d + 1)$-dimensional classical system. Some new techniques such as the loop algorithm (namely a new type of cluster algorithm) are reported here to overcome critical slowing down in quantum Monte Carlo simulations. Random fields and inhomogeneity effects are studied in quantum spin systems. The critical exponents of quantum spin chains are also evaluated numerically using the quantum transfer-matrix method. The Haldane gap is confirmed for spin $S = 1$ and 2 by numerical calculations.

The variational Monte Carlo method is also included here together with some recent applications of it to the Hubbard model, the t-J model and the Kondo lattice.

Some interesting applications of quantum Monte Carlo methods to fermion systems are also presented to investigate the role of strong correlations and fluctuations of electrons and to clarify the mechanism of the high-T_c superconductivity. Not only thermal properties but also quantum-mechanical ground-state properties have been studied by the projection technique using auxiliary fields. The algorithm in the infinite dimensional limit is very useful to solve models of strongly correlated systems and to clarify a transition to an incommensurate magnetic state in the Hubbard model. The Hubbard-Stratonovich transformation and the auxiliary-fields approach are also effectively used to interpret geometrically the negative-sign problem in terms of a Berry phase in the Hubbard and Heisenberg models. The maximum entropy method is applied to the degenerate single-impurity Anderson model. Many other interesting topics on fermion and boson systems are discussed in the present book.

Numerical calculations of transport coefficients and relaxation functions derived by Kubo's linear-response theory are also given. Thus, the quantum Monte Carlo method is of great use even for investigating dynamical quantum fluctuations in non-equilibrium systems.

I believe that the present book will be useful to active researchers in the frontier of condensed matter physics as well as to young graduate students who want to start learning the quantum Monte Carlo methods.

I would like to thank Dr. Dorota Lipowska for critical reading of some manuscripts by Japanese authors and also thank Dr. N. Hatano for his assistance in preparing the LaTeX style file.

Masuo Suzuki
Tokyo, September 1993

CONTENTS

Preface   v
  M. Suzuki

General Decomposition Theory of Exponential Operators   1
  M. Suzuki

Quantum Monte Carlo and Related Methods — Recent Developments   13
  N. Hatano and M. Suzuki

Monte Carlo Renormalization Group Study of the D=1 XXZ Model   49
  M. A. Novotny and H. G. Evertz

Overcoming Critical Slowing Down in Quantum Monte Carlo Simulations   65
  H. G. Evertz and M. Marcu

Quantum Manybody Spin Systems in Random Fields and Anisotropies   81
  P. Reed

Inhomogeneity Effects in Quantum Spin Systems   97
  S. Miyashita, J. Behre and S. Yamamoto

The Quantum Transfer Matrix and Its Application to Quantum Spin Chains   113
  K. Kubo

Transfer Matrices in Quantum Many-Body Systems   131
  T. Koma

Monte Carlo Calculations of Elementary Excitation   149
  M. Takahashi

The Decoupled Cell Method of Quantum Monte Carlo Calculation   163
  S. Homma

Decoupled Cell Monte Carlo Study of the Critical Properties of the
Spin-1/2 Ferromagnetic Heisenberg Model in Three Dimensions   179
  R. J. Creswick and C. J. Sisson

Variational Monte Carlo Studies of Correlated Electrons   193
    H. Shiba

Quantum Monte Carlo Simulation of Multiband Fermion Systems
and its Application to Superconductivity   205
    K. Kuroki and H. Aoki

Quantum Monte Carlo in the Infinite Dimensional Limit   221
    M. Jarrell, H. Akhlaghpour and Th. Pruschke

Aspects of the Sign Problem   235
    J. H. Samson

Quantum Simulations of the Degenerate Single-Impurity Anderson Model   251
    J. Bonča and J. E. Gubernatis

Quantum Monte Carlo Simulation by Auxiliary Fields   265
    S. Sorella

Ground-State Projection Using Auxiliary Fields   285
    S. Fahy

Fermion Simulations of Correlated Systems   299
    M. Imada

Dirty Bosons in 2D: Phases and Phase Transitions   317
    N. Trivedi and M. Makivić

Path-Integral Quantum Monte Carlo Studies of the Static and Time-
Dependent Thermodynamics of the Vibrational Properties of Crystals   339
    A. R. McGurn

Relaxation of Quantum Systems in Fluctuating Media   355
    M. Takasu

GENERAL DECOMPOSITION THEORY OF EXPONENTIAL OPERATORS

Masuo Suzuki

Department of Physics, University of Tokyo, Tokyo 113, Japan

Abstract

A general decomposition theory of exponential operators is reviewed, and its possible applications to quantum Monte Carlo simulations are discussed. The generalized Trotter formula, the improved Trotter-like formula, and higher-order decomposition formulas are presented, and the convergence of these formulas is also discussed.

1 Introduction

It is difficult to solve exactly a quantum system, because its Hamiltonian is too complicated to diagonalize. However, in many cases, the corresponding Hamiltonian $\mathcal{H}$ is composed of some partial Hamiltonians $\{\mathcal{H}_j\}$, and each $\mathcal{H}_j$ is easily diagonalized independently. In this situation, the following Trotter formula [1,2] is very useful [3-12] :

$$e^{x(A+B)} = \lim_{n\to\infty}(e^{\frac{x}{n}A}e^{\frac{x}{n}B})^n,\qquad(1.1)$$

or more generally

$$e^{-\beta\mathcal{H}} = e^{-\beta(\mathcal{H}_1+\cdots+\mathcal{H}_q)} = \lim_{n\to\infty}(e^{-\frac{\beta}{n}\mathcal{H}_1}\cdots e^{-\frac{\beta}{n}\mathcal{H}_q})^n.\qquad(1.2)$$

If we take the trace of (1.2) using some orthogonal complete sets, then the problem of calculating the partition fuction $Z = \text{Tr}\exp(-\beta\mathcal{H})$ for the d-dimensional quantum system $\mathcal{H}$ is reduced to that of calculating the partition function of the corresponding $(d+1)$-dimensional classical system [3-12]. This transformation is called the ST-transformation. Thus, quantum Monte Carlo simulations are reduced to the corresponding classical (say, Metropolis) Monte Carlo simulations, though some new types of many-body interactions appear and consequently the corresponding algorithms are much more complicated [4-12].

The above ST-transformation is also useful in calculating numerically the partition function of a quantum system using the (quantum) transfer matrix [13-18]. This method is based on the exchangeability theorem [12,13] of the two limits $N \to \infty$ and $n \to \infty$, where N denotes the number of sites of the relevent system (or system size).

As the way of the decomposition of the original Hamiltonian $\mathcal{H}$ into the sub-Hamiltonians $\{\mathcal{H}_j\}$ is arbitrary, it is possible to devise some efficient ways of decomposition according to the feature of the relevant system. One convenient way is to decompose the original Hamiltonian $\mathcal{H}$ into clusters [13,19].

2   Toward Higher-Order Decomposition

If we symmetrize the decomposition (1.1), we find the following second-order decomposition [10,12]

$$e^{x(A+B)} = e^{\frac{x}{2}A}e^{xB}e^{\frac{x}{2}A} + O(x^3),\tag{2.1}$$

as is well known. Then, the Trotter formula (1.1) is improved as

$$e^{x(A+B)} = (e^{\frac{x}{2n}A}e^{\frac{x}{n}B}e^{\frac{x}{2n}A})^n + O(\frac{x^3}{n^2}).\tag{2.2}$$

When the m-th order approximant $F_m(x)$ is known, we have [5]

$$e^{x(A_1+\cdots+A_q)} = (F_m(\frac{x}{n}))^n + O(\frac{x^{m+1}}{n^m}).\tag{2.3}$$

There are many ways [12, 20-53] of constructing the m-th order approximant $F_m(x)$:

$$e^{x(A_1+\cdots+A_q)} = F_m(x) + O(x^{m+1}).\tag{2.4}$$

Historically, the following fourth-order symmetric decomposition

$$F_4(x) = e^{\frac{x}{2}B}e^{\frac{x}{2}A}e^{\frac{x^3}{4}S_3}e^{\frac{x}{2}A}e^{\frac{x}{2}B}\tag{2.5}$$

was found [10, 54] for the first time in this field of quantum Monte Carlo simulations. Here, S_3 is given by

$$S_3 = \frac{1}{6}[2A + B,[B,A]].\tag{2.6}$$

Later, the above foumula (2.5) was extended to the following general form [12]

$$e^{2x(A_1+\cdots+A_q)} = e^{xA_1}\cdots e^{xA_q}e^{x^3S_3}\cdots e^{x^{2n+1}S_{2n+1}}\cdots e^{x^{2n+1}S_{2n+1}}\cdots e^{x^3S_3}e^{xA_q}\cdots e^{xA_1},\tag{2.7}$$

where

$$S_3 = \frac{1}{6}\sum_{k=2}^{q}\{2\sum_{i\neq j}\sum_{j\leq k-1}A_i^\times A_j^\times + \sum_{i<k}(A_i^\times)^2\}A_k\tag{2.8}$$

with Kubo's notation [55]

$$A^\times B \equiv [A,B] = AB - BA,\tag{2.9}$$

and

$$S_{2n+1} = \frac{1}{2(2n+1)!}[\frac{\partial^{2n+1}}{\partial x^{2n+1}}(e^{-x^{2n-1}S_{2n-1}} \cdots e^{-x^3 S_3} e^{-xA_q} \cdots e^{-xA_1} e^{2x(A_1 + \cdots + A_q)}$$
$$e^{-xA_1} \cdots e^{-xA_q} e^{-x^3 S_3} \cdots e^{-x^{2n-1}S_{2n-1}})]_{x=0}. \tag{2.10}$$

According to the Baker-Campbell-Hausdorff (BCH) theorem, all the operators $\{S_{2n+1}\}$ are linear combinations of the commutators of $\{A_j\}$. This kind of higher-order decomposition becomes more complicated as n increases, because it includes higher-order commutators for larger n. Consequently, the above higher-order decomposition (2.7) except the fourth-order one is not so useful from a practical point of view.

3   Improved Trotter-like Formula——Fourth-Order Decomposition

As far as the fourth-order decomposition of (2.5) is concerned, the trace of the combination of (2.3) and (2.5) is reduced to the following convenient formula [39]

$$\text{Tr} \exp[x(A + B)] = \text{Tr}[(e^{xA/2n} e^{xB/n} e^{xA/2n})^n \exp(\frac{x^3}{4n^2} S_3)] + O(1/n^4) \tag{3.1}$$

when $x^3 \ll n^2$. This is very close to the symmetrized Trotter formula except only single correction term $\exp[(x^3/4n^2)S_3]$. Without the trace in (3.1), such correction terms appear n times, and it is more complicated. Some applications of the above improved Trotter formula to quantum spin systems will be given in the near future [56].

4   General Theory of Exponential Perturbation Expansion

The ordinary perturbation expansion destroys the symmetry of the relevant system such as the unitarity and symplectic property. The following exponential perturbation expansion

$$e^{x(A_1 + \cdots + A_q)} = F_m(x) + O(x^{m+1}) \tag{4.1}$$

with

$$F_m(x) = Q(p_1 x)Q(p_2 x) \cdots Q(p_r x) \tag{4.2}$$

and

$$Q(x) = e^{xA_1} e^{xA_2} \cdots e^{xA_q} \tag{4.3}$$

is very useful, because it retains the above symmetry.

A general theory of higher-order decomposition of the operator $\exp[x(A_1 + A_2 + \cdots + A_q)]$ was proposed by the present author [21-25]. It is reviewed briefly here. For convenience, we consider the following decomposition based on the second-order symmetric approximant

$$S(x) = e^{\frac{x}{2}A_1} \cdots e^{\frac{x}{2}A_{q-1}} e^{xA_q} e^{\frac{x}{2}A_{q-1}} \cdots e^{\frac{x}{2}A_1}, \tag{4.4}$$

namely

$$F_m(x) = S(p_1x)S(p_2x)\cdots S(p_rx). \tag{4.5}$$

The parameters $\{p_j\}$ can be determined so that $F_m(x)$ may become of the m-th order as in (4.1). The trivial condition on $\{p_j\}$ is given by

$$p_1 + p_2 + \cdots + p_r = 1. \tag{4.6}$$

Then we assume that $S(x)$ is expressed as

$$S(x) = e^{x\mathcal{H}+x^3R_3+x^4R_4+\cdots}. \tag{4.7}$$

Here, $R_3, R_4 \cdots$ denote the third-order, fourth-order, $\cdots$ correction terms, respectively. Using the "time-ordering" operation P with respect to the subscript j, we rewrite $F_m(x)$ as

$$\begin{aligned} F_m(x) &= \prod_{j=1}^{r} \exp[(xp_j)\mathcal{H} + (xp_j)^3R_3 + \cdots] \\ &= \text{P} \exp(\sum_{j=1}^{r}[(xp_j)\mathcal{H} + (xp_j)^3R_3 + \cdots]) \\ &= \sum \frac{x^{n_1+3n_3+\cdots}}{n_1!n_3!\cdots}\text{PS}(Y_1^{n_1}Y_3^{n_3}\cdots). \end{aligned} \tag{4.8}$$

Here, the symbol S denotes Kubo's symmetrization operation [57] with respect to the operators $\{Y_n\}$, and

$$Y_1 = \sum_{j=1}^{r}(p_j\mathcal{H}) \text{ and } Y_n = \sum_{j=1}^{r}(p_j^n R_n). \tag{4.9}$$

As p_j and p_j^n in (4.9) indicate "time" of $\mathcal{H}$ and R_n, respectively, they should not be separated from $\mathcal{H}$ and R_n, respectively, until the time-ordering operation P is performed. As was shown in the previous papers [23-25], Eq. (4.8) can be easily rewritten as

$$F_m(x) = e^{x\mathcal{H}} + \sum_{n_1n_3\cdots}' \frac{x^{n_1+3n_3+\cdots}}{n_1!n_3!}\text{PS}(Y_1^{n_1}Y_3^{n_3}\cdots), \tag{4.10}$$

where the symbol $\sum'$ in (4.10) denotes the summation over $n_1, n_3 \cdots$ excluding the case $n_3 = n_5 = \cdots = 0$. Thus, the condition for $F_m(x)$ to be correct up to the m-th order is given by the following requirement

$$C_m(n_1, n_3, \cdots) \equiv \text{PS}(Y_1^{n_1}Y_3^{n_3}\cdots) = 0 \tag{4.11}$$

for all non-negative integers $n_1, n_3 \cdots$ under the condition that $n_1 + 3n_3 + \cdots \le m$, excluding the case $n_3 = n_5 = \cdots = 0$.

After some algebraic calculations, we obtain the following equations [23] for determining the symmetric parameters $\{p_j\}$ satisfying the relation $p_{r+1-j} = p_j$:

$$\text{Fourth} - \text{order} \; : \; \sum p_k = 1, \quad \sum p_k^3 = 0 \; ; \tag{4.13}$$

$$\text{Sixth} - \text{order} \; : \; \sum p_k = 1, \quad \sum p_k^3 = 0, \quad \sum p_k^5 = 0, \quad \sum p_k^3 a_{1k} b_{1k} = 0 \; ; \tag{4.14}$$

$$\text{Eighth} - \text{order} \; : \; \sum p_k = 1, \quad \sum p_k^{2n-1} = 0 \quad \text{for } n = 2, 3, 4$$
$$\sum p_k^{2n-1} a_{1k} b_{1k} = 0 \quad \text{for } n = 2, 3, \quad \sum p_k a_{3k} b_{3k} = 0,$$
$$\sum p_k^3 a_{1k}^2 b_{1k}^2 = 0 \; ; \tag{4.15}$$

$$\text{Tenth} - \text{order} \; : \; \sum p_k = 1, \quad \sum p_k^{2n-1} = 0 \quad \text{for } n = 2, 3, 4, 5,$$
$$\sum p_k^{2n-1} a_{1k} b_{1k} = 0 \quad \text{for } n = 2, 3, 4,$$
$$\sum p_k^{2n-1} a_{1k}^2 b_{1k}^2 = 0 \quad \text{for } n = 2, 3,$$
$$\sum p_k^3 a_{1k}^3 b_{1k}^3 = 0, \quad \sum p_k^3 a_{3k} b_{1k} = 0, \quad \sum p_k^3 a_{5k} b_{1k} = 0,$$
$$\sum p_k^3 a_{3k} b_{1k}^3 = 0, \quad \sum p_k a_{3k}^2 b_{1k}^2 = 0, \quad \sum p_k a_{3k} b_{3k} = 0 \; ; \tag{4.16}$$

$$\text{Twelvth} - \text{order} \; : \; \sum p_k = 1, \quad \sum p_k^{2n-1} = 0 \quad \text{for } n = 2, 3, 4, 5, 6,$$
$$\sum p_k^{2n-1} a_{1k} b_{1k} = 0 \quad \text{for } n = 2, 3, 4, 5,$$
$$\sum p_k^{2n-1} a_{1k}^2 b_{1k}^2 = 0 \quad \text{for } n = 2, 3, 4,$$
$$\sum p_k^{2n-1} a_{1k}^3 b_{1k}^3 = 0 \quad \text{for } n = 2, 3,$$
$$\sum p_k^{2n-1} a_{3k} b_{1k} = 0 \quad \text{for } n = 3, 4,$$
$$\sum p_k^{2n-1} a_{3k}^2 b_{1k}^2 = 0 \quad \text{for } n = 1, 2$$
$$\sum p_k^{2n-1} a_{5k} b_{1k} = 0 \quad \text{for } n = 2, 3, \quad \sum p_k^3 a_{7k} b_{1k} = 0,$$
$$\sum p_k^5 a_{3k} b_{3k} = 0, \quad \sum p_k^3 a_{5k} b_{1k}^3 = 0, \quad \sum p_k a_{3k}^3 b_{1k} = 0,$$
$$\sum p_k^3 a_{3k} b_{1k}^5 = 0, \quad \sum p_k a_{3k}^2 b_{1k}^4 = 0, \quad \sum a_{5k} a_{3k} p_k b_{1k}^2 = 0,$$
$$\sum a_{5k} a_{3k} a_{1k} p_k b_{1k} = 0, \quad \sum p_k a_{3k}^2 a_{1k} b_{1k}^3 = 0, \quad \sum p_k^3 a_{1k}^4 b_{1k}^4 = 0,$$
$$\sum p_k^{2n-1} a_{3k} b_{1k}^3 = 0 \quad \text{for } n = 2, 3, \quad \sum p_k a_{3k} b_{3k} = 0, \tag{4.17}$$

where

$$a_{nk} = \sum_{j<k} p_j^n + \frac{1}{2} p_k^n, \quad \text{and} \quad b_{nk} = \sum_{j>k} p_j^n + \frac{1}{2} p_k^n. \tag{4.18}$$

The minimal number of the above equations on $\{p_j\}$ in each order m, $M(m)$, is given by [23,58]

$$M(2) = 1, \; M(4) = 2, \; M(6) = 4, \; M(8) = 8, \; M(10) = 16, \; M(12) = 34, \; M(14) = 74, \cdots \tag{4.19}$$

5   General Convergence Theorem

As was discussed in the preceding section, we can construct any higher-order decomposition $F_m(x)$. Then, there arises a question whether $\{F_m(x)\}$ converge to the original exponential operator in the limit $m \to \infty$ or not. Now we have the following general theorem [38]. First we define an approximant $Q_s(x)$ of index s by

$$\| Q_s(x) - e^{x\mathcal{H}} \| \le K_s \mid x \mid^{s+1} \tag{5.1}$$

for sufficiently small $\mid x \mid$ with some positive constant K_s. Here, $\mathcal{H} = A_1 + A_2 + \cdots + A_q$.

Theorem : Let $Q_s(x)$ be an approximant of index s of $e^{x\mathcal{H}}$. As the m-th approximant one adopts

$$F_m(x) = Q_s(p_{m1}x)Q_s(p_{m2}x) \cdots Q_s(p_{mn}x), \tag{5.2}$$

where the number $n = n(m)$ of factors in the m-th product depends on $m(n(m) \to \infty$ as $m \to \infty)$, and p_{mj} satisfies Eq. (4.6) and the condition that

$$\mid p_{mj} + p_{mj+1} + \cdots + p_{mn} \mid \text{ is bounded uniformly for both } m \text{ and } j. \tag{5.3}$$

Then, the above series of approximants $\{F_m(x)\}$ with the operators $\{A_j\}$ in a Banach space converges to $\exp(x\mathcal{H})$, namely

$$\lim_{m \to \infty} \| F_m(x) - e^{x\mathcal{H}} \| = 0, \text{ i.e., } \lim_{m \to \infty} F_m(x) = e^{x\mathcal{H}} \tag{5.4}$$

for all $x \in C$ under the condition that

$$\lim_{m \to \infty} \sum_{j=1}^{n(m)} \mid p_{mj} \mid^{s+1} = 0. \tag{5.5}$$

The limit (5.4) is uniform provided $\mid x \mid \le \delta$ for any positive number δ, namely in any compact region of x.

The above theorem can be applied even to the convergence of the ordinary Trotter formula. The condition (5.5) with $s = 1$ for the ordinary Trotter formula in which the parameters $\{p_{mj}\}$ are given by $p_{mj} = 1/m$ is easily confirmed as

$$\lim_{m \to \infty} \sum_{j=1}^{n(m)} \mid p_{mj} \mid^2 = \lim_{m \to \infty} \frac{n(m)}{m^2} = \lim_{m \to \infty} \frac{1}{m} = 0. \tag{5.6}$$

For the above theorem, we find that the convergence rate of the limit (5.4) is proportial to the left-hand side of (5.5) or of higher order than it.

6 Some Higher-Order Decomposition Formulas

As is seen from the general theorem in Section 5, a simple criterion of error estimate may be given by ε

$$\varepsilon \equiv \sum_{j=1}^{n(m)} \mid p_{mj} \mid^{s+1} . \tag{6.1}$$

As $s = 2$ in the present symmetric decomposition (4.5), we have

$$\varepsilon \equiv \sum_{j=1}^{n(m)} \mid p_{mj} \mid^3 . \tag{6.2}$$

Clearly, if some $\{p_{mj}\}$ are larger than unity, the error estimate ε becomes larger than that with $\mid p_{mj} \mid < 1$ for all j. Thus, we show here some "typical" solutions [35, 59] of the equations (4.12)–(4.16) which satisfy the criterion that $\mid p_{mj} \mid < 1$ for all j.

a) Fourth-order formula [21] :

$$S_4^*(x) = S^2(p_2 x)S((1 - 4p_2)x)S^2(p_2 x) \tag{6.3}$$

with

$$p_2 = (4 - 4^{1/3})^{-1} = 0.41449077179437573714235406286 0 \cdots . \tag{6.4}$$

This decomposition is recommended in practical applications [34]. In fact, the "error estimate" ε for the decomposition $S_4^*(x)$ is given by $\varepsilon = 0.57$, which is much smaller than $\varepsilon' = 2.5$ for the minimal fourth-order decomposition

$$(S_4(\frac{x}{2}))^2 = (S(\frac{p}{2}x)S(\frac{1-2p}{2}x)S(\frac{p}{2}x))^2 \tag{6.5}$$

with

$$p = \frac{1}{2 - \sqrt[3]{2}} = 1.35120719195965 \cdots . \tag{6.6}$$

Here we have compared $S_4^*(x)$ with $[S_4(x/2)]^2$ instead of $S_4(x)$, because the number of factors is relevant to this comparison.

b) Sixth-order formula $(M(6) = 4)$:

$$S_6(x) = S^2(p_1 x)S^2(p_2 x)S^2(p_3 x)S^2(p_4 x)S^2(p_3 x)S^2(p_2 x)S^2(p_1 x), \tag{6.7}$$

where

$$\begin{array}{llll} p_1 &=& 0.3922568052387732, & p_3 &=& -0.5888399920894384, \\ p_2 &=& 0.1177866066796810, & p_4 &=& 0.6575931603419684. \end{array} \tag{6.8}$$

c) Eighth-order formula ($M(8) = 8$ and $p_{16-j} = p_j$ for all j's) :

$$\begin{aligned}S_8(x) &= S(p_1 x)S(p_2 x)S(p_3 x)S(p_4 x)S(p_5 x)S(p_6 x)S(p_7 x)\\&\times S(p_8 x)S(p_7 x)S(p_6 x)S(p_5 x)S(p_4 x)S(p_3 x)S(p_2 x)S(p_1 x),\end{aligned} \tag{6.9}$$

where

$$\begin{aligned}p_1 &= 0.741670364350612953448278, & p_5 &= 0.299064181303655923844 4635,\\p_2 &= -0.409100825800031593 9973001, & p_6 &= 0.3346249182452981837849580,\\p_3 &= 0.190754710296238379953 8763, & p_7 &= 0.3152930923967665966320567,\\p_4 &= -0.5738624711160822666563877, & p_8 &= -0.796887939353529163540197888.\end{aligned} \tag{6.10}$$

d) Tenth-order formula ($M(10) = 16$ and $p_{32-j} = p_j$ for all j's) :

$$S_{10}(x) = S(p_1 x)S(p_2 x)\cdots S(p_{16} x)\cdots S(p_2 x)S(p_1 x), \tag{6.11}$$

where

$$\begin{aligned}p_1 &= 0.240272136691376817 6465371, & p_9 &= 0.590296672913859538 0899091\\p_2 &= 0.1772781495704011320291293, & p_{10} &= -0.7672080102554930808311795\\p_3 &= 0.7806667596279710935602836, & p_{11} &= -0.5065099791027394826592456\\p_4 &= -0.5594672025734347825531856, & p_{12} &= 0.6386615662932831893883190\\p_5 &= 0.4406755465172401339812858, & p_{13} &= 0.1160613001706441872046272\\p_6 &= -0.3639685198702672791482016, & p_{14} &= -0.7267723266415894584069335\\p_7 &= -0.4374102252651395419946745, & p_{15} &= 0.1780400262708314376713936\\p_8 &= 0.3457387003846991851974124, & p_{16} &= 0.7072908105367138459351750.\end{aligned} \tag{6.12}$$

These formulas will be useful in practical applications.

7 Decomposition Scheme of Ordered Exponentials

In the present section, we discuss the decomposition scheme [53,59] of the ordered exponentials [57] defined by

$$\begin{aligned}U(t + \Delta t, t) &= T \exp \int_t^{t+\Delta t} \mathcal{H}(s)ds\\&= 1 + \int_t^{t+\Delta t} \mathcal{H}(s)ds + \int_t^{t+\Delta t} ds_1 \int_t^{s_1} ds_2 \mathcal{H}(s_1)\mathcal{H}(s_2) + \cdots\\&= U(t + \Delta t, s)U(s, t),\end{aligned} \tag{7.1}$$

where $\mathcal{H}(t) = A_1(t) + \cdots + A_q(t)$.

Concerning the decomposition of these ordered exponentials, we have the following scheme [53] :

Middle-point decomposition scheme : A typical m-th order decomposition formula of the ordered exponential (7.1) is obtained immediately from the corresponding m-th order decomposition formula for the ordinary exponential operator, by replacing $\{A_j\}$ in $Q(x)$ and $S(x)$ by the middle point operators $\{A_j(t_{j,\text{mid}})\}$ at each time separation for $Q(x)$ (or $\tilde{Q}(x) = Q^{-1}(x)$ and $S(x)$).

For the derivation of this general scheme, see Ref. 53.

Now we present some explicit formulas. The first-order formula is given by

$$U_1(t + \Delta t, t) = e^{\Delta t A(t+\frac{1}{2}\Delta t)} e^{\Delta t B(t+\frac{1}{2}\Delta t)}, \tag{7.3}$$

corresponding to the ordinary decomposition formula

$$e^{\Delta t(A+B)} = e^{\Delta t A} e^{\Delta t B} + O((\Delta t)^2). \tag{7.4}$$

Similarly the second-order decomposition $U_2(t + \Delta t, t)$ is given by

$$U_2(t + \Delta t, t) = e^{\frac{1}{2}\Delta t A(t+\frac{1}{2}\Delta t)} e^{\Delta t B(t+\frac{1}{2}\Delta t)} e^{\frac{1}{2}\Delta t A(t+\frac{1}{2}\Delta t)}, \tag{7.5}$$

corresponding to the ordinary symmetrized decomposition

$$e^{\Delta t(A+B)} = e^{\frac{1}{2}\Delta t A} e^{\Delta t B} e^{\frac{1}{2}\Delta t A} + O((\Delta t)^3). \tag{7.6}$$

The third-order decomposition $U_3(t + \Delta t, t)$ is also given by [44,53]

$$U_3(t + \Delta t, t) = Q(p_5\Delta t; t_5)\tilde{Q}(p_4\Delta t; t_4)Q(p_3\Delta t; t_3)\tilde{Q}(p_2\Delta t; t_2)Q(p_1\Delta t; t_1), \tag{7.7}$$

where

$$Q(x; t) = e^{xA_1(t)} \cdots e^{xA_q(t)} \quad \text{and} \quad \tilde{Q}(x; t) = e^{xA_q(t)} \cdots e^{xA_1(t)}, \tag{7.8}$$

and $\{t_j\}$ denote the middle points of each time separation, namely

$$
\begin{array}{lll}
t_1 &=& t + \frac{1}{2}p_1\Delta t, \quad t_2 = t + (p_1 + \frac{1}{2}p_2)\Delta t, \\
t_3 &=& t + \frac{1}{2}\Delta t, \quad t_4 = t + (1 - p_1 - \frac{1}{2}p_2)\Delta t, \quad t_5 = t + (1 - \frac{1}{2}p_1)\Delta t.
\end{array} \tag{7.9}
$$

The parameters $\{p_s\}$ are given by

$$
\begin{array}{ll}
p_1 = p_5 = 0.2683300957817599 \cdots, & p_3 = -0.8393230460347997 \cdots, \\
p_2 = p_4 = 0.651331427235699 \cdots.
\end{array} \tag{7.10}
$$

The simplest fourth-order decomposition of an ordered exponential is given by

$$U_4(t + \Delta t, t) = S(p\Delta t; t_3)S((1 - 2p)\Delta t; t_2)S(p\Delta t; t_1), \tag{7.11}$$

where

$$S(x;t) = e^{\frac{x}{2}A_1(t)} \cdots e^{\frac{x}{2}A_{r-1}(t)} e^{xA_r(t)} e^{\frac{x}{2}A_{r-1}(t)} \cdots e^{\frac{x}{2}A_1(t)} \qquad (7.12)$$

and

$$t_1 = t + \frac{1}{2}p\Delta t, \ t_2 = t + \frac{1}{2}\Delta t, \ t_3 = t + (1 - \frac{1}{2}p)\Delta t. \qquad (7.13)$$

In general, the m-th order decomposition is given by

$$U_m(t + \Delta t, t) = Q(p_r \Delta t; t_r) \cdots Q(p_2 \Delta t; t_2)Q(p_1 \Delta t; t_1) \qquad (7.14)$$

or

$$U_m^{(s)}(t + \Delta t, t) = S(p_r \Delta t; t_r) \cdots S(p_2 \Delta t; t_2)S(p_1 \Delta t; t_1) \ ; \ p_{r+1-j} = p_j, \qquad (7.15)$$

where $\{t_j\}$ denote the middle points of each time separation, namely

$$\begin{aligned} t_1 &= t + \tfrac{1}{2}p_1\Delta t, \ t_2 = t + (p_1 + \tfrac{1}{2}p_2)\Delta t, \ t_3 = t + (p_1 + p_2 + \tfrac{1}{2}p_3)\Delta t, \cdots, \\ t_j &= t + (p_1 + p_2 + \cdots p_{j-1} + \tfrac{1}{2}p_j)\Delta t, \cdots, t_r = t + (1 - \tfrac{1}{2}p_r)\Delta t. \end{aligned} \qquad (7.16)$$

In practical applications, we have to calculate $U(t',t)$ for a finite (not small) time difference $(t'-t)$. For this purpose, we iterate the following procedure of multiplication

$$U(t',t) = U(t', t + (n-1)\Delta t) \cdots U(t + 2\Delta t, t + \Delta t)U(t + \Delta t, t) \qquad (7.17)$$

with $n\Delta t = t' - t$.

The above decomposition scheme of ordered exponentials can be used in studying the time-dependent Schrödinger equation [41-44], and many other time-dependent many-body problems.

8   Variational Principle and Inequalities

It will be very convenient to find some inequalities concerning the approximants presented in the preceding sections. For the Trotter formula (1.1), we have the following trace inequality [5]

$$\mathrm{Tr}\,(e^{\frac{A}{n}}e^{\frac{B}{n}})^n \geq \mathrm{Tr}\,e^{A+B} \qquad (8.1)$$

for finite hermitian matrices A and B. This is a generalization of the following Golden-Symanzik-Thompson inequality

$$\mathrm{Tr}\,e^A e^B \geq \mathrm{Tr}\,e^{A+B}. \qquad (8.2)$$

By the way, it has been also proven [60] that $\mathrm{Tr}(e^{A/n}e^{B/n})^n$ is a decreasing function of positive n.

Concerning the Baker-Campbell-Hausdorff formula, the following inequality [5] holds :

$$\mathrm{Tr}\,(e^{\frac{A}{n}}e^{\frac{B}{n}})^n \geq | \,\mathrm{Tr}\,(e^{\frac{A}{2n}}e^{\frac{B}{2n}}\exp(\frac{1}{8n^2}[B,A]))^{2n} \,|. \qquad (8.3)$$

It is an open question to extend the above inequality (8.1) to the higher-order decomposition (4.5) and a general cluster expansion of the density matrix [61].

References

1. H.F. Trotter, Proc. Ann. Math. Soc. **10** (1959) 545.
2. M. Suzuki, Commun. Math. Phys. **51** (1976) 183.
3. M. Suzuki, Prog. Theor. Phys. **56** (1976) 1454.
4. M. Suzuki, S. Miyashita and A. Kuroda, Prog. Theor. Phys. **58** (1977) 1377.
5. M. Suzuki, J. Stat. Phys. **43** (1986) 883.
6. M. Suzuki, ed., *Quantum Monte Carlo Methods*. Solid State Sciences, vol. 74 (Springer, Berlin, 1986).
7. M. Suzuki, Physica **A194** (1993) 432.
8. M. Suzuki, ed., *Quantum Monte Carlo Methods in Condensed Matter Physics* (World Scientific, (1993) in press).
9. M.H. Kalos, ed., *Monte Carlo Methods in Quantum Problems* (Reidel, Boston, 1982).
10. H. De Raedt and B. De Raedt, Phys. Rev. **A28** (1983) 3575.
    H. De Raedt and Ad Lagendijk, Phys. Reports **127** (1985) 233.
    H. De raedt, Comput. Phys. Rep. **7** (1987) 1.
11. N. Hatano and M.Suzuki, in this volume.
12. M. Suzuki, J. Math. Phys. **26** (1985) 601.
13. M. Suzuki, Phys. Rev. **B31** (1985) 2957.
14. M. Suzuki and M. Inoue, Prog. Theor. Phys. **78** (1987) 787.
15. H. Betsuyaku, Prog. Theor. Phys. **73** (1985) 319 ; ibid **75** (1986) 808.
16. T. Delica, K. Kopinga, H. Leschke and K.K. Mon, Europhys. Lett. **15** (1991) 55.
17. K. Kubo, Phys. Rev. **B46** (1992) 866.
18. N. Hatano and M. Suzuki, J. Phys. Soc. Jpn. **62** (1993) 1346.
19. T. Tsuzuki, Prog. Theor. Phys. **72** (1984) 956.
20. M. Suzuki, Phys. Lett. **113A** (1985) 299.
21. M. Suzuki, Phys. Lett. **A146** (1990) 319.
22. M. Suzuki, J. Math. Phys. **32** (1991) 400.
23. M. Suzuki, Phys. Lett. A **A165** (1992) 387.
24. M. Suzuki, J. Phys. Soc. Jpn. **61** (1992) 3015.
25. M. Suzuki, in *Fractals and Disorder*, edited by A. Bunde (North-Holland, 1992), i.e., Physica **A191** (1992) 501.
26. R.D. Ruth, IEEE Trans. Nucl. Sci. NS-30 (1983) 2669, and F. Neri, preprint (1988).
27. E. Forest and R.D. Ruth, Physica **D43** (1990) 105.
28. E. Forest, J. Math. Phys. **31** (1990) 1133, and preprint.
29. H. Yoshida, Phys. Lett. **A150** (1990) 262.
30. A.D. Bandrauk and H. Shen, Chem. Phys. Lett. **176** (1991) 428.
31. J.A. Oteo and J. Ros, J. Phys. A Math. Gen. **24** (1991) 5751.
32. P. de Vries, *Trottering Through Quantum Physics* (Natuurkunding Laboratorium der Universiteit van Amsterdam, Valckenierstraat 65, 1018 XE Amsterdam, (1991)).

33. W. Janke and T. Sauer, Phys. Lett. **A165** (1992) 199.
34. N. Hatano and M. Suzuki, Prog. Theor. Phys. **85** (1991) 481.
35. M. Suzuki, in *Field Theory and Collective Phenomena* (World Scientific, 1993).
36. M. Suzuki and T. Yamauchi, submitted to J. Math. Phys. (1993).
37. M. Suzuki and K. Umeno, in the Proceedings of the Sixth Annual Workshop on *Recent Developments in Computer Simulation Studies in Condensed Matter Physics*, edited by D.P. Landau et al. (Springer).
38. M. Suzuki, Commun. Math. Phys. (in press).
39. M. Suzuki, Phys. Lett. **180A** (1993) 232. All the factors $8n^2$ and $8n^3$ in this paper should read $4n^2$ and $4n^3$, repectively.
40. T. Itoh and D-S. Cai, Phys. Lett. **A171** (1992) 189.
41. M. Glasner, D. Yevick and B. Hermansson, Mathl. Comput. Modelling **16** (1992) 177, and Appl. Math. Lett. **4** (1991) 85.
42. B. Hermansson and D. Yevick, Opt. Lett. **36** (1991) 354.
43. M. Glasner, D. Yevick and B. Hermansson, Electronics Lett. **27** (1991) 475.
44. A. Terai and Y. Ono, to be published.
45. J. Candy and W. Rozmus, J. Comp. Phys. **92** (1991) 230.
46. Q. Sheng, IMA Journ. Numerical Analysis **9** (1989) 199.
47. A.N. Drozdov, Physica **A196** (1993) 283.
48. Z. Mei-Qing, Phys. Lett. A (1993).
49. S.T. Kuroda and T. Suzuki, Japan J. Appl. Math. **7** (1990) 231.
50. S.T. Kuroda, *Differential Equations with Applications to Mathematical Physics* (Academic Press, Inc. 1993).
51. K. Aomoto, preprint.
52. G. Gnudi and T. Watanabe, J. Phys. Soc. Jpn. **62** (1993) 3492.
53. M. Suzuki, Proc. Japan Acad., **69** Ser. B, No.7 (1993) 161.
54. J.E. Hirsch, R.L. Sugar, D.J. Scalapino, and R. Blankenbecler, Phys. Rev. **B26** (1982) 5033.
55. R. Kubo, J. Phys. Soc. Jpn. **12** (1957) 570.
56. H. Kobayashi, N. Hatano and M. Suzuki, in preparation.
57. R. Kubo, J. Phys. Soc. Jpn. **17** (1962) 1100.
58. The values $r_{min}(14)$ and $r_{min}(16)$ in Ref. 23 are incorrect, as was pointed out by R. McLachlan (preprint).
59. M. Suzuki, Physica A (1993).
60. H. Araki, Lett. Math. Phys. **19** (1990) 167.
61. R. Kubo, J. Chem. Phys. **20** (1952) 770.

QUANTUM MONTE CARLO AND RELATED METHODS
— RECENT DEVELOPMENTS —

Naomichi Hatano and Masuo Suzuki
Department of Physics, University of Tokyo
Hongo 7-3-1, Bunkyo-ku, Tokyo 113, Japan

Abstract

Quantum Monte Carlo methods are reviewed with emphasis on their methodological aspects. Methods based on the Suzuki-Trotter approximation, namely the world-line approach and the auxiliary-field approach, are mainly described. The transfer-matrix method, which is closely related to the quantum Monte Carlo methods, is also reviewed. Recently developed arguments on the negative-sign problem are presented.

1 Introduction

In this article we review quantum Monte Carlo methods emphasizing their methodological aspects.

Let us discuss two quantum mechanical models, namely the Heisenberg model and the Hubbard model. The quantum Heisenberg model is described by the Hamiltonian

$$\mathcal{H} = -J \sum_{\langle i,j \rangle} \boldsymbol{\sigma}_i \cdot \boldsymbol{\sigma}_j, \tag{1}$$

where $\boldsymbol{\sigma}$ denotes the Pauli operators σ^x, σ^y and σ^z. The other model is defined by the Hubbard Hamiltonian [1], which describes strongly correlated electrons on a lattice:

$$\mathcal{H} \equiv K + V. \tag{2}$$

Here K denotes the kinetic energy of tight-binding electrons,

$$K \equiv -t \sum_{\langle i,j \rangle} \sum_{\sigma=\uparrow,\downarrow} \left(c_{i\sigma}^{\dagger} c_{j\sigma} + c_{j\sigma}^{\dagger} c_{i\sigma} \right), \tag{3}$$

and V is the on-site Coulomb energy plus the chemical potential,

$$V \equiv U \sum_i \left(n_{i\uparrow} - \frac{1}{2} \right) \left(n_{i\downarrow} - \frac{1}{2} \right) - \mu \sum_i \sum_{\sigma=\uparrow,\downarrow} n_{i\sigma}. \tag{4}$$

13

We encounter an essential difficulty if we naively apply the Metropolis algorithm to quantum systems. In quantum statistical mechanics we calculate the expectation value of a quantity in the form

$$\langle Q \rangle \equiv \frac{\mathrm{Tr}\, Q e^{-\beta\mathcal{H}}}{\mathrm{Tr}\, e^{-\beta\mathcal{H}}} = \sum_{\psi} \langle \psi | Q e^{-\beta\mathcal{H}} | \psi \rangle \Big/ \sum_{\psi} \langle \psi | e^{-\beta\mathcal{H}} | \psi \rangle. \qquad (5)$$

In order to simulate the system as it is, we have to calculate the ratio

$$R \equiv \langle \psi' | e^{-\beta\mathcal{H}} | \psi' \rangle \Big/ \langle \psi | e^{-\beta\mathcal{H}} | \psi \rangle \qquad (6)$$

for a flip trial $\psi \longrightarrow \psi'$. Hence we have to diagonalize the matrix $\mathcal{H}$, which is practically impossible for large systems.

This difficulty is inevitable in quantum statistical mechanics. In classical Monte Carlo dynamics it is easy to calculate the ratio (6) because of the locality of the interaction. In a quantum mechanical system, however, the quantum coherence ranges over the whole system. Even if the interaction term in the Hamiltonian is local, the effect of the term is not local.

In the following sections we review several methods of studying quantum systems. The main subject is how to calculate the density matrix $e^{-\beta\mathcal{H}}$ in order to simulate the system. The Suzuki-Trotter (ST) decomposition, which is adopted in most methods reviewed here, is described in Section 2. The decomposition has been applied in the world-line approach (Section 3), the transfer-matrix method (Section 5), the Monte Carlo power method (Section 6) and the auxiliary-field approach (Section 7). The negative-sign problem, which appears in some cases, is explained in Section 4.

Modifications of some approaches, performed for the purpose of studying ground-state properties, are described in Section 8. The diffusion Monte Carlo methods, which were also devised for studying the ground state, are discussed in Section 9.

Handscomb's method (Section 10) and the decoupled-cell method (Section 11), in which the ST approximation is not utilized, are also presented.

2   ST decomposition

Here we describe the method of calculating the density matrix $e^{-\beta\mathcal{H}}$ approximately. The method is utilized in many approaches presented in the following sections. We also explain the construction of higher-order approximants.

Generally speaking, when we find an mth order approximant $f_m(x)$ of the exponential operator $e^{-x\mathcal{H}}$,

$$f_m(x) = e^{-x\mathcal{H}} + O(x^{m+1}), \qquad (7)$$

the density matrix is approximated by

$$e^{-\beta\mathcal{H}} = \left(e^{-\beta\mathcal{H}/n}\right)^n = \left(f_m\left(\beta/n\right)\right)^n + O\left(\beta^{m+1}/n^m\right). \qquad (8)$$

The extrapolation $n \to \infty$ reproduces the original density matrix. The problem is how to find the approximant f_m. It is preferable that the approximant retains some symmetry properties of the original exponential operator.

In the Feynman path-integral approach [2] the expansion approximant

$$e^{-\beta \mathcal{H}/n} = 1 - \beta \mathcal{H}/n + O\left(\beta^2/n^2\right) \qquad (9)$$

has been usually used. The ST decomposition [3, 4, 5], on the other hand, takes the form

$$e^{-\beta \mathcal{H}/n} \simeq e^{-\beta p_1 A_1/n} e^{-\beta p_2 A_2/n} \cdots, \qquad (10)$$

where we decompose the total Hamiltonian into partial Hamiltonians $\{A_l\}$ so that it may be easier to diagonalize each partial Hamiltonian. Approximants of this type, when they imitate time-evolution operators, retain the unitarity. Moreover, it is easy to obtain higher-order approximants as is shown below. In the present section we concentrate on the approximants of type (10).

The simplest example of the decomposition is [6, 3]

$$Q_i(x) \equiv \prod_{i=1}^{q} e^{-x \mathcal{H}_i} = e^{-x \mathcal{H}} + O\left(x^2\right) \qquad (11)$$

with $\mathcal{H} = \sum_{i=1}^{q} \mathcal{H}_i$, and $[\mathcal{H}_i, \mathcal{H}_j] \neq 0$ for $i \neq j$. Here the symbol $\prod_{i=1}^{q}$ indicates that operators are multiplied in the ascending order $i = 1, 2, \ldots, q$.

When we extrapolate the original operator $e^{-x \mathcal{H}}$ using a series of approximants $[f_m(\beta/n)]^n$ in the limit $n \to \infty$, the correction vanishes as n^{-m}. On the other hand, time necessary to calculate the approximant $[f_m(\beta/n)]^n$, in general, increases proportionally to the Trotter number n. Thus approximants with large m can improve the procedure of the extrapolation $n \to \infty$.

Suzuki showed the following fact [7, 8]: if an approximant S_{2m-1}, correct up to the power of x^{2m-1}, satisfies the relation

$$S_{2m-1}(x) S_{2m-1}(-x) = 1, \qquad (12)$$

then the approximant is actually correct up to the power of x^{2m}. An approximant Q_{2m-1}, if it does not satisfy (12), thus produces the approximant S_{2m} as follows (the symmetrization) [8]:

$$S_m(x) = Q_{m-1}(x/2) Q_{m-1}^{-1}(-x/2) = e^{-x \mathcal{H}} + O(x^{m+1}). \qquad (13)$$

Moreover, Suzuki [8, 9, 10] discovered a new way to construct approximants correct up to arbitrarily higher orders, namely the fractal decomposition. The main idea of the fractal decomposition is its recursive way of construction; an appropriate combination of lower-order approximants yields a higher-order approximant in such a way that the lowest-order corrections of the former cancel out with each other.

$\log C_m$

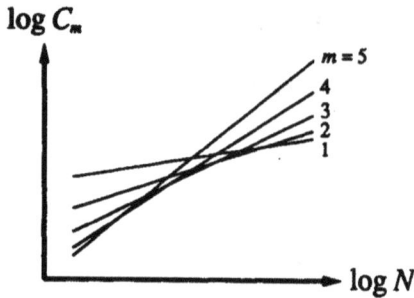

Figure 1: The magnitude of correction as a function of the number of multiplications.

$\log N$

Let us concentrate only on symmetrized approximants, that is, approximants which satisfy (12). An approximant S_{2m} is constructed of S_{2m-2} in the form:

$$S_{2m}(x) \equiv \prod_{j=1}^{2r-1} S_{2m-2}(p_{2m,j}x) \qquad (14)$$

with the coefficients $\{p_{2m,j}\}$ satisfying (i) the normalization condition, $\sum_{j=1}^{2r-1} p_{2m,j} = 1$, (ii) the cancellation condition, $\sum_{j=1}^{2r-1} p_{2m,j}^{2m-1} = 0$ and (iii) the symmetrization condition, $p_{2m,j} = p_{2m,2r-j}$ for $j = 1, 2, \ldots, r - 1$. The cancellation condition ensures that the approximant is correct up to the power of x^{2m-1}, and the symmetrization condition ensures its correctness up to the power of x^{2m}.

Which decomposition is the most advantageous of all the approximants $\{S_{2m}\}$? Let us leave out their application to Monte Carlo simulations for a while, and discuss the direct calculation of the products of the exponential operators (See Section 5). Comparing the fractal decompositions $\{S_{2m}\}$, we have to consider the following two factors. When we use a higher-order decomposition, (i) the magnitude of the correction decreases as β^{2m+1}/n^{2m}, while (ii) the computational time increases as $n(2r)^m$. (Here we assume that the coefficient of the correction term β^{2m+1}/n^{2m} does not grow rapidly as m increases.)

Suppose that we can afford the computational time to execute the multiplications N times. We can attain the Trotter number n given by $n \simeq N(2r)^{-m}$. The magnitude of the correction term of (8) is then of the order of

$$\beta^{2m+1} \big/ n^{2m} \simeq \beta^{2m+1}(2r)^{2m^2} \big/ N^{2m} \equiv C_m(N). \qquad (15)$$

We schematically show the function $C_m(N)$ in Fig. 1. When we fix the number of the multiplications, N, the most precise is the m_1th order decomposition, where m_1 is defined by

$$\frac{\partial}{\partial m}\log C_m(N)\bigg|_{m=m_1} = 0, \quad \text{or} \quad m_1 = \frac{\log(N/\beta)}{2\log(2r)}. \qquad (16)$$

Conversely, when we fix, to a value C, the magnitude of correction which we can allow, the most time-saving is the m_2th decomposition, where m_2 is given by

$$\frac{\partial}{\partial m} \log N_m(C)\bigg|_{m=m_2} = 0, \quad \text{or} \quad m_2 = \sqrt{\frac{\log(\beta/C)}{\log(2r)}}. \tag{17}$$

Here $N_m(C)$ is the inverse function of $C_m(N)$. Thus we can optimize the choice of a fractal decomposition. When it is optimized, the correction decreases rapidly as

$$C_{m_1}(N) = \beta^{m_1+1} \big/ N^{m_1} \sim N^{-\alpha \log N}, \tag{18}$$

where α is an appropriate constant.

As for the application of the fractal decomposition to Monte Carlo calculations, we have to take into account still another factor, namely the negative-sign problem. (The problem is reviewed in Section 4.) The negative-sign problem is more severe in the fractal decomposition than in the second-order decomposition, because some of the coefficients $\{p_{2m,j}\}$ are negative for the fractal decompositions [11]. Since the statistics deteriorates owing to the negative-sign problem, Monte Carlo simulations with the use of the fractal decompositions are critical at low temperatures.

Recently Suzuki has developed a general scheme for constructing higher-order approximants [12, 13]. Using these approximants, we can reduce the number of the multiplications of the exponential operators. As far as we calculate the approximants directly as in Section 5, this general scheme is quite useful. Unfortunately, the negative coefficients of $\{p_{2m,j}\}$ cannot be excluded [9], and hence the application of the decompositions to Monte Carlo simulations is still critical at low temperatures.

3   World-line approach

In this section we present one of the quantum Monte Carlo methods which utilize the ST decomposition reviewed above.

In the following we consider, as an example, the one-dimensional $S = 1/2$ Heisenberg ferromagnet:

$$\mathcal{H} = -\sum_{i=0}^{N} \sigma_i \cdot \sigma_{i+1}. \tag{19}$$

Let us describe how to obtain, by the ST decomposition, the average of a quantity Q, namely,

$$\langle Q \rangle_q \equiv \frac{1}{Z_q} \text{Tr}\, Q e^{-\beta \mathcal{H}} \quad \text{with} \quad Z_q \equiv \text{Tr}\, e^{-\beta \mathcal{H}}. \tag{20}$$

Here we use the simplest approximant

$$e^{-\beta \mathcal{H}} \simeq \left(e^{-\beta A/n} e^{-\beta B/n} \right)^n, \tag{21}$$

with the Hamiltonian (19) divided into two parts: $\mathcal{H} = A + B$ with

$$A \equiv - \sum_{i=0}^{N/2-1} \sigma_{2i} \cdot \sigma_{2i+1} \quad \text{and} \quad B \equiv - \sum_{i=0}^{N/2-1} \sigma_{2i+1} \cdot \sigma_{2i+2}. \tag{22}$$

Though there are many alternatives to this approximant as was mentioned in the previous section, we concentrate on it here for simplicity; it is straightforward to generalize the following argument to the other alternatives.

We thus approximately describe the partition function (20) of the quantum system $\mathcal{H}$ as follows [14, 15, 16, 17]:

$$Z_q \simeq Z_n \equiv \sum_{\{\sigma\}} W\{\sigma\}, \tag{23}$$

where

$$W\{\sigma\} \equiv \prod_{l=0}^{n-1} \left\langle \{\sigma\}_{2l} \left| e^{-\beta A/n} \right| \{\sigma\}_{2l+1} \right\rangle \left\langle \{\sigma\}_{2l+1} \left| e^{-\beta B/n} \right| \{\sigma\}_{2l+2} \right\rangle \tag{24}$$

with the periodic boundary condition $\{\sigma\}_{2n} \equiv \{\sigma\}_0$. Here every state vector $|\{\sigma\}_l\rangle$ is a direct product of eigenstates of σ_i^z:

$$|\{\sigma\}_l\rangle \equiv |\sigma_{0,l}\rangle |\sigma_{1,l}\rangle \cdots |\sigma_{N,l}\rangle, \tag{25}$$

where $\sigma_i^z |\sigma_{i,l}\rangle = \sigma_{i,l} |\sigma_{i,l}\rangle$ with $\sigma_{i,l} = \pm 1$. The summation in (23) runs over all the configurations of the $2n(N+1)$ variables $\sigma_{i,l}$.

We identify each summand of (23) with one of the Boltzmann factors of a classical system. The partition function of the quantum system Z_q is approximated by the partition function of the classical system Z_n.

Suzuki [14] pointed out the following fact: when a d-dimensional quantum Hamiltonian comprises only short-range interactions, we can always arrange the ST decomposition so that the resulting classical system is a $(d+1)$-dimensional system with *short-range* interactions. Thus we can simulate the classical system easily. We call the additional dimensional axis the Trotter direction.

The above general statement actually holds in the case of (23). Since each summand of A (or B) in (22) is commutable with others, we easily obtain their matrix elements in the following form:

$$\left\langle \{\sigma\}_{2l} \left| e^{-\beta A/n} \right| \{\sigma\}_{2l+1} \right\rangle = \prod_{i=0}^{N/2-1} w(2i, 2l)$$

$$\left\langle \{\sigma\}_{2l+1} \left| e^{-\beta B/n} \right| \{\sigma\}_{2l+2} \right\rangle = \prod_{i=0}^{N/2-1} w(2i+1, 2l+1) \tag{26}$$

with

$$w(i, l) \equiv \left\langle \sigma_{i,l}, \sigma_{i+1,l} \left| \exp\left(\frac{\beta}{n} \sigma_i \cdot \sigma_{i+1} \right) \right| \sigma_{i,l+1}, \sigma_{i+1,l+1} \right\rangle. \tag{27}$$

We can interpret this [14] as the Boltzmann factor of a two-dimensional spin system with four-spin interactions w. (The decomposition (22) is called the checkerboard decomposition [18, 19].) Monte Carlo calculation with the importance-sampling method [20] is applicable to the system if $W\{\sigma\} \geq 0$ for all $\{\sigma\}$.

Note that we have the following conservation rule similar to that of the six-vertex model:

$$\sigma_{i,l} + \sigma_{i+1,l} = \sigma_{i,l+1} + \sigma_{i+1,l+1}. \tag{28}$$

Hence we can draw lines connecting the corresponding down spins in the Trotter direction [18]. The lines are called *world lines*. There is a one-to-one correspondence between an allowed spin configuration and a world-line configuration, provided that we require the world lines not to intersect. Thus the summation in (23) runs only over the possible world-line configurations after all. We sample some of the world-line configurations in Monte Carlo simulations, creating, annihilating and moving world lines according to the Boltzmann weights (24).

Computational time necessary in the world-line Monte Carlo simulation is proportional to the number of spins of the classical system (23), namely to Nn. It is shorter than in the Monte Carlo simulation with the auxiliary-field approach described in Section 7. This merit comes from the fact that the classical system consists of short-range interactions.

The method has been successfully applied to quantum ferromagnets [21, 22, 23, 24, 25, 26, 27, 28, 29], to quantum antiferromagnets on bipartite lattices [30, 31, 32, 33, 34], and to one-dimensional fermion systems [35, 18, 19, 36, 37, 38]. On the other hand, for frustrated antiferromagnets and higher-dimensional fermion systems a problem arises, namely the negative-sign problem, which is described in the next section.

Once we obtain a set of data of the classical systems with a fixed temperature and various Trotter numbers, the Trotter extrapolation to the limit $n \to \infty$ yields data of the quantum system. We have to note that Trotter numbers must be large enough to ensure a sufficient convergence.

In order to improve the convergence for $n \to \infty$, the following method has been employed sometimes. The correction term depends on the way we decompose the Hamiltonian. We expect that the coefficient of the correction term $1/n^2$ of the approximant (21) is smaller in the case of the decomposition

$$A \equiv - \sum_{i=0}^{N/2m-1} \sum_{j=0}^{m-1} \sigma_{mi+j} \cdot \sigma_{mi+j+1}$$

$$\text{and} \quad B \equiv - \sum_{i=0}^{N/2m-1} \sum_{j=m}^{2m-1} \sigma_{mi+j} \cdot \sigma_{mi+j+1} \tag{29}$$

than in the case of the decomposition (22). We thus exactly treat quantum fluctuations inside the cluster of size m. This approach was first introduced in the quantum transfer-matrix method [39, 40, 41], which will be described in Section 5. We refer to this approach as the cluster-transfer-matrix method. Okabe and Kikuchi [42] first applied a decomposition of this type to a Monte Carlo study of the one-dimensional Heisenberg model. Nomura [43, 44] also studied $S = 1$ models by simulations based on this approach.

4   Negative-sign problem

In this section we review the negative-sign problem, which is one of the most serious problems in the quantum Monte Carlo method.

The first difficulty comes from the fact that the Boltzmann weight (24) becomes negative for some configurations $\{\sigma\}$ when we treat frustrated antiferromagnets or fermion systems. We can solve this difficulty using the reweighting method as is shown below. If it is solved, however, the statistics of the obtained Monte Carlo data deteriorates at low temperatures; this is the second difficulty of the negative-sign problem, which has not been resolved so far.

The problem appears when the spin system is frustrated, that is, when the number of antiferromagnetic bonds around a plaquette of the lattice is odd.

Consider, as an example, the Heisenberg antiferromagnet on a triangle: $\mathcal{H} = \sigma_1 \cdot \sigma_2 + \sigma_2 \cdot \sigma_3 + \sigma_3 \cdot \sigma_1$. We use the approximant

$$e^{-\beta\mathcal{H}} \simeq \left(e^{-\beta\sigma_1\cdot\sigma_2/n} e^{-\beta\sigma_2\cdot\sigma_3/n} e^{-\beta\sigma_3\cdot\sigma_1/n} \right)^n. \tag{30}$$

The procedure explained in the previous section gives the partition function (23) with

$$W\{\sigma\} \equiv \prod_{l=0}^{n} w(1, 3l)w(2, 3l+1)w(3, 3l+2). \tag{31}$$

In the present antiferromagnetic case, some local weights become negative: $w < 0$. These weights correspond to the configurations where a world line goes across a square of the four-body interaction. The total Boltzmann weight W can be negative; when $n = 1$, the two configurations depicted in Fig. 2 have the negative weights $W < 0$.

This is the first difficulty in the negative-sign problem; the importance sampling method is not applicable to a system if some of the Boltzmann weights of the system are negative. The Boltzmann weight of each configuration is proportional to the number of times when the configuration is generated in simulations, and needs to be positive.

The problem does not appear when the lattice of the spin system is bipartite, for example, square, honeycomb, or cubic. Local configurations which give negative Boltzmann weights always appear an even number of times, and the total weight is thus positive. This fact is confirmed as follows [33]. Consider the antiferromagnetic Hamiltonian on a bipartite lattice, $\mathcal{H} = \sum_{\langle i,j \rangle} \sigma_i \cdot \sigma_j$, and consider a decomposition similar to (8)-(10). We define the following unitary transformation U: for spins on a sublattice A, we rotate the spin space around the z axis by angle π, while we leave

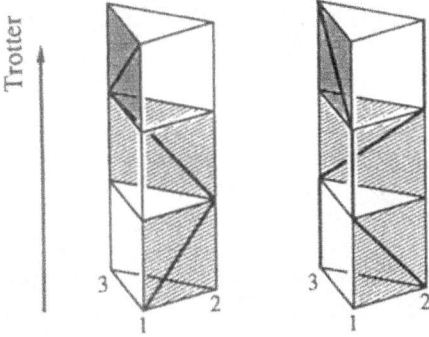

Figure 2: World-line configurations of the triangular antiferromagnet. These configurations have negative Boltzmann weights.

spins on the other sublattice B as they are. After we apply this unitary transformation, the local Boltzmann factor is always positive, and hence the total Boltzmann weight is also positive. It is impossible to define such a unitary transformation on frustrated lattices.

The negative-sign problem also appears when we apply the world-line approach to the two- and higher-dimensional Hubbard models. The following matrix element, for example, gives a negative Boltzmann factor owing to the anti-commutation relation of fermions:

$$\left\langle \uparrow,0,\uparrow,0 \left| e^{\beta t(c_{1\uparrow}^\dagger c_{2\uparrow}+c_{2\uparrow}^\dagger c_{1\uparrow})/n} e^{\beta t(c_{3\uparrow}^\dagger c_{4\uparrow}+c_{4\uparrow}^\dagger c_{3\uparrow})/n} \right| 0,\uparrow,0,\uparrow \right\rangle < 0, \tag{32}$$

where the base kets are defined by $|\uparrow,0,\uparrow,0\rangle \equiv c_{1\uparrow}^\dagger c_{3\uparrow}^\dagger |0\rangle$ and $|0,\uparrow,0,\uparrow\rangle \equiv c_{2\uparrow}^\dagger c_{4\uparrow}^\dagger |0\rangle$ with the vacuum state $|0\rangle$. The negative sign of the weights, in this case, results from the anti-commutation relation of fermion operators.

The naive solution of the negative-sign problem is to use the following identity [19, 45]:

$$\frac{\sum_{\{\sigma\}} Q\{\sigma\}W\{\sigma\}}{\sum_{\{\sigma\}} W\{\sigma\}} = \frac{\sum_{\{\sigma\}} Q\{\sigma\}R\{\sigma\}W'}{\sum_{\{\sigma\}} W'} \left/ \frac{\sum_{\{\sigma\}} R\{\sigma\}W'}{\sum_{\{\sigma\}} W'} \right. . \tag{33}$$

Here each $W'\{\sigma\}$ is the absolute value of the Boltzmann weight $W\{\sigma\}$, $W'\{\sigma\} \equiv |W\{\sigma\}|$, and $R\{\sigma\}$ is the sign of it, namely

$$R\{\sigma\} \equiv \text{sign}(W\{\sigma\}). \tag{34}$$

Hence we have $W = RW'$, which yields (33).

A quantity Q is thereby measured in simulations of the system defined by the new Boltzmann weight W':

$$\langle Q \rangle = \langle QR \rangle'_{\text{MCS}} \left/ \langle R \rangle'_{\text{MCS}}, \right. \tag{35}$$

where $\langle\cdots\rangle'_{\text{MCS}}$ stands for the Monte Carlo average in simulations of the system W'. The denominator of (35),

$$\langle R\rangle'_{\text{MCS}} = \frac{\sum_{\{\sigma\}} W}{\sum_{\{\sigma\}} |W|}, \tag{36}$$

is often called the negative-sign ratio.

The first difficulty of the negative-sign problem is thereby solved. Nevertheless this solution is exactly the point from which the second difficulty of the problem comes; a statistical error of the negative-sign ratio grows rapidly at low temperatures or for large systems.

This fact is can be explained as follows. The numerator of (36) is, in the limit $n \to \infty$, the partition function of the quantum system, Z_q, while the denominator of (36) converges to the partition function of another quantum system, Z'_q [46]. Thus we have $Z_q = e^{-\beta F}$ and $Z'_q = e^{-\beta F'}$, where F (F') stands for the free energy of the system Z_q (Z'_q). The negative-sign ratio is given by

$$\langle R\rangle'_{\text{MCS}} = e^{-\beta(F-F')} \qquad \text{as} \quad n \to \infty. \tag{37}$$

The value $\langle R\rangle'_{\text{MCS}}$ is bounded from above by unity, and therefore we have $F > F'$.

First, at low temperatures we can replace F and F' by the ground-state energies E_g and E'_g of the systems Z_q and Z'_q, respectively, and hence [47, 48, 11, 46]

$$\langle R\rangle'_{\text{MCS}} \simeq e^{-\beta(E_g-E'_g)} \qquad \text{as} \quad \beta \to \infty. \tag{38}$$

Second, assuming the extensivity of the free energy, we have [19]

$$\langle R\rangle'_{\text{MCS}} \simeq e^{-\beta(f-f')N} \qquad \text{as} \quad N \to \infty, \tag{39}$$

where N is the system size and f (f') is the free-energy density of the system Z_q (Z'_q).

The statistical error of the negative-sign ratio is given by

$$\Delta\langle R\rangle'_{\text{MCS}} \propto M^{-1/2}\sqrt{\langle R^2\rangle'_{\text{MCS}} - \langle R\rangle'^2_{\text{MCS}}} = M^{-1/2}\sqrt{1 - \langle R\rangle'^2_{\text{MCS}}}, \tag{40}$$

where M denotes the number of Monte Carlo steps. The relative error of the negative-sign ratio behaves as

$$\frac{\Delta\langle R\rangle'_{\text{MCS}}}{\langle R\rangle'_{\text{MCS}}} = M^{-1/2}\sqrt{\langle R\rangle'^{-2}_{\text{MCS}} - 1} \simeq M^{-1/2} \times \begin{cases} e^{\beta(E_g-E'_g)} \gg 1 & \text{as} \quad \beta \to \infty, \\ e^{\beta(f-f')N} \gg 1 & \text{as} \quad N \to \infty. \end{cases} \tag{41}$$

At low temperatures or for large systems, the statistical error grows rapidly, and hence we cannot precisely estimate the negative-sign ratio and physical quantities (35) by Monte Carlo simulations.

The negative-sign problem has been the main obstacle to Monte Carlo studies of frustrated quantum spin systems, e.g., of antiferromagnets on triangular lattices and on square lattices with next-nearest-neighbor interactions (the J_1-J_2 model).

Though the problem has not been completely resolved, an approach relieving the difficulty has been proposed [49, 50, 51] with the use of the so-called reweighting method, which has been employed in simulations of classical systems. The formula (33) holds for any set of W' when we change the definition of R from (34) to

$$R\{\sigma\} \equiv W\{\sigma\} \big/ W'\{\sigma\}. \tag{42}$$

We thus can take arbitrary W' instead of the specific choice $W' = |W|$ as long as the phase space of W' is equal to or larger than that of W.

From the viewpoint of the reweighting method, the negative-sign problem is explained as follows. Although the formula (33) is correct in the limit of an infinite number of Monte Carlo steps, it is known from simulations of classical systems that the formula may yield systematic and statistical errors in its actual use [52]. The errors may appear when the distribution of W in the phase space is considerably different from that of W'. This is actually the case in the negative-sign problem. Configurations sampled in the simulations of $W' = |W|$ are counted almost in vain.

We anticipate that the growth rate of the error, (41), is reduced when we take W' whose structure is more similar to that of W than to that of $|W|$. The method has been applied to the J_1-J_2 model [49, 50, 51] and, indeed, the data are improved at least in the region where frustration is not too strong.

In another approach to the negative-sign problem, the location of the nodal surface of the Boltzmann weight $W\{\sigma\}$ in the phase space $\{\sigma\}$ is assumed [53, 54]. This approach may be related to the fixed-node approximation in the diffusion Monte Carlo methods [55, 56].

Recently, the measurement of quantities in the ground state using the behavior (38) has been proposed. The ground-state energy of the positive-weight system, E'_g, can be measured without any significant statistical errors. On the other hand, owing to the behavior $\langle R \rangle'_{\text{MCS}} \propto \exp[-\beta(E_g - E'_g)]$, we may obtain the energy difference $(E_g - E'_g)$ analyzing the temperature dependence of the negative-sign ratio; see Fig. 3. Though the statistical error of each data point may be large, the combination of them may yield an accurate estimate of $E_g - E'_g$. Thus we can obtain the estimate of E_g [57].

To measure other quantities, the correction-ratio method may be applicable [58]. We can estimate rather precisely the quantity $\langle Q \rangle'_{\text{MCS}}$, which is measured with the factor R neglected. We can correct the estimate to obtain $\langle Q \rangle_{\text{MCS}}$ by the correction-ratio method.

The "ratio correction" is defined by

$$Q_{\text{RC}} \equiv \frac{\sum QW}{\sum QW'} = \frac{\sum QRW'}{\sum W'} \Big/ \frac{\sum QW'}{\sum W'} = \frac{\langle QR \rangle'_{\text{MCS}}}{\langle Q \rangle'_{\text{MCS}}}. \tag{43}$$

Comparing (43) with (33), we have

$$\langle Q \rangle_{\text{MCS}} = \langle Q \rangle'_{\text{MCS}} \frac{Q_{\text{RC}}}{\langle R \rangle'_{\text{MCS}}}. \tag{44}$$

$$\log \langle R \rangle'_{\text{MCS}} \, , \, \log Q_{\text{RC}}$$

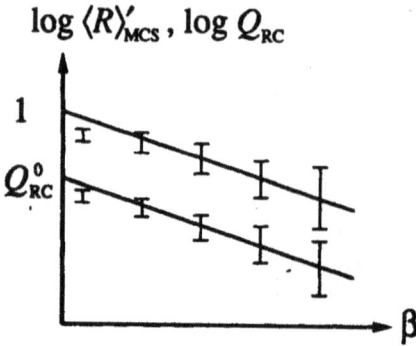

Figure 3: The temperature dependence of the negative-sign ratio, (38), and the ratio correction, (45).

If both of the quantities $\langle Q \rangle_{\text{MCS}}$ and $\langle Q \rangle'_{\text{MCS}}$ are finite at the zero temperature, the ratio correction must behave as

$$Q_{\text{RC}} \simeq Q_{\text{RC}}^0 e^{-\beta(E_{\text{t}} - E'_{\text{t}})} \qquad \text{as} \quad \beta \to \infty \tag{45}$$

to cancel the behavior of the denominator of (44), $\langle R \rangle'_{\text{MCS}}$; see Fig. 3. Hence we have

$$\langle Q \rangle_{\text{MCS}} \simeq \langle Q \rangle'_{\text{MCS}} \, Q_{\text{RC}}^0 \qquad \text{as} \quad \beta \to \infty. \tag{46}$$

If the asymptotic behavior (45) can be observed even at high temperatures (which is actually the case in [58]), we can avoid simulations at low temperatures, and hence suppress the statistical error.

The combination of the reweighting method and the correction-ratio method may be quite useful [51].

5   Quantum transfer-matrix method

In this section we describe the quantum transfer-matrix method which is based on the ST decomposition.

In (23) we rewrote the density matrix of a quantum system $e^{-\beta \mathcal{H}}$ as a product of matrices, using the ST decomposition. In Section 3 we described a method of multiplying the matrices by a Monte Carlo simulation. If the linear dimension of the matrices is small enough for a computer memory, their multiplication can be performed as it is [59, 60].

In other words, we express the partition function of the classical system in Fig. 4 in terms of a transfer matrix as

$$Z_n = \text{Tr}\,(T_1 T_2)^n, \tag{47}$$

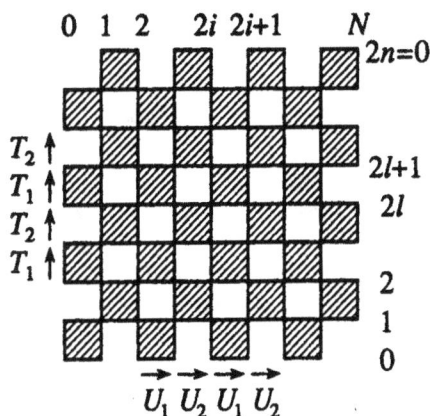

Figure 4: The real-space transfer matrices T_1 and T_2, and the virtual-space transfer matrices U_1 and U_2.

where T_1 and T_2 denote the matrices

$$T_1 = \left\langle \{\sigma\} \left| e^{-\beta A/n} \right| \{\sigma\}' \right\rangle \quad \text{and} \quad T_2 = \left\langle \{\sigma\} \left| e^{-\beta B/n} \right| \{\sigma\}' \right\rangle. \tag{48}$$

The matrices transfer a configuration of spins in a row to the one in the next row, namely in the Trotter direction. The linear dimension of the matrices is equal to the dimension of the Hilbert space; 2^{N+1} in the case of (19).

An important feature is aging the locality of the interaction as was already mentioned below the formula (25). Because of the locality, the transfer matrices are expressed as direct products of small matrices as in (26). We apply the matrices (27) to a vector successively; hence we do not have to store all matrix elements of T_1 and T_2. The size of the vector, when we treat an $S = 1/2$ system with 25 spins, is about 32×10^6. If we prepare an 8-byte real number for each component of the vector, the size of the memory storage is 256 megabytes, which is actually manageable on recent large-scale computers.

Note that the transfer matrix is explicitly block-diagonalized owing to the conservation law (28). Each sector is labeled with the total magnetization of spins in a row, or in other words, with a number of world lines. A state in a sector does not mix with states in the other sectors. Therefore the application of the transfer matrix to a state vector can be carried out in each sector separately. This aspect of the matrix is useful in reducing the size of the memory storage and computational time.

The maximum size manageable by this method above (which is called the real-space transfer-matrix method by contrast with the method described below) is less than 30 spins so far. An interesting reformulation of the transfer-matrix method was proposed [60, 61, 62] in order to treat much larger systems. The method is called the virtual-space transfer-matrix method. We describe the system in Fig. 4 in terms of matrices which transfer a configuration on a column to the one on the next column,

i. e. in the space direction:

$$Z_n(N) = \text{Tr}\,(U_1 U_2)^{N/2}. \tag{49}$$

Here we have assumed the periodic boundary condition $\sigma_N = \sigma_0$. We call U_1 and U_2 virtual-space transfer matrices; see Fig. 4.

Each transfer matrix is again expressed as a direct product of small matrices. We derive the matrices from the matrix (27) changing the ordering of the matrix elements.

The transfer matrices U_1 and U_2 are explicitly block-diagonalized into sectors. This fact can be understood as follows. If we redefine the signs of the spins on the $(l+1)$th column [63, 64], that is,

$$\sigma_{i,l+1} \longrightarrow \sigma'_{i,l+1} \equiv -\sigma_{i,l+1} \qquad \text{for all } \ i, \tag{50}$$

then the matrices U_1 and U_2 conserve the staggered magnetization. This conservation rule is derived from the original conservation law (28) as follows:

$$\sigma_{i,l} - \sigma_{i,l+1} = -\sigma_{i+1,l} + \sigma_{i+1,l+1} = \sigma'_{i+1,l} - \sigma'_{i+1,l+1}. \tag{51}$$

Hence each sector of transfer matrices is labeled with the total staggered magnetization of spins on a column.

The virtual-space transfer-matrix method is convenient especially in studying thermodynamic properties of long chains. Computational time grows only proportionally to the system size, and the size of a necessary memory storage does not depend on the system size. (The size of the storage now depends on the Trotter number. We have to store a configuration of spins on a column.) Many one-dimensional systems have been thus studied. Takada and Kubo investigated critical properties of $S = 1/2$ XXZ chains [65] and $S = 1$ XXZ chains [66] extensively. Many other studies of the latter model have been reported [67, 68, 69] in connection with the Haldane problem [70, 71].

A remarkable feature of the virtual-space transfer-matrix method emerges when we consider the thermodynamic limit of the quantum system. Several eigenvalues of the virtual-space transfer matrix describe the thermodynamic behavior of the system.

The feature comes from the following interchangeability theorem [60, 72]. The free-energy density of the quantum N-spin system in the thermodynamic limit is written in the form

$$f_q = -\lim_{N \to \infty} \lim_{n \to \infty} (N\beta)^{-1} \log \text{Tr}\,(U_1 U_2)^{N/2}. \tag{52}$$

The interchangeability theorem states that we can exchange the order of the two limit operations $N \to \infty$ and $n \to \infty$. In the limit $N \to \infty$, the partition function converges to

$$\lim_{N \to \infty} \text{Tr}\,(U_1 U_2)^{N/2} = \Lambda_1^N, \tag{53}$$

where the maximum eigenvalue of the matrix U_1U_2 is defined as Λ_1^2. The free-energy density (53) is thus written as

$$f_q = -\beta^{-1} \lim_{n\to\infty} \log \Lambda_1. \tag{54}$$

It is remarkable that a single eigenvalue gives a thermodynamic quantity.

The virtual-space transfer-matrix method yields still another useful formula, namely for the correlation length. A spin-spin correlation function $\langle \sigma_i^\alpha \sigma_{i+L}^\alpha \rangle$ in the thermodynamic limit is written in terms of the virtual-space transfer matrix as follows:

$$\left\langle \sigma_i^\alpha \sigma_{i+L}^\alpha \right\rangle = \lim_{N\to\infty} \lim_{n\to\infty} \frac{1}{Z_n(N)} \text{Tr}\, \sigma^\alpha (U_1U_2)^{L/2} \sigma^\alpha (U_1U_2)^{(N-L)/2}. \tag{55}$$

After exchanging the limit operations, the correlation length is given by

$$\xi^{-1} \equiv -\lim_{L\to\infty} L^{-1} \log \left\langle \sigma_i^\alpha \sigma_{i+L}^\alpha \right\rangle = \lim_{n\to\infty} \log (\Lambda_1/\Lambda_2), \tag{56}$$

where the second-largest eigenvalue of the matrix U_1U_2 is defined as Λ_2^2.

Suzuki and Inoue [72, 73] first pointed out that it is possible to treat quantum systems analytically on the basis of the formulae (54) and (56). This approach has been further developed as the thermal Bethe-ansatz method by several researchers [63, 64, 74, 75, 76]. They obtained thermal Bethe-ansatz equations analytically for one-dimensional spin-1/2 models and for one-dimensional Hubbard models with a finite Trotter number, and solved them numerically. As for the XYZ model without a magnetic field and the XXZ model with a magnetic field, Takahashi [77, 78] obtained the equations in the infinite-Trotter-number limit, and thereby calculated the temperature dependence of the free energy and the correlation length numerically rigorously.

6 Monte Carlo power method

In this section we describe a Monte Carlo power method, by which we can numerically obtain the maximum eigenvalue of a transfer matrix.

Analytic solutions of the virtual-space transfer matrix, which are mentioned in the preceeding section, are not available for higher-spin chains. Hence we have to calculate the eigenvalues of the matrix numerically.

The most straightforward method of obtaining the maximum eigenvalue of an asymmetric matrix T is the power method; we calculate the Rayleigh quotient,

$$\lambda(N) \equiv \left\langle \phi \left| T^{N+1} \right| \phi \right\rangle \Big/ \left\langle \phi \left| T^N \right| \phi \right\rangle, \tag{57}$$

and then take the limit $N \to \infty$. If the test vector $|\phi\rangle$ and the eigenvector belonging to the maximum eigenvalue Λ_1 have an overlap, the eigenvalue emerges in the form

$\lambda(N) = \Lambda_1 + O(e^{-N/\xi})$. The correction term disappears rapidly. Thus the free-energy density (54) is given by

$$f_{\mathrm{q}} = -\frac{1}{\beta}\lim_{n\to\infty}\lim_{N\to\infty}\log\frac{\left\langle\phi\left|(U_1U_2)^{N/2+1}\right|\phi\right\rangle}{\left\langle\phi\left|(U_1U_2)^{N/2}\right|\phi\right\rangle}. \tag{58}$$

The problem here is the following: when we increase the Trotter number, the size of a necessary memory storage grows rapidly especially in higher-spin cases. Hence the Trotter number attainable so far is less than 10 even for $S = 1$. At temperatures below $T \leq 0.1$, the convergence of the numerical data to the infinite-Trotter-number limit may not be satisfactory.

A solution of this problem is to use the Monte Carlo power method [79] described below, by which we can estimate the Rayleigh quotient. The essential idea is a special choice of the test vector $|\phi\rangle$:

$$|\phi\rangle = \sum_{\{\sigma\}_0}|\{\sigma\}_0\rangle, \tag{59}$$

where the symbol $\{\sigma\}_i$ denotes a configuration of $2n$ spins on the ith column, that is, $\{\sigma\}_i \equiv \{\sigma_{i,0},\ldots,\sigma_{i,2n-1}\}$. In other words, we define the test vector as a superposition of all the members of the orthonormal set.

Employing the choice (59), we interpret the denominator of the Rayleigh quotient as

$$\left\langle\phi\left|(U_1U_2)^{N/2}\right|\phi\right\rangle = \sum_{\{\sigma\}}\langle\{\sigma\}_0|U_1|\{\sigma\}_1\rangle\cdots\left\langle\{\sigma\}_{N-1}|U_2|\{\sigma\}_N\right\rangle \equiv \sum_{\{\sigma\}}W\{\sigma\}, \tag{60}$$

where each vector denotes a configuration of spins on a column. We thus identify (60) with the partition function of the classical system in Fig. 4. On the boundaries of the system, any configurations $\{\sigma\}_0$ and $\{\sigma\}_N$ can appear because of the choice (59). Hence, the *free boundary conditions* are required for $i = 0$ and $i = N$.

The numerator of the Rayleigh quotient is interpreted as follows:

$$\left\langle\phi\left|(U_1U_2)^{N/2+1}\right|\phi\right\rangle = \sum_{\{\sigma\}}W\{\sigma\}Q_{\mathrm{eig}}\{\sigma\}, \tag{61}$$

where

$$Q_{\mathrm{eig}}\{\sigma\} \equiv \sum_{\substack{\{\sigma\}_{N+1}\\\{\sigma\}_{N+2}}}\left\langle\{\sigma\}_N|U_1|\{\sigma\}_{N+1}\right\rangle\left\langle\{\sigma\}_{N+1}|U_2|\{\sigma\}_{N+2}\right\rangle. \tag{62}$$

The Rayleigh quotient hence can be written in the form

$$\lambda(N) = \sum_{\{\sigma\}}Q_{\mathrm{eig}}W\Big/\sum_{\{\sigma\}}W = \langle Q_{\mathrm{eig}}\rangle_{\mathrm{MCS}}. \tag{63}$$

In a Monte Carlo simulation of the system $W\{\sigma\}$, a direct measurement of the quantity Q_{eig} at the end $i = N$ yields an estimate of the Rayleigh quotient. We can improve statistics measuring a similar quantity at the other end $i = 0$.

It is notable that we can calculate the free energy of the system; the ordinary Monte Carlo methods with importance sampling do not yield the free energy.

When the second-largest eigenvalue of a transfer matrix is the largest eigenvalue of a block-diagonalized sector, the restriction of the summation in (59) enables us to obtain the eigenvalue. Thus we can estimate the correlation length (56) as well. This feature has been utilized in a study of the $S = 2$ spin chain [80].

7 Auxiliary-field approach

In this section we review the auxiliary-field approach, which has been developed particularly for lattice fermion problems; for example the Hubbard model (2).

The auxiliary-field approach has been devised in order to circumvent the difficulty of the negative-sign problem of the Hubbard model, which we mentioned in Section 4. Particularly in the half-filled-band case on a bipartite lattice, we can resolve the difficulty by means of the auxiliary-field approach.

The auxiliary-field approach, in particular the grand-canonical Monte Carlo method, consists of the following four steps [81, 82]. (i) We approximate the density matrix with the Trotter approximation decomposing the Hamiltonian into K and V. (ii) We introduce an auxiliary Bose field to break up the Coulomb interaction between fermions (the Stratonovich-Hubbard transformation). (iii) We trace out the fermion degrees of freedom. The partition function becomes a sum with respect to the Bose-field degrees of freedom. (iv) We replace the sum by Monte Carlo summation using the importance sampling.

In the first step we approximate the density matrix with the Trotter approximant decomposing the Hamiltonian into two parts, K and V:

$$e^{-\beta \mathcal{H}} \simeq \left(e^{-\beta K/n} e^{-\beta V/n}\right)^n. \tag{64}$$

The second step is to make use of the Stratonovich-Hubbard transformation. In the original form, the transformation is nothing but the inverse of the Gaussian integral [83, 84]. Hirsch devised another version of the transformation using a discrete Bose field, or an Ising variable [85]:

$$\exp\left[-xU\left(n_\uparrow - \frac{1}{2}\right)\left(n_\downarrow - \frac{1}{2}\right) + \frac{xU}{4}\right] = \frac{1}{2}\sum_{s=\pm1} e^{xJs(n_\uparrow - n_\downarrow)} = \frac{1}{2}\sum_{s=\pm1}\prod_{\sigma=\uparrow,\downarrow} e^{xJ\sigma s n_\sigma}, \tag{65}$$

where the coefficient x is defined by $\cosh xJ = e^{xU/2}$. The readers can confirm the identity (65) easily by checking it in the four cases $(n_\uparrow, n_\downarrow) = (0,0),(0,1),(1,0),(1,1)$.

The auxiliary Ising field s plays a role of a mean field. Consider the case that an upward-spin electron occupies a site, $n_\uparrow = 1$. Then the system favors the state $s = 1$ because of the coupling $sn_\uparrow$, and hence tends to refuse occupation of the site by a downward-spin electron because of the coupling $-sn_\downarrow$. The field s is likely to point in the same direction as the spin of an electron at the site.

Applying the transformation (65) to (64), we express the partition function of the system in the form:

$$Z_q \simeq Z_n = A^{-1} \sum_{\{s\}} W_\uparrow\{s\} W_\downarrow\{s\},$$ (66)

where

$$W_\sigma\{s\} \equiv \mathrm{Tr} \prod_{l=1}^{n} \left(e^{-\beta K_\sigma/n} e^{-\beta \tilde{V}_\sigma\{s\}_l/n} \right)$$ (67)

with a new potential energy

$$\tilde{V}_\sigma\{s\}_l \equiv -\sum_i (J\sigma s_{il} + \mu) n_{i\sigma},$$ (68)

and a coefficient $A \equiv 2^{Nn} e^{\beta UN/4}$. Note that $\{s\}$ consists of Nn pieces of Ising variables, that is, $\{s_{il}\}$ for $i = 1, 2, \ldots, N$ and $l = 1, 2, \ldots, n$.

We have described the system as the one where boson fields and fermions are interacting. A Monte Carlo method of studying such systems has been already proposed [86, 87, 88]. We have only to apply the formalism to the present problem.

The third step is to carry out the trace operation in (67). This is possible because the operators K_σ and $\tilde{V}_\sigma$ are written in quadratic forms with respect to fermion operators:

$$- \beta K/n = c_i^\dagger \overline{K}_{ij} c_j \quad \text{and} \quad - \beta \tilde{V}/n = c_i^\dagger \overline{V}_{ij} c_j,$$ (69)

where $\overline{K}$ and $\overline{V}$ are $N \times N$ matrices defined so as to reproduce K and (68). Summation over repeated indices is assumed in (69). (For the time being we leave out the subscript σ.) Note that the potential energy $\tilde{V}$ depends on the auxiliary fields $\{s\}_l$, and hence $\overline{V}$ also depends on them. (Hereafter the bar over a symbol indicates an $N \times N$ matrix.)

The trace operation yields [87, 82, 89]

$$W = \mathrm{Tr}\, e^{c_i^\dagger \overline{H}_{ij} c_j} = \det \left[\overline{I} + e^{\overline{H}} \right],$$ (70)

where $\overline{I}$ denotes the $N \times N$ identity matrix, and the $N \times N$ matrix $\overline{H}$ is defined by

$$\prod_{l=1}^{n} \left(e^{\overline{K}} e^{\overline{V}} \right) = e^{\overline{H}}.$$ (71)

After the third step we regard (66) as the partition function of a classical system with Nn Ising spins interacting through certain long-range interactions.

The fourth step is to carry out the Monte Carlo simulation. Consider a flip of a spin on the nth layer: $s_{jn} \longrightarrow s'_{jn} = -s_{jn}$. In order to perform a simulation, we have to calculate the ratio $R = W'/W$, where W' denotes the weight calculated for a new spin configuration. After some algebra, we have

$$R = 1 + \gamma \left(1 - \overline{G}_{jj} \right).$$ (72)

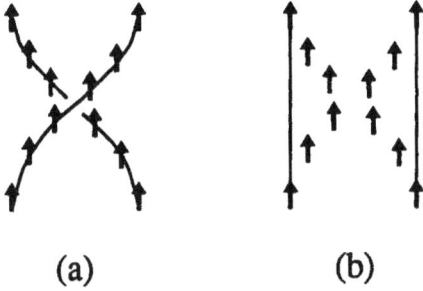

Figure 5: The auxiliary-field configurations (arrows) and the fermion world-line configurations (thick lines). (a) This configuration has a negative Boltzmann weight. (b) This configuration has a positive Boltzmann weight. The contributions of both of the configurations are summed up when we trace out the fermion degrees of freedom.

Here $\overline{G}$ denotes the equal-time Green's function, which is expressed by the following simple formula [87, 82]:

$$\overline{G} = \frac{1}{W}\mathrm{Tr}\, c_k c_l^\dagger e^{c_i^\dagger \overline{H}_{ij} c_j} = \left(\overline{I} + e^{\overline{H}}\right)^{-1}. \tag{73}$$

The parameter γ in (72) is defined by

$$\gamma + 1 \equiv \exp\left(\frac{-2\beta J \sigma s_{jn}}{n}\right). \tag{74}$$

The probability that s_{jn} is flipped in a step of the simulation is given by $R/(1+R)$. Note that, except in the half-filled case, the value R can be negative; we must replace R by $|R|$ as was done in Section 4.

We notice that it is convenient to keep the values $\{\overline{G}\}$ in a memory storage. After we update a spin s_{jn}, we have to update the values of $\overline{G}$ as well: $\overline{G}\{s\} \longrightarrow \overline{G}' \equiv \overline{G}\{s\}'$. We can construct $\overline{G}'$ from $\overline{G}$ in a form $\overline{G}' = \overline{C}\,\overline{G}$ with a sparse matrix $\overline{C}$. This completes the algorithm.

We can expect to ease the negative-sign problem in this scheme because of the following reason. Consider the auxiliary-field configuration shown in Fig. 5. As was discussed above, an Ising field at a site and the spin of an electron at the site tend to point in the same direction. When two upward-spin electrons exist, the most favored world-line configuration is the one shown in Fig. 5 (a) [90]. This configuration gives a negative weight as was described in (32). In the auxiliary-field approach, however, the fermion degrees of freedom are traced out in (67); in other words, many other world-line configurations are summed up. For example, the world-line configuration in Fig. 5 (b), with the auxiliary-fields fixed, gives a positive weight. We can expect that cancellation among all the world-line configurations results in reduction of the negative weight.

In the half-filled case, in particular, the negative-sign problem disappears. The signs of the weight factors $W_\uparrow$ and $W_\downarrow$ are the same, and hence the product of them is always non-negative [81, 82].

In addition, there is no restriction on measurable quantities in the present method. In the world-line approach, on the other hand, there are some quantities that are hard to measure [19].

The computational time for one Monte Carlo step depends on the size of the system as $N^3 n$, which comes from the fact that the interaction between Ising fields is of long range. This size-dependence is a drawback of the method: compare it with the dependence proportional to Nn in the world-line approach.

At low temperatures, besides the negative-sign problem, we face another problem, namely the numerical instabilities. Calculating the matrix e^H in (71), we have to multiply many exponential operators. At low temperatures, each operator has a wide spectrum. As we multiply the operators, we have to add very small numbers to very large numbers. Such a summation often results in a numerical error.

This problem has been overcome by repeating the orthogonalization [91, 92]. We execute the Gram-Schmidt orthogonalization of the exponential operators; thus we can sum up the small-number contributions separating them from the large-number contributions.

The auxiliary-field approach has been successfully applied to the half-filled Hubbard model [81, 82, 93, 94, 95], and to the attractive-interaction Hubbard model [96]. In both cases the negative-sign problem does not appear.

The approach has been applied even to the non-half-filled Hubbard model [97, 92, 98]. The argument for the temperature dependence of the negative-sign problem (38) has been developed in the studies of the model [48, 57, 99].

The method is also extended to such an algorithm [100, 101, 102] as is applicable to the t-J model.

8   Ground-state algorithms

We have already introduced Monte Carlo algorithms for the studies of finite-temperature properties. In the present section we review algorithms for the studies of ground-state properties.

The study of ground-state phase transitions driven by quantum fluctuation is now one of the most interesting problems in condensed-matter physics. In this problem, we define the critical points in a parameter space of the relevant model as those at which an energy gap from the ground state vanishes. At these critical points, correlation functions of the system show a power-law behavior instead of the exponential decay, and the correlation length diverges.

When we investigate ground-state phase transitions using quantum Monte Carlo methods, however, there arises a problem that it is difficult to take the zero-temperature limit. We examine the relevant system first at finite temperatures, and take the zero-temperature limit to extract a ground-state property. This can be performed practically at temperatures low enough to satisfy the inequality $T \ll \Delta E$. Here

ΔE denotes the energy gap just above the ground state. Near phase boundaries the energy gap is narrow, and hence it is difficult to satisfy the inequality.

Moreover, when we treat frustrated spin systems or fermion systems, the negative-sign problem becomes quite serious at low temperatures.

Some attempts to circumvent the problems have been proposed [103, 104, 105, 106, 107]. They are essentially based on the following formula:

$$e^{-\beta \mathcal{H}}|\psi\rangle \longrightarrow e^{-\beta E_g}|\psi_g\rangle \qquad \text{as} \quad \beta \to \infty, \tag{75}$$

where $|\psi_g\rangle$ is the ground-state vector, E_g is the ground-state energy, and $|\psi\rangle$ is a trial vector which is not orthogonal to $|\psi_g\rangle$. If the trial vector is orthogonal to the first excited state, the extraction of the ground state may be performed even at rather high temperatures (though the variable $T = 1/\beta$ now loses its meaning as the temperature). A quantity in the ground state is given by

$$\langle Q \rangle_g \equiv \langle \psi_g |Q| \psi_g \rangle = \lim_{\beta \to \infty} \langle\!\langle Q \rangle\!\rangle_\beta \tag{76}$$

with

$$\langle\!\langle Q \rangle\!\rangle_\beta \equiv \left\langle \psi \left| e^{-\beta \mathcal{H}} Q e^{-\beta \mathcal{H}} \right| \psi \right\rangle \Big/ \left\langle \psi \left| e^{-2\beta \mathcal{H}} \right| \psi \right\rangle. \tag{77}$$

In some methods the limit procedure $\beta \to \infty$ itself is simulated, by interpreting multiplications of a transfer matrix as a Markov process. We describe these methods in Section 9. In the present section we fix the parameter β to estimate quantities by a simulation. Analyzing the data we evaluate the quantities in the ground state. We describe an application of the world-line approach (Section 3) and the auxiliary-field approach (Section 7) to the calculation of (76).

First we describe the application of the world-line approach [107].

Following the standard procedure [21, 16, 17], we decompose the density matrix. We should keep the Trotter number n larger than β to eliminate the correction term. We express operators in terms of the orthonormal set of states which diagonalizes $\{\sigma_i^z\}$. At the same time we expand the trial function with respect to the base as

$$|\psi\rangle = \sum F(\{\sigma\}) |\{\sigma\}\rangle. \tag{78}$$

Thus we interpret the denominator of (77) as the partition function of an Ising-spin system of the checkerboard type [19]. The partition function is given by [107]

$$\left\langle \psi \left| e^{-2\beta \mathcal{H}} \right| \psi \right\rangle \simeq Z_n \equiv \sum F(\{\sigma\}_0) W(\{\sigma\}_0, \{\sigma\}_1, \ldots, \{\sigma\}_{4n+1}) F(\{\sigma\}_{4n+1}). \tag{79}$$

where

$$
\begin{aligned}
& W(\{\sigma\}_0, \{\sigma\}_1, \{\sigma\}_2, \ldots, \{\sigma\}_{4n+1}) \\
& \equiv \left\langle \{\sigma\}_0 \left| e^{-\beta A/(2n)} \right| \{\sigma\}_1 \right\rangle \cdots \left\langle \{\sigma\}_{4n} \left| e^{-\beta A/(2n)} \right| \{\sigma\}_{4n+1} \right\rangle.
\end{aligned}
\tag{80}
$$

On the other hand, in the finite-temperature algorithm described in Section 3, we transform the partition function of the quantum system into that of the corresponding Ising system, (23). This system differs from the system (79) in boundary conditions. In the finite-temperature algorithm we require the periodic-boundary conditions $\{\sigma\}_{2n} = \{\sigma\}_0$. In the formulation (79) we require constrained-boundary conditions on the spins $\{\sigma\}_0$ and $\{\sigma\}_{4n+1}$; a spin configuration $\{\sigma\}_0$ appears with a rate determined by the Boltzmann factor $W(\{\sigma\}_0)$ times the extra factor $F(\{\sigma\}_0)$.

When we perform the importance sampling of (79), we have to flip the boundary spins $\{\sigma\}_0$ and $\{\sigma\}_{4n+1}$, $\{\sigma\} \longrightarrow \{\sigma\}'$, according to the ratio $R = R_1 R_2$, where

$$R_1 \equiv \frac{W(\{\sigma\}'_0)}{W(\{\sigma\}_0)} \quad \text{and} \quad R_2 \equiv \frac{F(\{\sigma\}'_0)}{F(\{\sigma\}_0)}. \tag{81}$$

The factor R_1 can be calculated easily as is shown in Section 3. The factor R_2 can also be calculated for a wide class of trial functions [107]. The other spins are flipped only according to the factor R_1.

Next we describe the application of the auxiliary-field approach to the calculation of (77).

We utilize the Trotter approximation combined with the Stratonovich-Hubbard transformation (65). At the same time we define the trial function as a direct product of one-particle states [106, 97, 91, 92, 98, 108]:

$$|\psi\rangle \equiv \bigotimes_{m=1}^{M} \left(\sum_{k=1}^{N} F_{km} |k\rangle \right), \tag{82}$$

where $|k\rangle \equiv c_k^\dagger |0\rangle$, and M is the number of the fermions.

We thus express (77) in a form similar to (66), that is,

$$\left\langle \psi \left| e^{-2\beta\mathcal{H}} \right| \psi \right\rangle \simeq A^{-1} \sum_{\{s\}} W_\uparrow\{s\} W_\downarrow\{s\}. \tag{83}$$

The Boltzmann factor W_σ reduces to

$$W_\sigma\{s\} = \bigotimes_{m_1=1}^{M} \bigotimes_{m_2=1}^{M} \left(\langle k|\, {}^tF_{m_1 k} \right) \left(\left(e^{\overline{H}} F \right)_{lm_2} |l\rangle \right) = \det\left[{}^tFe^{\overline{H}}F \right], \tag{84}$$

instead of (70) [98]. The product ${}^tF\overline{H}F$ is an $M \times M$ matrix. The equal-time Green's function is given by the following $N \times N$ matrix,

$$\overline{G} = F \left({}^tFe^{\overline{H}}F \right)^{-1} {}^tFe^{\overline{H}}, \tag{85}$$

instead of (73) [98]. We obtain the flip ratio of Boltzmann weights, $R = W'/W$, in the form

$$R = 1 + (\gamma + 1)\overline{G}_{jj}, \tag{86}$$

instead of (72). Thus we can use the algorithm given in Section 7 as it is.

As was discussed at the beginning of the present section, if we optimize the trial vector, we may achieve the zero-temperature limit even at rather high temperatures. This property is useful in studying ground-state phase transitions [107], in which the energy gap becomes narrow. It is also useful in studying the ground state of frustrated spin systems and fermion systems [109], in which the negative-sign problem appears at low temperatures.

9 Diffusion Monte Carlo method

In this section we review the diffusion Monte Carlo method, which has been proposed under the names of the Green's-function Monte Carlo method and the projector Monte Carlo method.

The aim of the Green's-function Monte Carlo method [110, 111, 112, 113, 89] and the projector Monte Carlo method [104, 105] is to study ground-state properties of quantum systems. The former was proposed at first for studying on quantum gases and liquids in continuous space; the latter was proposed for quantum lattice problems. There is no essential difference between them. The diffusion Monte Carlo method has been generalized to the thermofield quantum Monte Carlo method [114, 115, 116, 117], by which we can study thermodynamic properties.

The starting point is the power method, which was also utilized in the previous section. The ground state emerges after successive applications of a matrix to a test vector:

$$|\psi_g\rangle = \lim_{n\to\infty} T^n |\psi\rangle , \qquad (87)$$

where the test vector is an appropriate superposition of the base kets. In the Green's-function Monte Carlo method the matrix T is defined by $T \equiv I - \Delta\tau\mathcal{H}$ with $\Delta\tau < 1/n$, while in the projector Monte Carlo method $T \equiv e^{-\Delta\tau A}e^{-\Delta\tau B}$ is used in association with the decomposition of the Hamiltonian $\mathcal{H} = A + B$. These definitions are equivalent to each other up to the first power of $\Delta\tau$.

Although the starting point is quite similar to the one in the ground-state algorithm described in the previous section, the sampling methods are considerably different in both cases. In the diffusion Monte Carlo method, the limit procedure $n \to \infty$ itself is simulated by interpreting multiplications of the matrix as a Markov process.

The diffusion Monte Carlo method is based on the similarity between (87) and a stochastic process. The convergence to a stationary distribution of a Markov process is described by

$$P_{stat} = \lim_{n\to\infty} L^n P_{init}. \qquad (88)$$

Here L denotes a stochastic matrix whose element $L(i,j)$ is the probability that a random walker hops from the jth state to the ith state in a time interval $\Delta\tau$. The stochastic matrix should satisfy

$$\sum_i L(i,j) = \text{constant} \qquad \text{for all } j. \tag{89}$$

We simulate the diffusion of a walker in the Hilbert space, moving the walker from a state $\{\sigma\}$ to another state $\{\sigma\}'$. In the case of (87), however, the matrix T is not stochastic, that is, the condition (89) does not hold. We decompose the matrix T into the product of a stochastic matrix D and a weight w:

$$T(\{\sigma\}', \{\sigma\}) = w(\{\sigma\}')D(\{\sigma\}', \{\sigma\}), \tag{90}$$

where D satisfies the condition (89). Simulating the diffusion, we accumulate a weight of every state $\{\sigma\}(t)$ along a diffusion path Γ: $W(\Gamma) \equiv \prod_\Gamma w(\{\sigma(t)\})$. After M trials we obtain a set of final bases $\{|\{\sigma(n)\}_m\rangle\}$ $(m = 1, 2, \ldots, M)$ and a set of their weights $\{W(\Gamma_m)\}$. An estimate of the ground-state wave function is given by

$$|\psi_\text{g}\rangle \simeq \frac{1}{M} \sum_{m=1}^M W(\Gamma_m) |\{\sigma(n)\}_m\rangle. \tag{91}$$

Hetherington pointed out [118] that a problem may appear when the number n is large. Since the matrix T is not a stochastic matrix, the random walkers do not generally reproduce the ground-state distribution; in other words, the formula (91) is not correct. As n increases, the distribution of the random walkers systematically deviates further from the ground-state distribution; see [118] for a concrete example.

There appears a dilemma. The above difficulty forces us to terminate the simulation at a small number n. When n is too small, on the other hand, the convergence to the ground state is insufficient.

In order to remedy the situation we introduce a guidance function f_G, and modify the process as follows [103]:

$$|\widetilde{\{\sigma\}}\rangle \equiv f_\text{G}(\{\sigma\}) |\{\sigma\}\rangle \qquad \text{and} \qquad \tilde{T}(\{\sigma\}', \{\sigma\}) \equiv \frac{f_\text{G}(\{\sigma\}')}{f_\text{G}(\{\sigma\})} T(\{\sigma\}', \{\sigma\}). \tag{92}$$

It is known from some examples that the systematic error is reduced when the guidance function is similar to the ground-state function. When not even a rough estimate of the ground-state function is known, we cannot avoid making some assumption to choose the guidance function. There is no general way to know the systematic error in that case.

We can point out the following advantages of the diffusion Monte Carlo method over the world-line approach: (i) Samples $\{\{\sigma(n)\}_m\}$ are independent of each other instead of being auto-correlated. (ii) Once a state $\{\sigma(l)\}$ is generated, memory of the previous state $\{\sigma(l-1)\}$ can be discarded.

On the other hand, it would be desirable to overcome the disadvantage concerning the systematic error.

When we apply the method to frustrated spin systems or fermion systems, the negative-sign problem appears. Some elements of the matrix T for these models are negative, and hence some w in (90) are negative. Cancellation occurs in the summation (91).

In order to circumvent the cancellation, the fixed-node approximation has been devised [55, 119]. (i) We hypothesize about the location of the nodal surfaces (where the sign of the ground-state wave function changes) and divide the Hilbert space into several regions at the assumed surfaces. (ii) We carry out the diffusion of states within each region individually. It is known [56] that the ground-state energy estimated by the approximation gives an upper bound of the true ground-state energy. (iii) We optimize the assumed location of the nodal surfaces. The approximation may be convenient as a variational approach.

An application of the Green's-function Monte Carlo method to quantum lattice problems has emerged quite recently. The one-dimensional $S = 1$ Heisenberg antiferromagnet [120, 121, 122] and the two-dimensional Heisenberg antiferromagnet [123, 124, 125, 126, 127, 128, 129, 130] have been studied in this method. The two-dimensional Hubbard model has been also studied [131] with the use of the fixed-node approximation.

10 Handscomb's method

In this section we review Handscomb's method. It is quite different from the methods described above. Here we sample expansion terms of the high-temperature expansion scheme performing a Monte Carlo simulation.

Consider the ferromagnetic Heisenberg model:

$$\mathcal{H} \equiv -\frac{J}{2} \sum_{\langle i,j \rangle} \sigma_i \cdot \sigma_j. \tag{93}$$

The summand of $\mathcal{H}$ can be rewritten in the form

$$\sigma_i \cdot \sigma_j = 2P_b - 1, \tag{94}$$

where b denotes the nearest-neighbor bond $\langle i, j \rangle$. The operator P_b is the exchange operator, which causes a permutation between states of the sites i and j; for example, $P_b |\uparrow\downarrow\rangle = |\downarrow\uparrow\rangle$. Leaving out a constant term $JN/2$, we can express $\mathcal{H}$ as the sum of P_b.

Utilizing this property, we find the following high-temperature expansion of the model:

$$Z_q = \sum_{n=0}^{\infty} \sum_{C_n} \frac{(\beta J)^n}{n!} W(C_n), \tag{95}$$

where $W(C_n) \equiv \mathrm{Tr}\, P_{b_1} P_{b_2} \cdots P_{b_n}$. Here C_n denotes a sequence of nearest-neighbor bonds: $C_n \equiv \{b_1, b_2, \ldots, b_n\}$.

Handscomb proposed to replace the summations $\sum_n$ and $\sum_{C_n}$ by a Monte Carlo summation [132, 133, 134, 135, 136, 137]. We successively generate sequences C_n using the importance-sampling technique. We define the following Monte Carlo "flips": (i) Addition of a bond to the sequence C_n, (ii) Deletion of a bond from the sequence C_n, and (iii) Cyclic permutation of the sequence C_n.

It is essential for the applicability of the method that the ratio of the weight $W(C_{n+1})/W(C_n)$ can be calculated quickly. As for the isotropic Heisenberg model, we can obtain the weights analytically. When we can decompose a sequence C_n into k strings disconnected from each other, the weight of the sequence is given by $W(C_n) = 2^k$; we can estimate the ratio $W(C_{n+1})/W(C_n)$ easily [133, 134, 136].

An advantage of the present method is that we directly evaluate the quantum-statistical average. In the methods based on the ST decomposition we introduce the Trotter number n and analyze the data to obtain the quantum limit $n \to \infty$. In the present method no extrapolation procedures are necessary.

The method has been applied to the ferromagnetic Heisenberg model [135, 136]. The result, however, was inconsistent with the conclusion which is widely accepted now [138, 139]. The method has been also applied to the exchange-interaction models with higher S [140]; the models are defined in terms of the summation of the exchange operators for an arbitrary S, which is generally a polynomial of $S_i \cdot S_j$ [141].

In the antiferromagnetic case $J < 0$, however, the expansion (95) is an alternating series because a sign comes from the coefficient $(\beta J)^n$ for odd integers n. Numerical evaluation of a summation over an alternating series is generally difficult because of cancellation on a large scale. Another series expansion has been proposed to remedy the situation using the expression [142, 143, 144]

$$- \sigma_i \cdot \sigma_j = 2(h_b^2 - h_b) - 1, \qquad (96)$$

instead of the expression (94). Here the operator h_b is given by $h_b \equiv \sigma_i^+ \sigma_j^- + \sigma_i^- \sigma_j^+$. The partition function is expanded with respect to h_b in a form similar to (95). The summand does not vanish only when the operators $\{h_b\}$ form a closed circle on the lattice. For bipartite lattices the number of the edges of a closed circle is even, and hence the summand is always positive.

The method, applied to the antiferromagnetic Heisenberg model on a square lattice, has yielded results [143, 145] which can be compared to the results obtained by the world-line Monte Carlo method [32, 33, 34]; see Ref. [146] for a review.

Applicability of the present method strongly depends on the computational time necessary to evaluate $W(C_n)$. In the Heisenberg model, the evaluation is easy because of the $SU(2)$ symmetry of the model. As for other models, it may not be the case; we may have to expand the weight W further to obtain a tractable expression [147, 148].

Kadowaki and Ueda [149, 150, 151] have performed simulations by the simple-sampling method in order to evaluate the very expansion coefficients. Since less than ten coefficients have been estimated, they have had to employ a kind of Padé approximation. Though a systematic error may be hidden in the results, their estimates seem to agree with other results.

11 Decoupled-cell method

In this section we discuss the decoupled-cell method. This method is rather different from the methods mentioned so far.

As is explained in Section 1, it is troublesome to apply naively the classical Metropolis algorithm to quantum-mechanical systems because it is difficult to calculate the ratio of the Boltzmann factors. In the decoupled-cell method the Boltzmann factor of finite-size cells approximates the Boltzmann factor of the whole system.

Consider a flip trial of a spin at the ith site, that is, $\sigma_i \longrightarrow -\sigma_i$; we leave other spins as they are. When we naively adopt the Metropolis algorithm, we have to calculate the ratio

$$R = \left\langle -\sigma_i \left| e^{-\beta \mathcal{H}} \right| -\sigma_i \right\rangle \Big/ \left\langle \sigma_i \left| e^{-\beta \mathcal{H}} \right| \sigma_i \right\rangle. \tag{97}$$

Since it is difficult to do it as was described in Section 1, we employ the following approximation [152, 153]:

$$R \sim \left\langle -\sigma_i \left| e^{-\beta \mathcal{H}_i^{(\nu)}} \right| -\sigma_i \right\rangle \Big/ \left\langle \sigma_i \left| e^{-\beta \mathcal{H}_i^{(\nu)}} \right| \sigma_i \right\rangle, \tag{98}$$

where $\mathcal{H}_i^{(\nu)}$ denotes the Hamiltonian of a cell whose origin is the ith site and whose radius is ν. Here the cell should be small enough to be easily diagonalized. Then we perform a simulation using the approximate ratio (98). Physical quantities are measured in a suitable spin representation; when we employ the x-direction as the quantization axis, we obtain the transverse magnetization and transverse correlations [154].

In the approximation (98) we neglect quantum-mechanical coherence between the inside and the outside of the cell $\mathcal{H}^{(\nu)}$. This neglect may be justified when the coherence length is smaller than ν. Hence, at least at high temperatures, we expect that the method works properly [155].

An advantage of the present method over the ST approach is the disappearance of the negative-sign problem. Since we use only the diagonal elements of the cell density matrix in (98), the ratio $r(\sigma)$ is necessarily positive. Thus the method looks applicable even to frustrated spin systems at low temperatures [156, 157]. At low temperatures, however, the coherence effect appears distinctly, and may exceed the size of the decoupled cell. In this case the applicability of the method is questionable. In fact we observe a systematic error in the results for the one-dimensional XY model [153].

The error decreases when we treat larger cells. Concerning the convergence in the limit $\nu \to \infty$ at low temperatures, only an empirical discussion is available so far

[157]. In contrast, the convergence in the infinite-Trotter-number limit in the case of the Suzuki-Trotter approach is well known, as was mentioned in Section 2. Scaling of the cell size ν with the temperature should be discussed in the future.

12   Dynamics and the maximum-entropy method

In this section we describe a method of studying dynamics. There has been no efficient Monte Carlo algorithm for the direct measurement of functions which describe dynamic properties; for example, the real-time Green's function. Having Obtained Monte Carlo data of thermodynamic quantities, we have to carry out the analytic continuation from the imaginary time $i\tau$ to the real time t. However, the analytic continuation based on noisy numerical data is difficult in general.

Recently the maximum-entropy method developed in the field of the information theory has been applied to the present problem [158, 159, 160]. We describe the method assuming that Monte Carlo data of thermodynamic quantities are available.

Hereafter we concentrate on fermion systems. The linear response of a system at finite temperatures to a time-dependent external field is described by the real-time Green's function [161]. The retarded Green's function and the advanced Green's function are defined as follows:

$$G^{R}(t) = \begin{cases} i\left\langle\left\{c(t),c^{\dagger}\right\}\right\rangle & \text{for } t > 0, \\ 0 & \text{for } t < 0, \end{cases}$$

$$G^{A}(t) = \begin{cases} 0 & \text{for } t > 0, \\ -i\left\langle\left\{c(t),c^{\dagger}\right\}\right\rangle & \text{for } t < 0, \end{cases} \tag{99}$$

(see [162] for reference). Here $c(t)$ denotes the Heisenberg representation of the operator c, $c(t) \equiv e^{i\mathcal{H}t}ce^{-i\mathcal{H}t}$. Note that the symbol $\langle\cdots\rangle$ denotes the thermal average as it did above. Hence the Green's functions also depend on the temperature, though we have left out the argument β.

The Matsubara Green's function, on the other hand, is defined by

$$\mathcal{G}(\tau) \equiv \left\langle c(\tau)c^{\dagger}\right\rangle \tag{100}$$

with $c(\tau) \equiv e^{\tau\mathcal{H}}ce^{-\tau\mathcal{H}}$. The function is defined in the domain $[-\beta, \beta]$.

The Matsubara Green's function can be estimated by means of Monte Carlo simulations, while the real-time Green's function cannot. There is a relation between their Fourier components, that is,

$$\mathcal{G}(\omega_n) = \begin{cases} G^{R}(i\omega_n) & \text{for } \omega_n > 0, \\ G^{A}(i\omega_n) & \text{for } \omega_n < 0, \end{cases} \tag{101}$$

where $\omega_n \equiv (2n+1)\pi/\beta$ in fermion systems (the Matsubara frequencies). The functions are thus related through the analytic continuation.

The analytic continuation is equivalent to the inverse-Laplace transformation. Introducing the spectral density in the Lehmann representation,

$$\rho(\omega) \equiv (2\pi i)^{-1} \left(G^{R}(\omega + i0) - G^{A}(\omega - i0) \right), \tag{102}$$

we obtain the following relation similar to the Laplace transformation:

$$\mathcal{G}(\tau) = \int_{-\infty}^{\infty} \frac{e^{-\tau\omega}}{1 + e^{-\beta\omega}} \rho(\omega) d\omega. \tag{103}$$

Now the problem is how to carry out the inverse transformation of (103), or how to obtain a set of data of the spectral density $\{\rho_i\} \equiv \{\rho(\omega_i)\}$ from a set of noisy data of the Matsubara Green's function $\{\mathcal{G}_{\text{data}}(\tau_j) \pm \sigma_j\}$. As a naive method of the inverse transformation of (103) we may apply the least-squares method; we determine the set $\{\rho_i\}$ so that the set $\{\mathcal{G}_{\text{fit}}(\tau)\}$ given by

$$\mathcal{G}_{\text{fit}}(\tau) \equiv \sum_i \frac{e^{-\tau\omega_i}}{1 + e^{-\beta\omega_i}} \rho(\omega_i) \Delta\omega \tag{104}$$

may minimize the value

$$\chi^2 \equiv \sum_j \frac{(\mathcal{G}_{\text{data}}(\tau_j) - \mathcal{G}_{\text{fit}}(\tau_j))^2}{\sigma_j^2}. \tag{105}$$

The least-squares method, however, does not work in general. The naive application of the least-squares method yields a guess $\{\rho_i\}$ which generally does not satisfy physical conditions. We have some information about the spectral density $\rho(\omega)$ prior to the data analysis; that is, we know its positivity, $\rho(\omega) \geq 0$ for all ω, and the sum rule $\int_{-\infty}^{\infty} \rho(\omega) d\omega = 1$. If the guess does not satisfy these conditions, it might be nonsense from a physical point of view.

Data-analysis problems of this type frequently appear in many fields of science. The maximum-entropy method is one of the standard methods of solving such problems; see [163] for a review. A common feature of the difficulty in these problems is the competition between measured data and prior knowledge; when we insist on the measured data, we might make a guess which contradicts the prior knowledge, and vice versa. By means of the maximum-entropy method we control the competition.

The precision of measured data may be indicated by χ^2. On the other hand, in order to measure how a guess $\{\rho_i\}$ is likely in comparison with the prior knowledge, we use the Shannon entropy,

$$S[\rho] = \sum_i \left[\rho(\omega_i) - m_i - \rho(\omega_i) \ln\left(\frac{\rho(\omega_i)}{m_i}\right) \right] \Delta\omega. \tag{106}$$

Here we have made a proposal $\{m_i\}$, which is called a "default model". The default model should be chosen so as to satisfy the prior knowledge. The Shanon entropy indicates the difference between the guess and the default model.

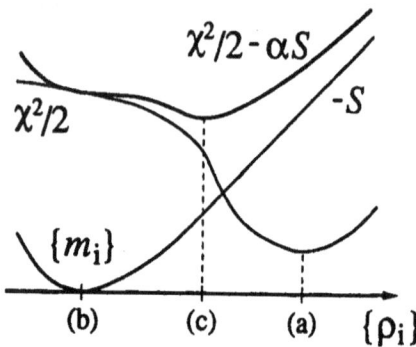

Figure 6: The value χ^2, (105), and the entropy S, (106). The points (a), (b) and (c) on the abscissa correspond to the maximum of S, the minimum of $\chi^2/2$ and the minimum of (107), respectively.

In the maximum entropy method, we arrive at a guess $\{\rho\}$ by minimizing the quantity

$$\chi^2/2 - \alpha S. \qquad (107)$$

Thereby we arrange a compromise between measured data and prior knowledge; see Fig. 6. The maximum of S (or the minimum of $-S$) corresponds to the guess of the default model $\rho(\omega_i) = m_i$, while the minimum of $\chi^2/2$ corresponds to the guess by the least-squares method.

An algorithm of determining the parameter α as well as the one of estimating the appropriateness of a default model have been given [160].

Although the estimation of possible systematic errors may not be easy, the method has been successfully applied to the Anderson model [164, 165], dilute magnetic alloys [166, 167], one- and two-dimensional Heisenberg antiferromagnets [168, 169] and the one-dimensional t-J model [170].

Acknowledgment

We are grateful to Dr. Dorota Lipowska for her critical reading of the manuscript.

References

1. J. Hubbard, Proc. Roy. Soc. A **276** (1963) 238-257.
2. R. P. Feynman and A. R. Hibbs, *Quantum Mechanics and Path Integrals* (McGraw-Hill, New York, 1965).
3. M. Suzuki, Commun. Math. Phys. **51** (1976) 183-190.
4. M. Suzuki, Commun. Math. Phys. **57** (1977) 193-200.
5. M. Suzuki, J. Math. Phys. **26** (1985) 601-612.
6. H. F. Trotter, Proc. Am. Math. Soc. **10** (1959) 545-551.

7. M. Suzuki, Phys. Lett. A **113** (1985) 299-300.
8. M. Suzuki, Phys. Lett. A **146** (1990) 319-323.
9. M. Suzuki, J. Math. Phys. **32** (1991) 400-407.
10. H. Yoshida, Phys. Lett. A **150** (1990) 262-268.
11. N. Hatano and M. Suzuki, Prog. Theor. Phys. (1991) 481-491.
12. M. Suzuki, Phys. Lett. A **165** (1992) 387-395.
13. M. Suzuki, J. Phys. Soc. Jpn. **61** (1992) 3015-3019.
14. M. Suzuki, Prog. Theor. Phys. **56** (1976) 1454-1469.
15. M. Suzuki, Prog. Theor. Phys. **58** (1977) 755-766.
16. M. Suzuki, J. Stat. Phys. **43** (1986) 883-909.
17. M. Suzuki, ed., *Quantum Monte Carlo Methods*, Solid-State Sciences **74** (Springer, Berlin, 1987).
18. J. E. Hirsch, D. J. Scalapino, R. L. Sugar and R. Blankenbecler, Phys. Rev. Lett. **47** (1981) 1628-1631.
19. J. E. Hirsch, R. L. Sugar, D. J. Scalapino and R. Blankenbecler, Phys. Rev. B **26** (1982) 5033-5055.
20. N. C. Metropolis, A. W. Rosenbluth, M. N. Rosenbluth, A. H. Teller and E. Teller, J. Chem. Phys. **21** (1953) 1087-1092.
21. M. Suzuki, S. Miyashita and A. Kuroda, Prog. Theor. Phys. **58** (1977) 1377-1387.
22. J. J. Cullen and D. P. Landau, Phys. Rev. B **27** (1983) 297-313.
23. H. De Raedt, B. De Raedt, J. Fives and A. Lagendijk, Phys. Lett. **104** (1984) 430-434.
24. H. De Raedt, B. De Raedt and A. Lagendijk, Z. Phys. B – Condensed Matter **57** (1984) 209-220.
25. E. Loh, Jr., D. J. Scalapino and P. M. Grant, Phys. Rev. B **31** (1985) 4712-4714.
26. E. Loh, Jr. and D. J. Scalapino, Physica Scripta **32** (1985) 327-333.
27. H.-Q. Ding and M. S. Makivić, Phys. Rev. B **42** (1990) 6827-6830.
28. H.-Q. Ding, Phys. Rev. B **45** (1992) 230-242.
29. M. S. Makivić, Phys. Rev. B **46** (1992) 3167-3170.
30. S. Miyashita, J. Phys. Soc. Jpn. **57** (1988) 1934-1946.
31. J. D. Reger and A. P. Young, Phys. Rev. B **37** (1988) 5978-5981.
32. H.-Q. Ding and M. S. Makivić, Phys. Rev. Lett. **64** (1990) 1449-1452.
33. M. S. Makivić and H.-Q. Ding, Phys. Rev. B **43** (1991) 3562-3574.
34. H.-Q. Ding, Phys. Lett. A **159** (1991) 355-357.
35. H. De Raedt and A. Lagendijk, Phys. Rev. Lett. **46** (1981) 77-80.
36. J. E. Hirsch and D. J. Scalapino, Phys. Rev. B **27** (1983) 7169-7185.
37. J. E. Hirsch and D. J. Scalapino, Phys. Rev. B **29** (1984) 5554-5561.
38. J. E. Hirsch, Phys. Rev. Lett. **53** (1984) 2327-2330.
39. T. Tsuzuki, Prog. Theor. Phys. **73** (1985) 1352-1368.
40. T. Tsuzuki, Prog. Theor. Phys. **75** (1986) 225-242.
41. H. Betsuyaku, Prog. Theor. Phys. **75** (1986) 774-789.
42. Y. Okabe and M. Kikuchi, J. Phys. Soc. Jpn. **56** (1987) 1963-1973.

43. K. Nomura, Phys. Rev. B **40** (1989) 2421-2425.
44. K. Nomura, Phys. Rev. B **40** (1989) 9142-9146.
45. M. Takasu, S. Miyashita and M. Suzuki, Prog. Theor. Phys. **75** (1986) 1254-1257.
46. N. Hatano and M. Suzuki, Phys. Lett. A **163** (1992) 246-249.
47. I. Morgenstern, Z. Phys. B – Condensed Matter **77** (1989) 267-273.
48. E. Y. Loh Jr., J. E. Gubernatis, R. T. Scalettar, S. R. White, D. J. Scalapino and R. L. Sugar, Phys. Rev. B **41** (1990) 9301-9307.
49. T. Nakamura, N. Hatano and H. Nishimori, J. Phys. Soc. Jpn. **61** (1992) 3494-3502.
50. T. Nakamura, N. Hatano and H. Nishimori, in *Computer Aided Innovation of New Materials II*, ed. by M. Doyama, J. Kihara, M. Takada and R. Yamamoto (Elsevier, Amsterdam).
51. T. Nakamura and N. Hatano, J. Phys. Soc. Jpn. **62** (1993) 3062-3070.
52. E. P. Münger and M. A. Novotny, Phys. Rev. B **43** (1991) 5773-5783.
53. S. B. Fahy and D. R. Hamann, Phys. Rev. Lett. **65** (1990) 3437-3440.
54. S. Fahy and D. R. Hamann, Phys. Rev. B **43** (1991) 765-779.
55. D. M. Ceperley and B. J. Alder, Phys. Rev. Lett. **45** (1980) 566-569.
56. J. W. Moskowitz, K. E. Schmidt, M. A. Lee and M. H. Kalos, J. Chem. Phys. **77** (1982) 349-355.
57. D. R. Hamann and S. B. Fahy, Phys. Rev. B **41** (1990) 11352-11363.
58. N. Furukawa and M. Imada, J. Phys. Soc. Jpn. B **60** (1991) 810-824.
59. H. Betsuyaku, Phys. Rev. Lett. **53** (1984) 629-632.
60. M. Suzuki, Phys. Rev. B **31** (1985) 2957-2965.
61. H. Betsuyaku, Prog. Theor. Phys. **73** (1985) 319-331.
62. T. Yokota and H. Betsuyaku, Prog. Theor. Phys. **75** (1986) 46-58.
63. T. Koma, Prog. Theor. Phys. **78** (1987) 1213-1218.
64. T. Koma, Prog. Theor. Phys. **81** (1989) 783-809.
65. S. Takada and K. Kubo, J. Phys. Soc. Jpn. **55** (1986) 1671-1685.
66. K. Kubo and S. Takada, J. Phys. Soc. Jpn. **55** (1986) 438-441.
67. H. Betsuyaku and T. Yokota, Prog. Theor. Phys. **75** (1986) 808-827.
68. T. Delica, K. Kopinga, H. Leschke and K. K. Mon, Europhys. Lett. **15** (1991) 55-61.
69. K. Kubo, Phys. Rev. B **46** (1992) 866-873.
70. F. D. M. Haldane, Phys. Rev. Lett. **50** (1983) 1153-1156.
71. F. D. M. Haldane, Phys. Lett. A **93** (1983) 464-468.
72. M. Suzuki and M. Inoue, Prog. Theor. Phys. **78** (1987) 787-799.
73. M. Inoue and M. Suzuki, Prog. Theor. Phys. **78** (1988) 645-664.
74. T. Koma, Prog. Theor. Phys. **83** (1990) 655-659.
75. M. Yamada, J. Phys. Soc. Jpn. **59** (1990) 848-856.
76. H. Tsunetsugu, J. Phys. Soc. Jpn. **60** (1991) 1460-1463.
77. M. Takahashi, Phys. Rev. B **43** (1991) 5788-5797; Erratum, Phys. Rev. B **44** (1991) 5397.

78. M. Takahashi, Phys. Rev. B **44** (1991) 12382-12394.
79. T. Koma, J. Stat. Phys. **71** (1993) 269-297.
80. N. Hatano and M. Suzuki, J. Phys. Soc. Jpn. **62** (1993) 1346-1353.
81. J. E. Hirsch, Phys. Rev. Lett. **51** (1983) 1900-1903.
82. J. E. Hirsch, Phys. Rev. B **31** (1985) 4403-4419.
83. R. L. Stratonovich, Doklady Akad. Nauk SSSR **115** (1957) 1097-1100 [in Russian], Soviet Phys. Doklady **2** (1958) 416-419 [translated into English].
84. J. Hubbard, Phys. Rev. Lett. **3** (1959) 77-78.
85. J. E. Hirsch, Phys. Rev. B **28** (1983) 4059-4061.
86. D. J. Scalapino and R. L. Sugar, Phys. Rev. Lett. **46** (1981) 519-521.
87. R. Blankenbecler, D. J. Scalapino and R. L. Sugar, Phys. Rev. D **24** (1981) 2278-2286.
88. D. J. Scalapino and R. L. Sugar, Phys. Rev. B **24** (1981) 4295-4308.
89. H. De Raedt and W. von der Linden, in *The Monte Carlo Methods in Condensed Matter Physics*, ed. by K. Binder, Topics in Applied Physics **71** (Springer, Berlin, 1992) 249-284. [Note that "the projector Monte Carlo method" of their terminology corresponds to the ground-state algorithm of the auxiliary-field approach described in Section 8, not to the projector Monte Carlo method described in Section 9.]
90. J. E. Hirsch, Phys. Rev. B **34** (1986) 3216-3220.
91. S. Sorella, S. Baroni, R. Car and M. Parrinello, Europhys. Lett. **8** (1989) 663-668.
92. S. R. White, D. J. Scalapino, R. L. Sugar, E. Y. Loh, J. E. Gubernatis and R. T. Scalettar, Phys. Rev. B **40** (1989) 506-516.
93. J. E. Gubernatis, D. J. Scalapino, R. L. Sugar and W. D. Toussaint, Phys. Rev. B **32** (1985) 103-116.
94. J. E. Hirsch, Phys. Rev. **35** (1987) 1851-1859.
95. J. E. Hirsch and S. Tang, Phys. Rev. Lett. **62** (1989) 591-594.
96. J. E. Hirsch, Phys. Rev. Lett. **54** (1985) 1317-1320.
97. S. Sorella, E. Tosatti, S. Baroni, R. Car and M. Parrinello, Int. J. Mod. Phys. B **1** (1988) 993-1003.
98. M. Imada and Y. Hatsugai, J. Phys. Soc. Jpn. **58** (1989) 3752-3780.
99. S. Sorella, Int. J. Mod. Phys. B **5** (1991) 937-976.
100. X. Y. Zhang, E. Abrahams and G. Kotliar, Phys. Rev. Lett. **66** (1991) 1236-1239.
101. X. Y. Zhang, Mod. Phys. Lett. B **5** (1991) 1255-1265.
102. X. Y. Zhang, R. L. Sugar and R. T. Scalettar, Phys. Rev. B **45** (1992) 471-472.
103. M. H. Kalos, D. Levesque and L. Verlet, Phys. Rev. A **9** (1974) 2178-2195.
104. J. Kuti, Phys. Rev. Lett. **49** (1982) 183-186.
105. R. Blankenbecler and R. L. Sugar, Phys. Rev. D **27** (1983) 1304-1311.
106. G. Sugiyama and S. E. Koonin, Annals of Physics **168** (1986) 1-26.
107. N. Hatano and M. Suzuki, J. Phys. Soc. Jpn. **62** (1993) 847-850.
108. W. von der Linden, I. Morgenstern and H. de Raedt, Phys. Rev. B **41** (1990)

4669-4673.

109. N. Furukawa and M. Imada, J. Phys. Soc. Jpn. B **60** (1991) 3669-3674.

110. M. H. Kalos, Phys. Rev. **128** (1962) 1791-1795.

111. M. H. Kalos, ed., *Monte Carlo Methods in Quantum Problems*, NATO ASI Series C **125** (Reidel, Dordrecht, 1984).

112. K. E. Schmidt and M. H. Kalos, in *Application of the Monte Carlo Method in Statistical Physics*, ed. by K. Binder, Topics in Current Physics **36** (Springer, Berlin, 1984) 125-143; 2nd edition (Springer, Berlin, 1987).

113. K. E. Schmidt and D. M. Ceperley, in *The Monte Carlo Methods in Condensed Matter Physics*, ed. by K. Binder, Topics in Applied Physics **71** (Springer, Berlin, 1992) 205-248.

114. M. Suzuki, Phys. Lett. A **111** (1985) 440-444.

115. M. Suzuki, J. Stat. Phys. **42** (1986) 1047-1070.

116. M. Suzuki and H. Betsuyaku, Phys. Rev. B **34** (1986) 1829-1834.

117. M. Suzuki, S. Miyashita and M. Takasu, Phys. Rev. B **35** (1987) 3569-3575.

118. J. H. Hetherington, Phys. Rev. A **30** (1984) 2713-2719.

119. S. A. Vitiello and P. A. Whitlock, Phys. Rev. B **44** (1991) 7373-7377.

120. J. B. Parkinson, J. C. Bonner, G. Müller, M. P. Nightingale and H. W. J. Blöte, J. Appl. Phys. **57** (1985) 3319-3321.

121. M. P. Nightingale and H. W. J. Blöte, Phys. Rev. B **33** (1986) 659-661.

122. M. Takahashi, Phys. Rev. Lett. **62** (1989) 2313-2316.

123. T. Barnes and G. J. Daniell, Phys. Rev. **37** (1988) 3637-3651.

124. T. Barnes and E. S. Swanson, Phys. Rev. B **37** (1988) 9405-9409.

125. T. Barnes, D. Kotchan and E. S. Swanson, Phys. Rev. B **39** (1989) 4357-4362.

126. M. Gross, E. Sánchez-Velasco and E. Siggia, Phys. Rev. **39** (1989) 2484-2493.

127. J. Carlson, Phys. Rev. B **40** (1989) 846-849.

128. N. Trivedi and D. M. Ceperley, Phys. Rev. B **40** (1989) 2737-2740.

129. N. Trivedi and D. M. Ceperley, Phys. Rev. B **41** (1990) 4552-4569.

130. K. J. Runge, Phys. Rev. B **45** (1992) 7229-7236.

131. G. An and J. M. van Leeuwen, Phys. Rev. B **44** (1991) 9410-9417.

132. D. C. Handscomb, Proc. Camb. Phil. Soc. **58** (1962) 594-598.

133. D. C. Handscomb, Proc. Camb. Phil. Soc. **60** (1964) 115-122.

134. J. W. Lyklema, Phys. Rev. Lett. **49** (1982) 88-90.

135. J. W. Lyklema, Phys. Rev. B **27** (1983) 3108-3110.

136. J. W. Lyklema, in *Monte Carlo Methods in Quantum Problems*, ed. by M. H. Kalos, NATO ASI Series C **125** (Reidel, Dordrecht, 1984) 145-155.

137. Y. C. Chen, H. H. Chen and F. Lee, Phys. Rev. B **43** (1991) 11082-11087.

138. P. Schlottmann, Phys. Rev. Lett. **54** (1985) 2131-2134.

139. M. Takahashi and M. Yamada, J. Phys. Soc. Jpn. **54** (1985) 2808-2811.

140. Y. C. Chen, H. H. Chen and F. Lee, Phys. Lett. A **130** (1988) 257-130.

141. H. H. Chen and R. L. Joseph, J. Math. Phys. **13** (1972) 725-739.

142. D. H. Lee, J. D. Joannopoulos and J. W. Negele, Phys. Rev. B **30** (1984) 1599-1602.

143. E. Manousakis and R. Salvador, Phys. Rev. Lett. **60** (1988) 840-843.
144. G. Gomez-Santos, J. D. Joannopoulos and J. W. Negele, Phys. Rev. B **39** (1989) 4435-4443.
145. E. Manousakis and R. Salvador, Phys. Rev. B **39** (1989) 575-585.
146. E. Manousakis, Rev. Mod. Phys. **63** (1991) 1-62.
147. A. W. Sandvik and J. Kurkijärvi, Phys. Rev. B **43** (1991) 5950-5961.
148. A. W. Sandvik, J. Phys. A: Math. Gen. **25** (1992) 3667-3682.
149. S. Kadowaki and A. Ueda, Prog. Theor. Phys. **75** (1986) 451-454.
150. S. Kadowaki and A. Ueda, Prog. Theor. Phys. **78** (1987) 224-236.
151. S. Kadowaki and A. Ueda, Prog. Theor. Phys. **82** (1989) 493-506.
152. S. Homma, H. Matsuda and N. Ogita, Prog. Theor. Phys. **72** (1984) 1245-1247.
153. S. Homma, H. Matsuda and N. Ogita, Prog. Theor. Phys. **75** (1986) 1058-1065.
154. T. Horiki, S. Homma, H. Matsuda and N. Ogita, Prog. Theor. Phys. **82** (1989) 507-513.
155. H. Matsuda, K. Ishii, S. Homma and N. Ogita, Prog. Theor. Phys. **80** (1988) 583-587.
156. S. Homma, K. Sano, H. Matsuda and N. Ogita, Prog. Theor. Phys. **87** (1986) 127-138.
157. C. Zeng and V. Elser, Phys. Rev. B **42** (1990) 8436-8444.
158. R. N. Silver, D. S. Sivia and J. E. Gubernatis, in *Quantum Simulations of Condensed Matter Phenomena*, ed. by J. D. Doll and J. E. Gubernatis, (World Scientific, Singapore, 1990) 340-354.
159. R. N. Silver, D. S. Sivia and J. E. Gubernatis, Phys. Rev. B **41** (1990) 2380-2389.
160. J. E. Gubernatis, M. Jarrell, R. N. Silver and D. S. Sivia, Phys. Rev. B **44** (1991) 6011-6029.
161. R. Kubo, J. Phys. Soc. Jpn. **12** (1957) 570-586.
162. A. A. Abrikosov, L. P. Gorkov and I. E. Dzyaloshinski, translated by R. A. Silverman, *Methods of Quantum Field Theory in Statistical Physics*, the Dover edition (Dover Publications, New York, 1975).
163. J. Skilling, ed., *Maximum Entropy and Baysian Methods*, Fundamental Theories of Physics, (Kluwer Academic Publishers, Dordrecht, 1989).
164. R. N. Silver, J. E. Gubernatis, D. S. Sivia and M. Jarrell, Phys. Rev. Lett. **65** (1990) 496-499.
165. M. Jarrell, J. E. Gubernatis and R. N. Silver, Phys. Rev. B **44** (1991) 5347-5350.
166. M. Jarrell, D. S. Sivia and B. Patton, Phys. Rev. B **42** (1990) 4804-4807.
167. M. Jarrell, J. Gubernatis, R. N. Silver and D. S. Sivia, Phys. Rev. B **43** (1991) 1206-1209.
168. J. Deisz, M. Jarrell and D. L. Cox, Phys. Rev. B **42** (1990) 4869-4872.
169. M. Makivić and M. Jarrell, Phys. Rev. Lett. **68** (1992) 1770-1773.
170. J. Deisz, K.-H. Luk, M. Jarrell and D. L. Cox, Phys. Rev. B **46** (1992) 3410-3419.

MONTE CARLO RENORMALIZATION GROUP STUDY OF THE D=1 XXZ MODEL

M. A. Novotny[1] and H. G. Evertz[1,2]

[1] Supercomputer Computations Research Institute,
Florida State University,
Tallahassee, FL 32306-4052 USA

[2] Department of Physics and Astronomy
and Center for Simulational Physics
University of Georgia,
Athens, GA 30602 USA

e-mail: novotny@scri.fsu.edu

Abstract

We report current progress on the synthesis of methods to alleviate two major difficulties in implementing a Monte Carlo Renormalization Group (MCRG) for quantum systems. In particular, we have utilized the loop-algorithm to reduce critical slowing down, and we have implemented an MCRG method in which the symmetries of the classical equivalent model need not be fully understood, since the Renormalization Group is given by the Monte Carlo simulation. We report preliminary results obtained when the resulting MCRG method is applied to the $d=1$ XXZ model. Our results are encouraging. However, since this model has a Kosterlitz-Thouless transition, it does not provide a full test of our MCRG method.

1 Introduction

Although the Monte Carlo method introduced by Metropolis *et al.*[1] has been a useful tool in the study of the critical behavior of classical systems, its application to the study of the critical behavior of quantum systems requires that many difficulties be circumvented. Some of these difficulties are already present in the study of classical systems, while others are present only in the study of quantum systems. In this paper we utilize present knowledge in overcoming the difficulties in such studies, and we apply the method to the study of the one-dimensional XXZ quantum spin model.

The first difficulty to be surmounted in the study of quantum critical behavior was the problem of how to apply the Monte Carlo method to quantum systems. This difficulty was circumvented when Suzuki[2, 3, 4] proposed to use the formula of

49

Trotter[5] to map any d-dimensional quantum system onto a $d+1$-dimensional classical equivalent. Convergence properties of the Suzuki-Trotter transformation[6] and of higher-order decompositions have been described[7]. Although this decomposition is applicable in general, it can lead to other difficulties not encountered in the study of most classical systems.

One of the problems often introduced by the application of the Suzuki-Trotter transformation is the minus-sign problem. The minus-sign problem occurs when the 'Boltzmann weights' of the classical equivalent of a quantum problem are not all positive. In the Monte Carlo method Boltzmann weights are interpreted as probabilities, and the minus-sign problem to date is a serious one and can only be decreased. This is done by performing a simulation with any chosen probability distribution, and reweighting to the Boltzmann weights of the classical equivalent[8, 9, 10, 11, 12, 13]. Unfortunately, this moves the difficulty from one of principle to one of acquiring adequate statistics. In general the desired quantities are ratios of two numbers, each of which comes from the difference between two large numbers with statistical errors. The minus-sign problem is therefore still largely unsolved. It is interesting to note that it is possible to make some classical statistical mechanical models have a minus-sign problem. For example, the simulation of the $d=2$ Ising ferromagnet has a minus-sign problem in a certain representation[9, 14].

A difficulty which occurs in the study of both classical and quantum critical behavior is the critical slowing down that occurs in the standard Monte Carlo method near the critical point. The difficulty is that standard Monte Carlo algorithms employ a *local* update procedure, and consequently 'information' diffuses through the lattice slowly in a random walk fashion. This difficulty was first mastered for the ferromagnetic Potts model by Swendsen and Wang[15] where the important realization was to perform *non-local* updates on clusters closely related to the critical clusters of the model. Similar procedures have been devised for a number of systems and have recently been reviewed [16, 17]. The classical equivalent obtained from the Suzuki-Trotter transformation often maps onto a *vertex* model, and it is only recently that algorithms to alleviate critical slowing down for vertex models have been devised [18, 19, 20].

In the study of phase transitions universal quantities, such as critical exponents, are desired. However, a phase transition occurs only in the thermodynamic limit — whereas Monte Carlo simulations by necessity can be done only on finite lattices. The best ways to obtain estimates of the critical exponents from Monte Carlo studies are to use finite-size scaling methods (for reviews see [21, 22]) or renormalization group (RG) methods[23, 24]. One marriage of Monte Carlo and RG methods was given by Swendsen[25, 26, 27]. In Swendsen's implementation of the Monte Carlo Renormalization Group (MCRG) one calculates only correlation functions (not coupling constants). To successfully apply this MCRG method, it is important that the RG chosen preserves the symmetries of the Hamiltonian[28]. However, this presents a problem for quantum MCRG calculations since typically after the application of the Suzuki-Trotter formula one does not generally know the underlying 'Hamiltonian' of the $d+1$-dimensional classical equivalent. If you do not know the 'Hamiltonian', how can you know its symmetries? In particular, if the minus-sign problem is present, the concept of a 'Hamiltonian' on the classical equivalent is not well defined. If the classical equivalent is a vertex model, the constraint due to the continuity of vertex loops is difficult to preserve in an RG treatment. This is the difficulty that has delayed progress on the application of MCRG procedures to quantum systems after the initial success of quantum MCRG methods to the transverse Ising model[29, 30, 31], for which the underlying classical equivalent Hamiltonian is known and has no global constraints. For a review of quantum MCRG see Sec. 8 of Ref. [32]. Similar difficulties have been encountered in applying Wilson's idea of real-space blocking[24] to exact-diagonalization studies, which has only recently been surmounted[33, 34, 35].

A method to overcome the difficulty of deciding what RG procedure to use in quantum studies was recently reported by Münger and Novotny[14]. The idea is to modify a momentum-space MCRG developed by Swendsen[27], but to let the Monte Carlo simulation itself decide which RG to use. In momentum space this leads to the RG procedure of systematically discarding the least important degrees of freedom as one renormalizes the lattice. This momentum-space MCRG procedure was successfully applied to the square lattice q-state Potts model with q not restricted to integer values. Thus the MCRG in Ref. [14] was performed on a staggered 6-vertex model which is obtained from a mapping of the q-state Potts model.

In this paper we apply this MCRG for the first time to a quantum system — the 6-vertex model, which is the classical equivalent of the $d=1$ spin-$\frac{1}{2}$ quantum XXZ model. The XXZ Hamiltonian is

$$\mathcal{H} = + \sum_{i=1}^{N} \left(\sigma_i^x \sigma_{i+1}^x + \sigma_i^y \sigma_{i+1}^y + \lambda \sigma_i^z \sigma_{i+1}^z \right), \qquad (1)$$

with partition function $Z = \exp(-\beta \mathcal{H})$. Periodic boundary conditions on the chain of N spins are used. Notice that $\mathcal{H}$ is invariant under $(\beta \to -\beta, \lambda \to -\lambda)$ by rotating every second spin through an angle of π, and is thus equivalent to the Hamiltonian $\mathcal{H} = -\sum_{i=1}^{N} \left(\sigma_i^x \sigma_{i+1}^x + \sigma_i^y \sigma_{i+1}^y - \lambda \sigma_i^z \sigma_{i+1}^z \right)$.

The case $\lambda = 1$ corresponds to the isotropic Heisenberg antiferromagnet, while the case $\lambda = 0$ corresponds to the quantum XY chain.

We have chosen this model since both the loop-algorithm of Ref. [18, 19, 20], and the MCRG program of Ref. [14] were written to study the 6-vertex model, which is the classical equivalent of the Hamiltonian in Eq. (1).

2 Monte Carlo Method

It is well known [2, 6, 36, 37], that the $d=1$ XXZ chain can be mapped to a $(1+1)$ dimensional classical spin system by Suzuki's method. The main steps are a breakup of the Hamiltonian for an N-spin chain into pieces living on even and odd bonds, an application of the Trotter-Suzuki formula, and an insertion of complete

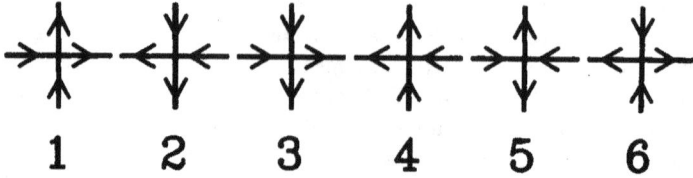

Figure 1: The 6 vertices which are allowed in the model are shown. For the XXZ model the weights are $w_5=w_6=1$ and $w_1=w_2=w_3=w_4$.

sets of intermediate states. The result is a system of classical spins $s_{i,j} = \pm 1$ on a periodic $N \times M$ chessboard lattice with the Euclidean time (Trotter) direction having extension M, and four-spin interactions on all "black" plaquettes of the chessboard [2, 6, 36, 37]. The mapping becomes exact as $M \to \infty$.

The four spins $(s_{i,j}; s_{i+1,j}; s_{i,j+1}; s_{i+1,j+1})$ bordering an interacting plaquette obey the continuity constraint

$$s_{i,j} + s_{i+1,j} = s_{i,j+1} + s_{i+1,j+1}. \qquad (2)$$

The *world lines* connecting sites with $s_{i,j} = 1$ are thus continuous.

This classical spin system is a vertex model: vertices are located at the center of interacting plaquettes; they are connected by lines. Each such line touches a lattice site (i,j). If $s_{i,j} = +1$ (-1), then we place an arrow that points upwards (downwards) in the Trotter direction on this line. Note that the resulting lattice of arrows and vertices has lines tilted 45 degrees w.r.t. the chessboard lattice.

The constraint of Eq. (2) is that of the 6-vertex model: each vertex has two arrows pointing into the vertex and two pointing away from the vertex. The six possible vertex configurations are shown in Fig. 1. Their weights w_i in the partition function are the weights of the interacting plaquettes, namely [37]

$$\begin{aligned} a &= w_1 = w_2 = e^{-\lambda\hat{\beta}}, \\ b &= w_3 = w_4 = e^{\lambda\hat{\beta}}\sinh 2\hat{\beta}, \\ c &= w_5 = w_6 = e^{\lambda\hat{\beta}}\cosh 2\hat{\beta}, \end{aligned} \qquad (3)$$

with $\hat{\beta} = \beta/(2M)$. The 6-vertex model is exactly solved [42]. It is governed by the parameter [42]

$$\Delta \equiv \frac{a^2 + b^2 - c^2}{2ab} = \frac{\sinh(-2\lambda\hat{\beta})}{\sinh 2\hat{\beta}} \xrightarrow{\hat{\beta}\to 0} -\lambda. \qquad (4)$$

Here we use periodic boundary conditions for the vertex lattice. The normal XXZ boundary conditions would be periodic in both the spin chain direction and the

Trotter direction, and correspond to periodic boundary conditions in the 45-degree directions for the vertex lattice.

Monte Carlo simulations of constrained systems like the 6-vertex model have suffered from severe critical slowing down in the past. The situation has been drastically improved by the advent of the loop cluster algorithm [18, 19, 20], which we now sketch very briefly. In terms of arrows, the constraint of Eq. (2) is a "divergence = zero" condition. All arrows therefore lie on directed closed loops (like magnetic field lines). The loop algorithm stochastically constructs such a loop and flips the direction of all arrows on the loop in a single Monte Carlo step while maintaining detailed balance. It can therefore perform large steps in phase space, and produce statistically independent configurations within a few Monte Carlo steps even at infinite correlation length on large lattices [18]. Note that the loop algorithm also easily produces loops that wind around the lattice. At the KT transition ($\frac{a}{c} = \frac{b}{c} = \frac{1}{2}$), for example, autocorrelation times are reduced by about two orders of magnitude on a 64^2 lattice with respect to a Metropolis simulation.

3 MCRG Method

Here we use the MCRG method described in detail in Ref. [14]. The idea of this MCRG method is to let the Monte Carlo simulation itself provide the renormalization group to be used. This is done by building on the MCRG method developed by Swendsen [25, 26] to calculate critical exponents. Rather than use a block spin procedure in real space, the RG is defined in momentum space in a similar fashion as the procedure introduced by Swendsen in Ref. [27]. Rather than using the Fourier transform of the configurations [27], we use the Fourier transform of each state individually [38],

$$\hat{v}_{\mathbf{k}}^{\alpha} = \sum_{\mathbf{r}} \delta_{\alpha, v_{\mathbf{r}}} \exp\left(i\mathbf{k} \cdot \mathbf{r}\right), \tag{5}$$

where α labels the vertex state v, and the summation is over the simulated lattice. In our case, α takes the values of one through six. An inverse Fourier transform over a restricted part of the momentum-space is then performed. The same restricted part is used for all generated configurations, and the inverse transform is given by

$$v_{\mathbf{s}}^{\prime\,\alpha} = \sum_{\mathbf{k}}{}' \hat{v}_{\mathbf{k}}^{\alpha} \exp\left(-i(\mathbf{k} - \mathbf{m}_{\alpha}) \cdot \mathbf{s}\right), \tag{6}$$

where $\mathbf{m}_{\alpha}$ is a shift in momentum space, and the prime on the summation indicates that the sum is over a restricted part of momentum space. Finally, on each site of the reduced lattice for each configuration the state α is chosen for which the real part, $\Re(v_{\mathbf{s}}^{\prime\,\alpha})$ is largest. Of course this choice is arbitrary, since for example the state with the largest modulus could have been chosen.

Both the restricted part of momentum space and the shift $\mathbf{m}_{\alpha}$ are the same for all generated configurations and are given by the Monte Carlo simulation itself. The shift m_α is chosen such that the important fluctuations, as determined by the peak in the momentum-space plot from the entire Monte Carlo simulation, are shifted to the center of the Brillouin zone. This transforms each state into a mainly 'ferromagnetic' state, and allows the use of normal ferromagnetic operators to obtain critical exponents from the linearized transformation matrix $\mathbf{T}^*$. The restricted region of the shifted momentum space is chosen to be the region in the center of the Brillouin zone that gives $N'=L'\times L'$ spins on transforming back to real space. Note that this MCRG procedure does not preserve the constraints of the vertex model (the continuity of loops). However, the renormalized variables should be considered to be composite variables, and the vertex constraints on the original lattice are taken into account during the RG procedure in a way dictated by the Monte Carlo simulation of the model.

In principle, one could avoid transforming back to real space, but at the expense of defining the 'majority rule' for the states in momentum space, and at the expense of defining operators for the linearized transformation matrix $\mathbf{T}^*$ in momentum space; neither of which has been attempted to the authors' knowledge. The transformation back to real space reduces the lattice size and gives the RG. However, a normalization is also required. This is done by multiplying the new real-space spins with a constant such that $\sum |v_r|^2/N = \sum' |v'_s|^2/N'$, where N' is the number of lattice sites on the reduced lattice, and the primed summation is over renormalized spins v'_s on the reduced lattice. Other normalization methods can be devised [14], but they should all lead to the same critical exponents from the MCRG.

The critical properties are obtained in the usual fashion from the linearized RG transformation matrix $\mathbf{T}^*$ from the elements

$$T^*_{\alpha\beta} = \frac{\partial K^{(n)}_\alpha}{\partial K^{(n+m)}_\beta}, \tag{7}$$

where $K^{(n)}_\alpha$ is the coupling constant corresponding to the operator S_α following the nth RG transformation. The K_α include factors of $1/k_B T$. From the chain rule,

$$\frac{\partial \langle S^{(n)}_\gamma \rangle}{\partial K^{(n+m)}_\beta} = \sum_\alpha T^*_{\alpha\beta} \frac{\partial \langle S^{(n)}_\gamma \rangle}{\partial K^{(n+m)}_\alpha}, \tag{8}$$

where the angular brackets indicate the thermal average of the operator. The partial derivatives can be easily calculated from the canonical ensemble:

$$\frac{\partial \langle S^{(n)}_\gamma \rangle}{\partial K^{(n+m)}_\alpha} = \langle S^{(n)}_\gamma S^{(n+m)}_\alpha \rangle - \langle S^{(n)}_\gamma \rangle \langle S^{(n+m)}_\alpha \rangle. \tag{9}$$

The thermal averages in Eq. (9) are calculated between the renormalized lattices after n and $n+m$ RG transformations. These averages can be calculated at the simulated set of parameters, but they can also be calculated with parameters other than the simulated ones using a single reweighting method [8, 10, 11, 12], or using data from multiple simulations for different parameters [12, 13, 14].

The number of operators in Eq. (8) is infinite. However, the MCRG estimate for the matrix $\mathbf{T}^*$ is found from Eq. (8) by first truncating all the matrices in Eq. (8) and then multiplying Eq. (8) by the inverse of the truncated matrix that multiplies $\mathbf{T}^*$ in Eq. (8). To first order in perturbation theory, it has been shown that this procedure takes into account the operators which have not been included in the truncation procedure [39]. Since we have performed the shift m_α for each state, we need to include only 'ferromagnetic' operators.

Once the largest eigenvalues, $\lambda_1 \geq \lambda_2 \geq \cdots$, of the linearized transformation matrix $\mathbf{T}^*_{\alpha\beta}$ are determined, the eigenvalues of the RG are given by

$$y_i = \ln(\lambda_i)/\ln(b), \tag{10}$$

where the scale factor b is given by the ratio of the lengths on the two lattice sizes used, $b = L'_{large}/L'_{small} = (N'_{large}/N'_{small})^{1/d}$, where d is the dimension of the model studied. The exponents y_i are irrelevant if they are less than zero, are marginal if $y_i = 0$, and are relevant and give universal critical exponents for models associated with the fixed point if $y_i > 0$.

For a second-order transition the two largest critical exponents of the RG (y_T from the operators in the thermal section and y_H from operators in the magnetic section) allow the critical exponents to be calculated in the normal fashion from $\nu = 1/y_T$ and $\eta = 2 + d - 2y_H$.

Special care must be taken in the case of a model with a Kosterlitz-Thouless (KT) transition[40]. The authors in Ref. [40] show that although the magnetic exponents are given by $\eta = 2 + d - 2y_H$ in the normal fashion, the values of ν for a KT transition must be obtained by fitting to the KT RG equations. We have not done such a fitting procedure, but have rather concentrated only on the magnetic exponents. In general one can also use the MCRG method to obtain the values of the critical couplings[40, 41]. However since the critical couplings for the $d=1$ XXZ model are known exactly, we have not tried to find them using MCRG.

4   Theoretical results for the $d=1$ XXZ Model

The $d=1$ spin-$\frac{1}{2}$ XYZ model has been studied using a wide variety of methods. These methods have ranged from exact methods[42, 43, 44], to universality and mapping methods[45, 46], to conformal invariance methods[47, 48], to series expansion methods[49], to exact diagonalization studies[50, 51, 52], to Monte Carlo methods[4, 53, 54]. The phase diagram which has emerged from these studies is illustrated in the 6-vertex representation in Fig. 2[42]. We have performed our MCRG calculations only on the F-model (where $w_1 = w_2 = w_3 = w_4$ and $w_5 = w_6$) at points illustrated by the open circles in Fig. 2. In Fig. 2 region I corresponds to a ferroelectric region where the lowest energy state is one with all vertices equal to 1 or 2. In this region the excited states give a negligible contribution to the partition function and the system is frozen in one of the two ground states. The situation is similar for

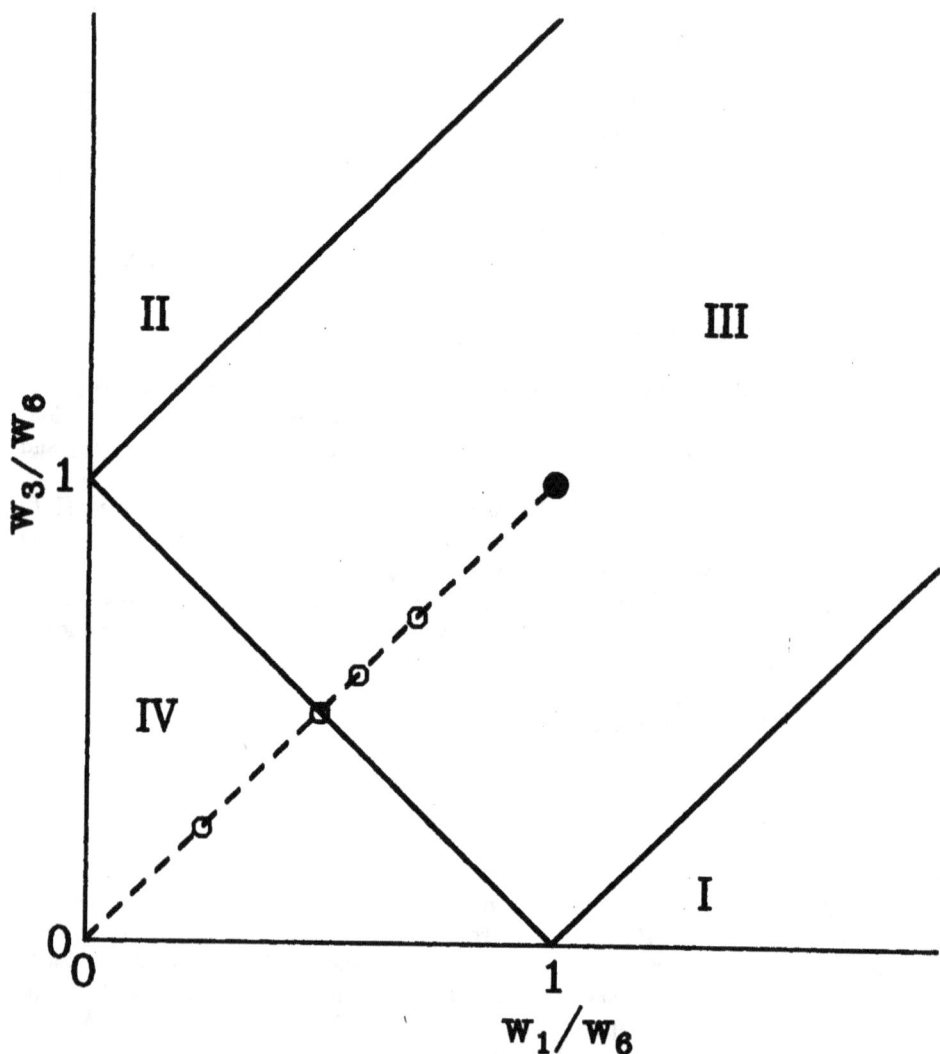

Figure 2: The phase diagram for the 6-vertex model is shown. The vertex weights satisfy $w_1=w_2$, $w_3=w_4$, and $w_5=w_6$. The dashed line corresponds to the F-model where $w_1=w_3$ as well, and this line terminates at the 'infinite temperature' point where $w_1=w_5$ (filled circle). The open circles represent the points where we have made MCRG calculations.

region II, where all vertices equal 3 or 4, and region IV where the two ordered states have vertices of 5 on one sublattice and 6 on the other sublattice. Region III is the disordered phase where there is no spontaneous order, nor interfacial tension, but the correlation length is infinite and correlations decay as an inverse power of distance rather than exponentially[42].

The F-model with $w_1 < w_5/2$ (which corresponds to the asymmetry parameter of Eq. (1), $\lambda > 1$) is in region IV. Since in this region the model is completely ordered, one expects to obtain a low-temperature discontinuity fixed point (the Nienhuis-Nauenberg criterion[55]) with the largest RG eigenvalue $y=d$.

There is a Kosterlitz-Thouless (KT) transition in the F-model when $w_1 = w_5/2$ (where $\lambda = 1$, and the model is the Heisenberg model). In the XXZ model there is a KT transition from the massless into the antiferroelectric ordered domain at $\lambda = 1$. This occurs through a Kosterlitz-Thouless mechanism caused by the excitations of an umklapp process[45, 46]. At $\lambda = 1$ the vortex operator (which comes from the critical exponent x_T^{8V} of the 8-vertex model) is marginal.

For the F-model in region III the critical exponents are those of the critical 8-vertex model[42] and are the same as those of the Luttinger model[45]. From conformal invariance[48] the entire operator content is known. Stringlike solutions of the Bethe-*Ansatz* equations yield excited states corresponding to operators $O_{i,j}$ with dimensions

$$x_{i,j} = i^2 x_p + j^2/4x_p \qquad i,j = 0,1,2,\cdots. \tag{11}$$

These operators are the analogs of the Gaussian-model operators composed of spin-wave excitations of index i and a "vortex" of vorticity j. In Eq. (11) $x_p = (\pi - \gamma)/2\pi$, where $\gamma = \arccos(\lambda)$. The operators $O_{1,0}$ and $O_{2,0}$ correspond respectively to the polarization and energy operators of the 8-vertex model. The operator $O_{0,1}$ is the crossover operator of the 8-vertex model, or equivalently the energy operator of the Ashkin-Teller model. From the dimensions $x_{i,j}$ the RG exponents are found through

$$y_{i,j} = d - x_{i,j}. \tag{12}$$

The exponents $\eta_{i,j} = 2 + d - 2y_{i,j} = 2x_{i,j}$ govern the behavior of correlations. For example, for $O_{0,1}$ one has $\eta_{0,1} = 2x_{0,1} = 2 + d - 2y_{0,1}$ and this critical exponent governs the correlation $\eta_z = \eta_{0,1}$ with

$$\langle S^z(0)S^z(r) \rangle \sim r^{-\eta_z} \tag{13}$$

where $S^z = \frac{1}{2}\sigma^z$.

5 MCRG results

We have applied the MCRG method described in Sec. 2 to the 6-vertex model on $L \times L$ lattices with periodic boundary conditions. The 11 operators which we included in the truncated linearized RG transformation matrix $\mathbf{T}^*$ are shown in Fig. 3. Note that because the momentum-space RG does not conserve the 6-vertex constraint on

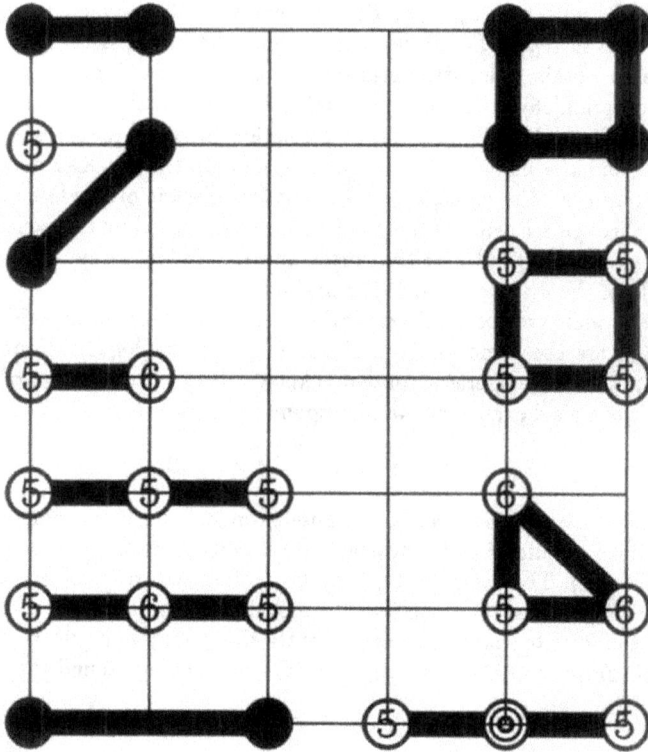

Figure 3: The 11 operators used for the truncated linearized RG transformation matrix $\mathbf{T}^*$ are shown. Filled circles represent a vertex in any state, circles with numbers inside represent a δ-function for that vertex state, and the 'bull's eye' in the last operator is any vertex which is not a 5 or 6. The square lattice is drawn with light lines, while heavy lines join the interacting sites. For the operator to be nonzero all filled-circle interacting vertices must have the same vertex state. All combinations which are related by translation, rotation, and reflection to those shown were included in the calculation.

n	m 1	2	3	4
	Largest RG Exponent			
1	1.9998	1.9996	1.9981	1.9864
2	2.0003	1.9980	1.9868	
3	1.9963	1.9840		
4	1.9811			
	Second RG exponent			
1	0.6449	1.0041	1.2214	1.1063
2	0.7962	1.1325	1.0128	
3	0.9313	0.7794		
4	0.9683			
	RG scale factor b			
1	1.080	1.286	1.588	1.800
2	1.191	1.471	1.667	
3	1.235	1.400		
4	1.133			

Table 1: The two largest exponents of the linearized transformation matrix, $\mathbf{T}^*$, are listed for the simulation in the ordered phase at $w_1/w_5 = \frac{1}{4}$. The lattice simulated was a 32×32 lattice, and the sizes of the renormalized $L' \times L'$ lattices were L'=27, 25, 21, 17, and 15. Listed are the exponents between renormalized lattices n and $n+m$ as well as the corresponding RG scale factors b. For example, the entry with $n=2$ and $m=3$ corresponds to an RG transformation between lattices with $L'=25$ and $L'=15$. The data should converge for large n, m, but also become susceptible to finite size effects. The expected largest eigenvalue is that of the zero-temperature fixed point since the system is ordered in Region IV of Figure 2. Consequently the largest exponent should be that of the discontinuity fixed point, $y=d=2$. Our data converge to this value extremely rapidly. The next largest eigenvalue would be the first analytic correction, which approaches unity.

the renormalized lattice, we have not included polarization-type operators in the 11 measured operators. Our calculations were done on a CRAY Y-MP432 supercomputer, and the Fourier transforms used were FFT's in the NAG library (the explicit routines used depended on the lattice sizes the FFT was applied to).

We first studied the F-model with $w_1 = w_5/4$ in the ordered region IV of Fig. 2, and the results are shown in Table 1. The lattice simulated for Table 1 was 32×32 and the renormalized lattices were $L' \times L'$ with L'=27, 25, 21, 17, and 15. The RG analysis was performed using 15000 configurations which were generated skipping 250 Monte Carlo cluster updates between configurations and with 250000 cluster updates for thermalization. The exponential autocorrelation time, $\tau_{\exp}$ in units of 'sweeps'[20] is given by the autocorrelation time in units of cluster updates times the average cluster size divided by $2L^2$. Here $\tau_{\exp}=43(5)$, and it is most clearly visible in the sublattice energy[20]. The average cluster size is 0.96 times $2L^2$. In principle one can obtain estimates for the statistical errors of the eigenvalues in Table 1, for example by analyzing J bins of the generated configurations. We have not yet performed such an analysis.

The values of Tables 1 and 2 should be read in a specific fashion to see the convergence of the exponents. In order to obtain good critical exponents from the MCRG one needs to be able to penetrate the *linear* region about the fixed point, since the exponents are calculated from the linearized transformation matrix $T^*_{\alpha\beta}$. It may take several iterations to be able to get close enough to the fixed point that a linear approximation is reasonable. Consequently, the first iterations (the ones starting from the largest lattices, and hence on the first lines in the tables) may need to be ignored, or at least to be used only to see how the convergence toward the linear region is proceeding. For this reason we do not show the iteration between the original lattice and the next largest lattice in the tables. The exponents should then converge for large values of n and m.

If n, m become too large, however, there are additional difficulties. Smaller scale factors b will have the larger statistical errors[14]. Also, in MCRG studies the larger statistical errors typically occur on the smallest lattices, which correspond to the last entry in each line of the tables. The smallest lattices will also have larger finite-size effects, which are particularly disruptive for the longer-range operators included in the calculation of $T^*_{\alpha\beta}$.

In Table 1 one sees extremely rapid convergence to the eigenvalue $y=d=2$ associated with the Nienhuis-Nauenberg criterion for a discontinuity fixed point [55]. The next-largest eigenvalue, while not having converged too well, seems to be approaching the value of unity, which would give the first analytical correction to the discontinuity fixed-point behavior.

Our other MCRG results are shown in Table 2. These are at the KT point ($\lambda=1$), the XY point (free-fermion case with $\lambda=1/\sqrt{2}$), and an intermediate point with $\lambda=1/2$. Table 2 lists the two largest eigenvalues obtained from our MCRG procedure. Each value of λ studied took approximately 23 hours of Y-MP time for the RG analysis of the generated configurations. The time required for the Monte Carlo portion of the runs depended on the value of λ since the cluster-size per volume depended on λ. For $\lambda=0, \frac{1}{2}$, and 1 the cluster-size was 0.15, 0.09, 0.038 times $2L^2$ and the Y-MP CPU time used corresponded to 3.9, 2.5, and 1.4 hours. The lattice simulated for Table 2 was 64×64, and the renormalized lattices were $L'\times L'$ with $L'=51, 41, 31, 21$, and 15. The RG analysis was again performed using 15000 configurations which were generated skipping 250 cluster updates between configurations and using 250000 cluster updates for thermalization. Consequently, a total of 4×10^6 cluster updates were generated. The exponential autocorrelation time, which is most clearly visible in the sublattice energy[20], for $\lambda=0, \frac{1}{2}$, and 1, is $\tau_{exp}=4.0(2), 8.1(9)$, and $5.7(6)$, respectively.

The convergence toward the linear region about the fixed point is expected to be particularly slow when a marginal eigenvalue is present, as is the case in Table 2. Consequently, in Table 2 convergence has not yet occured, but the trend toward convergence of the largest eigenvalue is evident and in agreement with the expected result.

In Table 2 we see only one nearly converged largest exponent y for each value of λ.

		$w_1/w_5 = 1/2$	$\lambda = 1$				$w_1/w_5 = 1/\sqrt{2}$	$\lambda = 0$	
n	m	1	2	3	4	m 1	2	3	4
		Largest RG Exponent				Largest RG Exponent			
1		2.022	1.809	1.707	1.667	1.902	1.477	1.342	1.299
2		1.676	1.624	1.605		1.098	1.145	1.165	
3		1.591`	1.581			0.975	1.070		
4		1.563				0.844			
		Second RG Exponent				Second RG Exponent			
1		1.466	1.068	0.835	0.763	1.185	0.493	0.175	0.198
2		0.636	0.585	0.576		−.520	−.234	−.005	
3		0.436	0.500			−.610	−.166		
4		0.358				−.931			

		$w_1/w_5 = 0.5825\cdots$	$\lambda = 1/2$		
n	m	1	2	3	4
		Largest RG Exponent			
1		2.050	1.715	1.581	1.524
2		1.501	1.463	1.437	
3		1.397	1.391		
4		1.305			
		Second RG Exponent			
1		1.429	0.908	0.616	0.537
2		0.292	0.307	0.327	
3		0.029	0.216		
4		−.137			
n	m	1	2	3	4
		RG scale factor b			
1		1.244	1.645	2.429	3.400
2		1.323	1.952	2.733	
3		1.476	2.067		
4		1.400			

Table 2: The two largest exponents of the linearized transformation matrix T^* are listed for the simulations at $\lambda=0$, $\frac{1}{2}$, and 1. The lattice simulated was 64×64, and the sizes of the renormalized $L'\times L'$ lattices were L'=51, 41, 31, 21, and 15. Listed are the exponents between renormalized lattices, n and $n+m$, as well as the corresponding RG scale factors b. For example, the entry with $n=3$ and $m=2$ corresponds to an RG transformation between lattices with L'=31 and L'=15. The data should converge for large n, m; but should also become susceptible to finite size effects. For $\lambda=0$, $\frac{1}{2}$, and 1, the largest eigenvalues $y_z=2-\eta_z$, corresponding to the operator $O_{0,1}$, are 1, $\frac{5}{4}$ and $\frac{3}{2}$, respectively.

This exponent can be associated with the operator $O_{0,1}$ corresponding to the crossover exponent of the 8-vertex model, and the exponent η_z can be associated with the S^z correlations. The reason operators such as $O_{1,0}$ may not be found in our MCRG calculation is because such operators correspond to the polarization operator of the 8-vertex model, and such asymmetric operators are not included in the operators shown in Fig. 3. Such operators are not easily included since the momentum-space RG procedure does not preserve the local vertex constraint.

6 Discussion and Conclusions

We have implemented for the first time the marriage of a loop algorithm to alleviate critical slowing down and an MCRG procedure that does not necessitate that one know the symmetries of the underlying classical equivalent beforehand since they are given by the Monte Carlo. We have applied the resulting MCRG method to the study of the $d=1$ spin-$\frac{1}{2}$ quantum XXZ chain. Although this model has a Kosterlitz-Thouless transition, and consequently the convergence toward the fixed point is fairly slow, we are able to obtain the critical exponent η_z associated with the S^z correlation function reasonably well. The exponent η_z comes from the operator $O_{0,1}$, which is associated with the crossover exponent of the 8-vertex model. We have pointed out the difficulties in our study that have prevented us from obtaining the other exponents of the XXZ model. Although this is a preliminary study, our MCRG method shows a great deal of promis, since it is easily generalized to the study of many other quantum models in one and higher dimensions.

Acknowledgments
The authors wish to thank J. Adler, M. Marcu, A. Moreo, and P. A. Rikvold for useful discussions. The MCRG program used was based on code partially written by E. P. Münger. This research was supported in part by the Florida State University Supercomputer Computations Research Institute which is partially funded through contract # DE-FC05-85ER25000 by the U.S. Department of Energy.

References

1. N. Metropolis, A. W. Rosenbluth, M. N. Rosenbluth, A. H. Teller, and E. Teller, J. Chem. Phys. **31**, 1087 (1953).
2. M. Suzuki, Commun. Math. Phys. **51**, 183 (1976); **57**, 193 (1977).
3. M. Suzuki, Prog. Theor. Phys. **56**, 1454 (1976).
4. M. Suzuki, S. Miyashita, and A. Kuroda, Prog. Theor. Phys. **58**, 1377 (1977).
5. H. F. Trotter, Proc. Am. Math. Soc. **10**, 545 (1959).

6. M. Suzuki in *Quantum Monte Carlo Methods in Equilibrium and Nonequilibrium Systems*, editor M. Suzuki, Springer Series in Solid-State Sciences 74, (Springer-Verlag, Berlin, 1987).

7. M. Suzuki, Phys. Lett. A **165**, 387 (1992).

8. J. P. Valleau and D. N. Card, J. Chem. Phys. **57**, 5457 (1972).

9. M. A. Novotny, J. Appl. Phys. **53**, 7997 (1982).

10. M. Falcioni, E. Marinari, M. L. Paciello, G. Parisi, and B. Taglienti, Phys. Lett. **108B**, 331 (1982).

11. A. M. Ferrenberg and R. H. Swendsen, Phys. Rev. Lett. **61**, 2635 (1988).

12. E. P. Münger and M. A. Novotny, Phys. Rev. B **43**, 5773 (1991).

13. A. M. Ferrenberg and R. H. Swendsen, Phys. Rev. Lett. **63**, 1195 (1989).

14. E. P. Münger and M. A. Novotny, Phys. Rev. B **44**, 4314 (1991).

15. R. H. Swendsen and J. S. Wang, Phys. Rev. Lett. **58**, 86 (1987).

16. U. Wolff in *Lattice '89*, Capri 1989, N. Cabbibo et al., editors, Nucl. Phys. B (Proc. Suppl.) **17**, 93 (1990).

17. A. D. Sokal, in *Lattice '90*, Tallahassee 1990, U. M. Heller et al., editors, Nucl. Phys. B (Proc. Suppl.) **20**, 55 (1991).

18. H. G. Evertz, M. Marcu and G. Lana, Phys. Rev. Lett. **70**, 875 (1993).

19. H. G. Evertz and M. Marcu, in *Lattice 92*, Amsterdam 1992, ed. J. Smit et al., Nucl. Phys. B (Proc. Suppl.) **30**, 277 (1993).

20. H. G. Evertz and M. Marcu, in *Quantum Monte Carlo Methods in Condensed Matter Physics*, ed. M. Suzuki (World Scientific 1993).

21. M. N. Barber in *Phase Transitions and Critical Phenomena*, Vol. 8, editors C. Domb and J. L. Lebowitz (Academic Press, London, 1983).

22. V. Privman in *Finite Size Scaling and Numerical Simulation of Statistical Systems*, ed. V. Privman, (World Scientific, Singapore, 1990).

23. K. G. Wilson and M. E. Fisher, Phys. Rev. Lett. **28**, 240 (1972).

24. K. G. Wilson, Rev. Mod. Phys. **47**, 773 (1975).

25. R. H. Swendsen, Phys. Rev. Lett. **42**, 859 (1979).

26. R. H. Swendsen, Phys. Rev. B **20**, 2080 (1979).

27. R. H. Swendsen, Phys. Rev. Lett. **47**, 1159 (1981).

28. M. A. Novotny, D. P. Landau, and R. H. Swendsen, Phys. Rev. B **26**, 330 (1982).

29. M. Kolb, Phys. Rev. Lett. **51**, 1696 (1983).

30. M. A. Novotny and D. P. Landau, Phys. Rev. B **31**, 1449 (1985).

31. M. Kolb in *Quantum Monte Carlo Methods in Equilibrium and Nonequilibrium Systems*, editor M. Suzuki, Springer Series in Solid-State Sciences 74, (Springer-Verlag, Berlin, 1987).

32. M. Suzuki, J. Stat. Phys. **43**, 883 (1986).

33. S. R. White, Phys. Rev. Lett. **69**, 2863 (1992).

34. S. R. White in *Computational Approaches in Condensed-Matter Physics*, Vol. 70 in *Proceedings in Physics*, edited by S. Miyashita, M. Imada, and H. Takayama, page 97, (Springer-Verlag, Berlin, 1992).

35. M.-B. Lepetit and E. Manousakis, Phys. Rev. B **48**, 1028 (1993).

36. M. Barma and B. S. Shastry, Phys. Rev. B **18** 3351 (1978).
37. M. Marcu and A. Wiesler, J. Phys. A **18** 2479 (1985).
38. P. A. Rikvold, unpublished.
39. R. Shankar, Phys. Rev. B **33**, 6515 (1986).
40. L. Biferale and R. Petronzio, Nucl. Phys. B **328**, 677 (1989).
41. C. F. Baillie, R. Gupta, K. A. Hawick, and G. S. Pawley, Phys. Rev. B **45**, 10438 (1992).
42. R. J. Baxter, *Exactly Solved Models in Statistical Mechanics* (Academic Press, London, 1982).
43. S. Katsura, Phys. Rev. **127**, B1508 (1962).
44. E. Lieb, T. Schultz, and D. Mattis, Ann. Phys. (N.Y.) **16**, 407 (1961).
45. M. P. M. den Nijs, Phys. Rev. B **23**, 6111 (1981).
46. J. L. Black and V. J. Emery, Phys. Rev. B **23**, 429 (1981).
47. J. L. Cardy in *Phase Transitions and Critical Phenomena*, Vol. 11, editors C. Domb and J. L. Lebowitz (Academic Press, London, 1987).
48. F. C. Alcaraz, M. N. Barber, and M. T. Batchelor, Phys. Rev. Lett. **58**, 771 (1987); Ann. Phys. (N.Y.) **182**, 280 (1988).
49. G. A. Baker, G. S. Rushbrooke, and M. E. Gilbert, Phys. Rev. **135**, A1272 (1964).
50. J. C. Bonner and M. E. Fisher, Phys. Rev. **135**, A640 (1964).
51. A. Moreo, Phys. Rev. B **36**, 8582 (1987).
52. F. C. Alcaraz and A. Moreo, Phys. Rev. B **46**, 2896 (1992).
53. J. J. Cullen and D. P. Landau, Phys. Rev. B **27**, 297 (1983).
54. J. W. Lyklema, Phys. Rev. Lett. **49**, 88 (1982).
55. B. Nienhuis and M. Nauenberg, Phys. Rev. Lett. **35**, 477 (1975).

OVERCOMING CRITICAL SLOWING DOWN IN QUANTUM MONTE CARLO SIMULATIONS

Hans Gerd Evertz[1,2] and Mihai Marcu[3]

[1] Supercomputer Computations Research Institute,
Florida State University,
Tallahassee, FL 32306-4052, USA
evertz@scri.fsu.edu

[2] Department of Physics and Astronomy
and Center for Simulational Physics
University of Georgia,
Athens, GA 30602, USA

[3] Racah Institute of Physics,
Hebrew University,
91904 Jerusalem, Israel
marcu@vax.huji.ac.il

Abstract

The classical $d+1$-dimensional spin systems used for the simulation of quantum spin systems in d dimensions are, quite generally, vertex models. Standard simulation methods for such models strongly suffer from critical slowing down. Recently, we developed the *loop algorithm*, a new type of cluster algorithm that to a large extent overcomes critical slowing down for vertex models. We present the basic ideas on the example of the F model, a special case of the 6-vertex model. Numerical results clearly demonstrate the effectiveness of the loop algorithm. Then, using the framework for cluster algorithms developed by Kandel and Domany, we explain how to adapt our algorithm to the cases of the 6-vertex model and the 8-vertex model, which are relevant for spin $\frac{1}{2}$ systems. The techniqes presented here can be applied without modification to 2-dimensional spin $\frac{1}{2}$ systems, provided that in the Suzuki-Trotter formula the Hamiltonian is broken up into 4 sums of link terms. Generalizations to more complicated situations (higher spins, different uses of the Suzuki-Trotter formula) are, at least in principle, straightforward.

1 Introduction

Simulations of quantum spin systems are based on mapping the problem to a sequence of simulations of an increasingly anisotropic classical spin system in one more dimension [1]. For example, the spin $\frac{1}{2}$ xxz chain is mapped onto a model with Ising-like spin variables, 4-spin interaction, and constraints. This model is identical to the 6-vertex model, when using eigenstates of S^z as complete sets of intermediate states, or to the 8-vertex model, when using eigenstates of S^x.

For simulations of many interesting physical situations, critical slowing down (CSD) is a major problem. Standard simulation algorithms employ *local* update procedures like e.g. the Metropolis and the heat bath algorithm. With local updates, "information travels slowly", like a random walk [2]. If the relevant length scale is the correlation length ξ, the number of updates necessary to decorrelate large regions, i.e. the autocorrelation time τ, grows like

$$\tau \propto \xi^z, \tag{1}$$

where $z \approx 2$ for local updates, as suggested by the random walk analogy. The dynamical critical exponent z is the quantitative measure of CSD.

A possible way out of this difficulty is to employ *nonlocal* updates, which decorrelate a configuration much more quickly. This is a constructive approach, since up to now no general recipe is known for devising efficient nonlocal moves. The hope that multigrid algorithms would be such a general procedure has unfortunately not been fulfilled. However, in recent years *cluster algorithms* have been successful in a variety of instances [3, 4, 5, 6, 7, 8, 9] (this is a nonexhaustive list of references).

The first cluster algorithm was invented by Swendsen and Wang (SW) [3] for the case of the Ising model. The basic idea is to perform moves that significantly change the Peierls contours characterizing a configuration. As the size of Peierls contours is, typically, anything up to the order of the correlation length, critical slowing down may be strongly reduced or even completely eliminated by this approach. The SW algorithm has been modified and generalized for other spin systems, mostly with two-spin interactions [4, 5, 7]. Notice that for these systems clusters are connected regions of spins, with the same dimensionality as the underlying lattice. A few generalizations along different lines were also done [6, 8, 9].

Recently [10, 11, 12] we introduced *cluster algorithms for vertex models and quantum spin systems*, which are the first cluster algorithms for models with constraints. While [10] is an adaptation of the valleys-to-mountains-reflections (VMR) algorithm [7], originally devised for solid-on-solid models, the *loop algorithm* introduced in [11, 12] does not resemble any existing scheme.

In vertex models the dynamical variables are localized on bonds, and the interaction, usually defined by giving the Boltzmann weights, is between all bonds meeting at a vertex. The possible bond variable values at a vertex are subject to constraints. The statistical weight of a configuration is given by the product over all vertices of the vertex weights.

Our scheme is devised such as to take into account the constraints automatically, and to allow a simple way to construct clusters. For usual spin systems most cluster algorithms start by "freezing" (also called "activating") or "deleting" bonds. Clusters are then sets of sites connected by frozen bonds. In the case of vertex models our idea

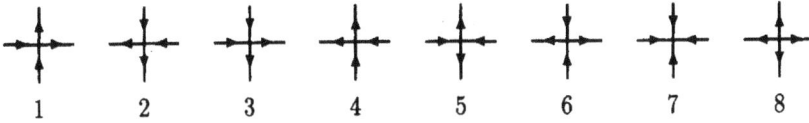

Figure 1: The 8 vertex configurations, $u = 1, ..., 8$, using the notations of [13, 14].

is to define clusters as *closed paths of bonds* ("loops"). To construct such clusters, we have to perform operations at vertices that generalize the freeze-delete procedure. In this context we introduce the concept of *break-up of a vertex*.

In this paper we discuss the loop algorithm in detail for the cases of the 6-vertex and the 8-vertex model [13, 14]. The 8-vertex model is defined on a square lattice. On each bond there is an Ising-like variable that is usually represented as an arrow. For example, arrow up or right means $+1$, arrow down or left means -1. At each vertex we have the constraint that there is an even number of incoming arrows (and consequenly an even number of outgoing ones). In fig. 1 we show the 8 possible configurations u at a vertex, numbered as in [13, 14]. The vertex weights $\rho(u)$ are symmetric under reversal of all arrows; in standard notation [13, 14], we have:

$$
\begin{aligned}
\rho(1) &= \rho(2) = a \,, \\
\rho(3) &= \rho(4) = b \,, \\
\rho(5) &= \rho(6) = c \,, \\
\rho(7) &= \rho(8) = d \,.
\end{aligned}
\tag{2}
$$

The 6-vertex model is defined by $d = 0$. That means that only those 6 vertex configurations are allowed, for which there are exactly two incoming and two outgoing arrows.

The 6-vertex model has two types of phase transitions: Kosterlitz-Thouless (KT) and KDP [13]. For the xxz chain at zero temperature, the former corresponds to the transition at the Heisenberg antiferromagnet point, while the latter corresponds to the transition at the Heisenberg ferromagnet. A submodel exhibiting the KT transition is the F model, defined by $c = 1$, $a = b = \exp(-K)$, $K \geq 0$. The coupling K plays the role of inverse temperature. The KT transition is at $K_c = \ln 2$. The correlation length is finite for $K > K_c$ and infinite for $K \leq K_c$. For the KDP transition, an example is the KDP model itself, defined by $a = 1$, $b = c = \exp(-K)$, $K \geq 0$.

In section 2 we shall describe the loop algorithm on the example of the F model. Besides describing the break-up of a vertex operation and explaining how clusters are constructed, we shall employ the principle of *minimal freezing* in order to optimize the algorithm with respect to a free parameter. In section 3 we analyze the performance of the loop algorithm for the F model. We investigate the exponential autocorrelation times at $K = K_c$ and at $K = K_c/2$. It turns out that the above mentioned optimization is crucial for reducing CSD. In the case of the optimal algorithm, we find a dynamical critical exponent of $z(K_c) = 0.71(5)$ and $z(K_c/2) = 0.19(2)$. In section 4 we review the general formalism for cluster algorithms developed by Kandel and Domany [6],

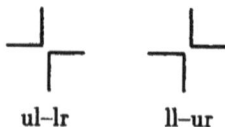

Figure 2: The two break-ups of a vertex used for the F model algorithm.

which we then use in section 5 to formulate the loop algorithm for the general arrow flip symmetric 6-vertex model. For this case too, we show how to find the optimal algorithm by minimizing freezing. In section 6 we develop the loop algorithm for the 8-vertex model. As opposed to the 6-vertex model, it is no longer possible to obtain a unique algorithm by minimizing freezing. We propose a way to overcome this problem. For further generalizations of our algorithm, this is an essential issue. While the discussion of the F model and the 6-vertex model was published before [11, 12], the algorithm for the 8-vertex model is published here for the first time. We present our conclusions in section 7. In particular, we briefly discuss the successful use of the loop algorithm for the study of the the spin $\frac{1}{2}$ Heisenberg antiferromagnet in two dimensions [15].

2   The Loop Algorithm for the F Model

If we regard the arrows on bonds as a vector field, the constraint at the vertices is a zero-divergence condition. Therefore every configuration change can be obtained as a sequence of *loop-flips*. By "loop" we denote an oriented, closed, non-branching (but possibly self-intersecting) path of bonds, such that all arrows along the path point in the direction of the path. A loop-flip reverses the direction of all arrows along the loop.

Our cluster algorithm performs precisely such operations, with appropriate probabilities. It constructs closed paths consisting of one loop or of several loops without common bonds. All loops in this path are flipped together.

We shall construct the path iteratively, following the direction of the arrows. Let bond b be the latest addition to the path. The arrow on b points to a new vertex v. There are two outgoing arrows at v, and what we need is a unique prescription for continuing the path through v. This is provided by a *break-up* of the vertex v. In addition to the break-up, we have to allow for *freezing* of v. By choosing suitable probabilities for break-up and freezing we shall satisfy detailed balance.

The *break-up* operation is defined by splitting v into two corners, as shown in fig. 2. We shall label the two possible break-ups of a vertex by ul–lr (upper-left–lower-right) and ll–ur (lower-left–upper-right). At any corner one of the arrows points towards v, while the other one points away from v. Thus we will not allow e.g. the ul–lr break-up for a vertex in the configuration 3. A "corner flip" is a flip of both arrows. For a given break-up, we only allow the configuration changes resulting from independent corner flips. This preserves the zero divergence condition at v. Notice that a single corner flip transforms a vertex of weight 1 into a vertex of weight e^{-K} and vice-versa. Detailed balance is satisfied with the following probabilities for choosing a given break-up (these probabilities depend on the current vertex configuration u):

$$
p_{\text{ul-lr}}(u) = \begin{cases} re^K & u = 1,2 \\ 0 & u = 3,4 \\ r & u = 5,6 \end{cases}, \qquad p_{\text{ll-ur}}(u) = \begin{cases} 0 & u = 1,2 \\ re^K & u = 3,4 \\ r & u = 5,6 \end{cases}; \qquad (3)
$$

Here r is a free parameter for now.

Freezing of a vertex means that its weight must not change. Since there are only two different vertex weights, we introduce two freezing probabilities. They are already determined by the requirement that for a given vertex configuration the sum of freezing and break-up probabilities must be one:

$$
p_{\text{freeze}}(u) = \begin{cases} 1 - re^K & u = 1,2,3,4 \\ 1 - 2r & u = 5,6 \end{cases}. \qquad (4)
$$

Notice that if we freeze a vertex of type 1, 2, 3 or 4, and we choose to flip the incoming arrow on the bond b, then we must also flip the outgoing arrow that lies on the straight-line continuation of b. We may in addition flip the remaining two arrows, but we do not have to. If on the other hand we freeze a vertex of type 5 or 6, and we choose to flip one of the incoming arrows, then we have to flip all remaining three arrows.

The range of possible values for r is obtained by requiring that all probabilities are between zero and one:

$$
0 \leq r \leq \min(\tfrac{1}{2}, e^{-K}). \qquad (5)
$$

Assume now that we have broken or frozen all vertices. Starting from a bond b_0, we proceed to construct a closed path by moving in the arrow direction. As we move from vertex to vertex, we always have a unique way to continue the path. If a vertex is broken, we enter and leave it along the same corner. If the vertex is frozen and of type 1, 2, 3 or 4, we pass through it on a straight line. At such vertices the path may be self-intersecting (like e.g. at the center of an "8"). Finally, if the latest bond b added to the cluster points to a frozen vertex v of type 5 or 6, the path continues both to the right and to the left of b. One of these directions can be considered as belonging to the loop we came from, the other one as belonging to a new loop, which also contains the second incoming arrow of v. The two loops have to be flipped together. Actually, the zero-divergence condition guarantees that all loops will eventually close.

The break-or-freeze decision for all vertices determines a unique partitioning of the lattice into closed paths that can be flipped independently. We choose to perform single cluster updates, i.e. we "grow" a *single path* from a random starting bond b_0, and flip it. The break-or-freeze decision is only needed for the vertices along the path. Thus the computer time for one path is proportional to the length of that path.

It is easy to see that our algorithm is correct. The proof of detailed balance is completely analogous to that for other cluster algorithms [5, 6]. The simplest formal proof is to show that the algorithm fits into the framework of Kandel and Domany [6]; we shall do this later, for the more general case of the 6-vertex model. The main ingredient here is that $p_{\text{ul-lr}}$ and $p_{\text{ll-ur}}$ already satisfy detailed balance locally. Furthermore, it is not difficult to see that any two allowed configurations can be connected by a finite number of cluster flips. Thus a finite power of the Markov matrix is ergodic.

How do we choose an optimal value for the parameter r ? We have seen that freezing of a vertex of type 5 or 6 forces us to flip two loops together. If we had broken up the vertex instead, we might have been allowed to flip the two loops independently. Thus more freezing leads to larger clusters. We conjecture that the *least possible freezing is optimal*. This is confirmed by numerical tests (see below). From eq. (4) we then obtain

$$r_{\text{opt}} = \begin{cases} \frac{1}{2} & K \leq K_c \\ e^{-K} & K \geq K_c \end{cases} . \tag{6}$$

By maximizing r we also minimize the freezing probability for vertices of type 1, 2, 3 and 4. Notice that if we choose $r = r_{\text{opt}}$, then for $K \leq K_c$ vertices of type 5 and 6 are never frozen, so every path consists of a single loop. For $K > K_c$ on the other hand, vertices of type 1, 2, 3 and 4 are never frozen, so we do not continue a path along a straight line through any vertex.

There are some distinct differences between our loop-clusters and more conventional spin-clusters. For spin-clusters, the elementary objects that can be flipped are spins; freezing binds them together into clusters. Our closed loops on the other hand may be viewed as a part of the *boundary* of spin-clusters (notice that the boundary of spin clusters may contain loops inside loops). It is reasonable to expect that, in typical cases, building a loop-cluster will cost less work than for a spin-cluster. This is an intrinsic advantage of the loop algorithm.

The last remark can be exemplified nicely for the F model, where a spin-cluster algorithm – the VMR algorithm [10] – is also available. At K_c one can see that if we use $r = r_{\text{opt}}$, loop-clusters are indeed parts of the boundary of VMR spin-clusters. Since flipping a loop-cluster is not the same as flipping a VMR cluster, we expect the two algorithms to have different performance. We found (see [10] and the next section) that in units of clusters, the VMR algorithm is more efficient, but in work units, which are basically units of CPU time, the loop algorithm wins. At $K_c/2$, where the loop-clusters are not related to the boundary of VMR clusters, we found the loop algorithm to be more efficient both in units of clusters and in work units, with a larger advantage in the latter. More details on the comparison between the VMR algorithm and the loop algorithm will be published elsewhere.

3   Numerical Study of the Algorithm's Performance

We tested our new algorithm on $L \times L$ square lattices with periodic boundary conditions[1], both at the transition point K_c and at $\frac{1}{2}K_c$ deep inside the massless phase. We carefully analyzed autocorrelation functions and determined the exponential autocorrelation time τ. At infinite correlation length, *critical slowing down* is quantified by the relation [5]

$$\tau \propto L^z . \tag{7}$$

Local algorithms are slow, with $z \approx 2$. To be on the safe side, we performed runs with a local algorithm that flips arrows around elementary plaquettes with Metropolis probability, and indeed found $z = 2.2(2)$ at $K = K_c$.

In order to make sure that we do observe the slowest mode of the Markov matrix, we measured a whole range of quantities and checked that they exhibit the same τ. As in [10], the slowest mode is strongly coupled to the sublattice energy[2]. The two sublattice energies [10] add up to the total energy. The constraints of the model cause them to be strongly anticorrelated. Within our precision the true value of τ is *not* visible from autocorrelations of the total energy, which decay very quickly. Only for the largest lattices do we see a small hint of a long tail in the autocorrelations. A similar situation occurred in [10], where, when decreasing the statistical errors, the decay governed by the true τ eventually became visible. Note that as a consequence of this situation, the so-called "integrated autocorrelation time" [5] is much smaller than τ, and it would be completely misleading to evaluate the algorithm based only on its values.

We shall quote autocorrelation times τ in units of "sweeps" [5]. We define a sweep such that on average each bond is updated once during a sweep. Thus, if τ^{cl} is the autocorrelation time in units of clusters, then

$$\tau = \tau^{\mathrm{cl}} \frac{< \text{cluster size} >}{2L^2} . \tag{8}$$

Each of our runs consisted of between 50000 and 200000 sweeps. Let us also define the exponent z^{cl} by $\tau^{\mathrm{cl}} \propto L^{z^{\mathrm{cl}}}$, and a cluster size exponent c by $< \text{cluster size} > \propto L^c$. We then have:

$$z = z^{\mathrm{cl}} - (2 - c) . \tag{9}$$

---

[1]In order to make contact with studies of the related BCSOS model [10], we did not allow loops that wind around the lattice. In [16] we allowed such loops and found somewhat smaller autocorrelation times.

[2]The sublattice energy is easily defined in the equivalent BCSOS model [10]. In vertex language, it is defined, up to irrelevant additive and multiplicative constants, as follows. Assign value $+1$ to arrows pointing in positive Cartesian directions, -1 otherwise. Label all vertices black or red in checkerboard fashion. Break black vertices ul–lr and red ones ll–ur (see fig. 2). Multiply the two arrow values at each resulting corner with each other, and, for black vertices only, with -1. Add all products. The other sublattice energy is obtained by interchanging black with red in the checkerboard assignment.

L	$K = K_c$	$K = \frac{1}{2}K_c$
8	1.8(1)	4.9(4)
16	3.0(2)	5.6(2)
32	4.9(4)	6.2(3)
64	7.2(7)	7.4(3)
128	15.5(1.5)	8.3(2)
256	20.5(2.0)	
z	0.71(5)	0.19(2)

Table 1: Autocorrelation time τ at $r=r_{opt}$, and the resulting dynamical exponent z.

Table 1 shows the autocorrelation time τ for the optimal choice $r = r_{opt}$. At $K = \frac{1}{2}K_c$, deep inside the massless phase, critical slowing down is almost completely absent. A fit according to eq. 7 gives $z = 0.19(2)$. The data are also consistent with $z = 0$ and only logarithmic growth. For the cluster size exponent c we obtained $c = 1.446(2)$, which points to the clusters being quite fractal. At the phase transition $K = K_c$ we obtained $z = 0.71(5)$, which is still small. The clusters seem to be less fractal: $c = 1.060(2)$.

We noted above that at $K = K_c$ and for the optimal choice of r, the loop-clusters are related to the VMR spin-clusters. In [10] we obtained for the VMR algorithm at $K = K_c$ the result $z^{cl} = 1.22(2)$, but we had $c = 1.985(4)$, which left us with $z = 1.20(2)$. In this case it is the smaller dimensionality of the clusters that make the loop algorithm more efficient.

As mentioned, no critical slowing down is visible for the integrated autocorrelation time of the total energy. At $K = K_c$, $\tau_{int}(E)$ is only 0.80(2) on the largest lattice, and we find $z_{int}(E) \approx 0.20(2)$. At $K = \frac{1}{2}K_c$, $\tau_{int}(E)$ is 1.1(1) on all lattice sizes, so $z_{int}(E)$ is zero.

What happens for non-optimal values of r? Table 2 shows our results on the dependence of z on r. z rapidly increases as r moves away from r_{opt}. This effect seems to be stronger at $\frac{1}{2}K_c$ than at K_c. We thus see that the optimal value of r indeed produces the best results, as conjectured from our principle of *least possible freezing*.

In the massive phase close to K_c, we expect that $z(K_c)$ will determine the behaviour of τ in a similar way as in ref. [10]. To confirm this, a finite size scaling analysis of τ for $K < K_c$ is required.

4   The Kandel-Domany framework

Cluster algorithms are conveniently described in the general framework of Kandel and Domany [6], which guarantees that we are using a correct Markov process. More important than convenience is the fact that this framework allows for a lot of flexibility

K	r	z
$\frac{1}{2}K_c$	0.500	0.19(2)
$\frac{1}{2}K_c$	0.450	1.90(5)
$\frac{1}{2}K_c$	0.400	$\geq 2.6(4)$
K_c	0.500	0.71(5)
K_c	0.475	0.77(6)
K_c	0.450	0.99(6)
K_c	0.400	$\geq 2.2(1)$

Table 2: Dependence of the dynamical critical exponent z on the parameter r. We use "$\geq$" where for our lattice sizes τ increases faster than a power of L.

in defining new algorithms.

Let us consider the system defined by the partition function

$$Z = \sum_u \rho(u)\,, \tag{10}$$

where u are the configurations to be summed over, and $\rho(u)$ is the Boltzmann weight of u. Let us define a *set* of new Boltzmann weight functions $\rho_i(\tilde{u})$. The index i numbers these "modified interactions". Usually $\rho(u)$ is a product over local Boltzmann weights $\rho^c(u)$, where c are cells in the lattice. In this case the index i will be a multiindex over the relevant set of cells.

Assume now that during a Monte Carlo simulation we arrived at a particular configuration u. We choose a new configuration in a two step procedure. The first step is to replace the weight function ρ with one of the functions ρ_i. The probability $p_i(u)$ of choosing a specific i only depends on the current configuration u. It has to satisfy the following equations:

$$p_i(u) = q_i \, \frac{\rho_i(u)}{\rho(u)}\,, \qquad \sum_i p_i(u) = 1\,, \tag{11}$$

where $q_i \geq 0$ are constants independent of u. The second step is to update the configuration by employing a procedure that satisfies detailed balance for the chosen Boltzmann weight ρ_i. Kandel and Domany have shown that this two-step procedure satisfies detailed balance for the original Boltzmann weight function ρ.

If ρ is a product over some set of lattice cells as described above, and we decide upon ρ_i's with a similar product structure, it is sufficient to choose for each component of the multiindex i (i.e. for each cell) the modified interaction independently. In this case we shall have to fulfill (11) independently for each cell. For vertex models, the cells will be the vertices themselves.

ll–ur   ul–lr   straight

Figure 3: All three possible break-ups of a vertex.

5   The Loop Algorithm for the Full 6-Vertex Model

For the purpose of the loop algorithm, the main difference between the F model and the full arrow flip symmetric 6-vertex model is that, for the Boltzmann weights of (2), $a \neq b$. Consequently, we have to replace the freezing operation for the vertex configurations 1, 2, 3, 4, by three different operations. Firstly there is the freezing of 1, 2 (i.e. the freezing of the weight a), secondly the freezing of 3, 4 (i.e. the freezing of the weight b). Thirdly there is a new break-up operation, that allows transitions between the groups of weights 1,2 and 3, 4. This new break-up is not a break-up into two corners, but into two straight lines, as shown in fig. 3. As with the ul–lr and ll–ur break-ups, the straight break-up splits a vertex into two pieces, each of which contains an incoming and an outgoing arrow. During a cluster flip, each piece may be flipped independently.

Clusters are contructed in a very similar fashion to the case of the F model. Starting from a random bond b_0, we go from vertex to vertex on a path in the direction of the arrows. If we encounter a vertex v for which one of the break-ups was chosen, there is a unique way to pass through it. If on the other hand v is frozen, we continue the path in the direction of one of the outgoing arrows, and mark the other two arrows as also belonging to the cluster. Two things can now happen. If the path returns at a later stage to the vertex v, this will automatically be along the two marked arrows. If the path closes without returning to v, we have to append to the cluster a new "elementary loop" starting with the two marked arrows. The constraint guarantees that all such loops will eventually close.

After having constructed the cluster, we flip it, as in the case of the F model. Thus we have completely characterized our algorithm, up to the definition of the probabilities to choose a break-up or freeze for a vertex. These probabilities are most conveniently described by using the framework of Kandel and Domany [6].

For a given vertex, which is in the current configuration u with the Boltzmann weight $\rho(u)$, we define 6 modified Boltzmann weights ρ_i, $i = 1, ..., 6$, corresponding to specific break-up and freeze operations. These definitions are contained in table 3. The labelling of the modified Boltzmann weights is completely arbitrary, and the fact that their number – six – is the same as the number of allowed vertex configurations is just a coincidence. As discussed in [6] (see also table 3), *freezing* is described by introducing one modified weight for each different value of $\rho(u)$. For example, to freeze the value a, we choose the interaction ρ_1 to be $\rho_1(\tilde{u}) = 1$ if $\rho(\tilde{u}) = a$, and $\rho_1(\tilde{u}) = 0$ otherwise. In other words, when ρ_1 is chosen, transitions between $\tilde{u} = 1$

i	action	$\rho_i(\tilde{u})$	$p_i(u)$
1	freeze 1,2	1, $\rho(\tilde{u}) = a$ 0, else	q_1/a, $\rho(u) = a$ 0, else
2	freeze 3,4	1, $\rho(\tilde{u}) = b$ 0, else	q_2/b, $\rho(u) = b$ 0, else
3	freeze 5,6	1, $\rho(\tilde{u}) = c$ 0, else	q_3/c, $\rho(u) = c$ 0, else
4	ll–ur	0, $\rho(\tilde{u}) = a$ 1, else	0, $\rho(u) = a$ $q_4/\rho(u)$, else
5	ul–lr	0, $\rho(\tilde{u}) = b$ 1, else	0, $\rho(u) = b$ $q_5/\rho(u)$, else
6	straight	0, $\rho(\tilde{u}) = c$ 1, else	0, $\rho(u) = c$ $q_6/\rho(u)$, else

Table 3: The modified Boltzmann weight functions ρ_i and the probabilities $p_i(u)$ to choose them at a vertex in current configuration u.

and $\tilde{u} = 2$ cost nothing, whereas the vertex configurations 3, 4, 5, and 6 are not allowed. Each *break-up* is also described by one modified weight. As an example take the ul–lr break-up. It is given by the modified weight number 5, with $\rho_5(\tilde{u}) = 1$ if $\rho(\tilde{u}) = a$ or c, and $\rho_5(\tilde{u}) = 0$ if $\rho(\tilde{u}) = b$. In other words, with the new interaction ρ_5, transitions between 1, 2, 5 and 6 cost nothing, while the vertex configurations 3 and 4 are not allowed. This corresponds precisely to allowing independent corner flips in a ul–lr break-up.

Notice that, in general, the break-ups correspond to allowing transitions between two groups of two configurations, each group being defined as the set of all configurations having the same given weight. By a slight abuse of language, we shall talk about the transition between the two weights. This point of view will be useful for the 8-vertex model.

In table 3 we also give the probabilities $p_i(u)$ to replace the original Boltzmann weight ρ by the modofied one ρ_i. Fulfillment of (11) ensures detailed balance and the proper normalization of probabilities. With the definitions in table 3, the first part of (11) is automatically fulfilled. Normalization of probabilities (i.e. the second part of (11)) implies constraints on the q_i:

$$q_1 + q_5 + q_6 = a \,,$$
$$q_2 + q_4 + q_6 = b \,, \tag{12}$$
$$q_3 + q_4 + q_5 = c \,.$$

We have seen that freezing forces loops to be flipped together. Previous experience with cluster algorithms suggests that it is advantageous to be able to flip loops independently, as far as possible. We therefore introduce the principle of *minimal freezing* as a guide for choosing the optimal values for the constants q_i: we shall minimize the freezing probabilities, given the constraints (12) and $q_i \geq 0$. From (12) we immediately see that it is possible to minimize q_1, q_2 and q_3 simultaneously (by increasing q_4, q_5 and q_6). But let us discuss in detail the optimal values of the q_i for all 4 phases of the model [13, 14].

Let us first look at phase IV, where $c > a + b$. To minimize the freezing of weight c, we have to minimize q_3. From (12), $q_3 = c - a - b + q_1 + q_2 + 2q_6$. With $q_i \geq 0$ this implies $q_{3,\min} = c - a - b$. The minimal value of q_3 can only be chosen if *at the same time* we set $q_1 = q_2 = 0$, i.e. minimize (in this case do not allow for) the freezing of the smaller weights a and b. The optimized parameters for phase IV are then:

$$q_1 = 0, \quad q_2 = 0, \quad q_3 = c - a - b,$$
$$q_4 = b, \quad q_5 = a, \quad q_6 = 0 . \tag{13}$$

In phase I the situation is technically similar. Here $a > b + c$, and we minimize freezing with $q_1 = a - b - c$ and $q_2 = q_3 = 0$. The same holds for phase II, $b > a + c$, where we obtain minimal freezing for $q_2 = b - a - c$ and $q_1 = q_3 = 0$.

Phase III (the massless phase) is characterized by $a, b, c < \frac{1}{2}(a + b + c)$. Here we can set all freezing probabilities to zero. Thus,

$$q_1 = 0, \quad 2q_4 = b + c - a ,$$
$$q_2 = 0, \quad 2q_5 = c + a - b ,$$
$$q_3 = 0, \quad 2q_6 = a + b - c . \tag{14}$$

We finish this section by showing how to obtain the F model algorithm described in section 2 from the more general 6-vertex algorithm. For the particular case of the F model, we only have phases III and IV, so we can set the freezing probabilities $q_1 = q_2 = 0$, i.e. we completely renounce these two freezing operations. Since $a = b$, (12) then implies $q_4 = q_5$. It is now straightforward to see that, with the identification of the straight break-up to the freezing of 1, 2, 3, 4, we have recovered the algorithm of section 2.

6   A Proposal for the 8-Vertex Model

For the 8-vertex model, the constraint at the vertices is no longer a zero-divergence condition. Therefore we cannot expect any more to have clusters that are made out of loops for which all the arrows point in the same direction. In the 6-vertex case this requirement was needed in order to preserve the constraints. The 8-vertex constraints on the other hand are already preserved if at each vertex the number of flipped arrows is even. This will allow us to devise an algorithm with clusters that are again collections of loops, but for which the orientation plays no role.

Let us consider the freeze and break-up operations for a given vertex v. Since we are dealing with the arrow flip symmetric 8-vertex model (see section 1), freezing would mean that the only allowed configuration change is the flip of all four arrows around v. This is similar to the 6-vertex case. Of course, we have here one extra weight that can be frozen, namely $d = \rho(7) = \rho(8)$. For the break-up the situation is only slightly more complicated. Consider as an example the ul–lr break-up. A corner flip performs transitions not only between the weights a and c, but also between the weights b and d. Within the framework of Kandel and Domany we can thus define two modified Boltzmann weights. The first, which we shall denote $\rho_{ac}(\tilde{u})$, obeys $\rho_{ac}(\tilde{u}) = 1$ for $\rho(\tilde{u}) = a$ or c, and $\rho_{ac}(\tilde{u}) = 0$ otherwise; it allows transitions without any cost between the vertex configurations 1, 2, 5, 6, and it prohibits the configurations 3, 4, 7, 8, from ever occuring. The second Boltzmann weight, which we shall denote $\rho_{bd}(\tilde{u})$, is obtained in the same way, by interchanging $a \leftrightarrow b$, $c \leftrightarrow d$, and $\{1,2,5,6\} \leftrightarrow \{3,4,7,8\}$. For the ll–ur and the straight break-ups we can also define two modified Boltzmann weights each in a completely analogous fashion.

The 10 modified Boltzmann weights can be described very elegantly using a symmetric matrix. To denote a weight, we use greek letters: α, β take values in the set $\{a, b, c, d\}$. The modified weights associated with freezing are denoted by $\rho_{\alpha\alpha}$, while those associated with the break-ups (two for each break-up) are denoted by $\rho_{\alpha\beta} \equiv \rho_{\beta\alpha}$, $\alpha \neq \beta$. They are fully characterized by:

$$\rho_{\alpha\beta}(\tilde{u}) = \begin{cases} 1 & \rho(\tilde{u}) = \alpha \quad \text{or} \quad \rho(\tilde{u}) = \beta \\ 0 & \text{else} \end{cases} . \tag{15}$$

The probabilities $p_{\alpha\beta}$ to choose one of these modified weights are then given by the first part of (11), while the second part of (11) implies the following 4 relations for the 10 constants $q_{\alpha\beta}$ ($\equiv q_{\beta\alpha}$):

$$\sum_{\beta} q_{\alpha\beta} = \alpha . \tag{16}$$

At this stage we again employ the principle of minimal freezing. In the case of the 6-vertex model, there were 3 freezing operations, hence 3 constants to minimize. Together with the 3 relations (12) this was enough for obtaining a unique optimal choice for the 6 q's. Here we have 4 freezing operations and 4 relations in (16). Thus even after minimizing freezing, we are left with two free parameters. We do not have any additional physical principle that allows us to fix them. Minimizing something else than freezing is certainly not good. It is also easy to check that no miraculous cancellations happen. The two undesired parameters are here to stay.

Let us indicate one possible solution to this problem. It is straightforward to prove that detailed balance is still fulfilled if the q's are chosen stochastically at each step, provided that this choice is independent of the current configuration u. Thus we propose the following procedure. Parametrize the q's such that freezing is minimal and (16) is fulfilled. For each vertex, then choose the two free parameters randomly, according to some fixed distribution.

After choosing a modified weight at each vertex, the clusters can be constructed in exactly the same way as for the 6-vertex model. If we start at a link b_0, "grow" a cluster and flip it, the amount of work will again be proportional to the cluster size. This completes the description of the loop algorithm for the 8-vertex model.

7   Conclusions and Outlook

We have presented a new type of cluster algorithm that considerably accelerates the simulation of vertex models. The clusters are closed paths of bonds, and the constraints at the vertices are automatically satisfied. We have successfully tested our algorithm for the F model and found remarkably small dynamical critical exponents.

The most important application of our algorithm seems to be the *critical acceleration of Quantum Monte Carlo simulations* This application is based on the fact that quantum spin systems in one and two dimensions can be mapped into vertex models in $1 + 1$ and $2 + 1$ dimensions via the Suzuki-Trotter formula and suitable splittings of the Hamiltonian [1].

The simplest example is the spin $\frac{1}{2}$ xxz quantum chain, which is mapped directly into the 6-vertex model or the 8-vertex model. For higher spins, more complicated vertex models result (e.g. 19-vertex model for spin one) [1].

For (2+1) dimensions, different splittings of the Hamiltonian lead to quite different vertex models, in particular on quite different lattice types [1]. For example, in the case of spin $\frac{1}{2}$ we can choose between simple 6-vertex or 8-vertex models on a quite complicated $2 + 1$ dimensional lattice (if the Hamiltonian is split into 4 sums of link terms), and models on a *bcc* lattice, with 8 bonds and a large number of configurations per vertex (if the Hamiltonian is split into 2 sums of plaquette terms).

For the simulation of the *2-dimensional Heisenberg antiferromagnet and ferromagnet* using the former splitting, all relevant formulas have been worked out in the present paper. Actually, the low temperature properties of the antiferromagnet have recently been investigated by Wiese and Ying [15] *using our algorithm*. Their calculation is, in our opinion, the first high quality verification of the magnon picture for the low lying excitations. In particular, this excludes to a much higher degree of confidence than before the speculation (some years ago widespread) that the model had a nonzero mass gap.

Notice that, similar to other cluster algorithms [5], it is straightforward to define improved observables. The investigation [15] in fact uses them.

Let us also remark that the loop algorithm can easily change global properties like the number of world lines or the winding number (see [1]). Thus it is well suited for simulations in the grand canonical ensemble.

Last, but not least, the loop algorithm just might open up a new avenue for taming the notorious fermion sign problem.

References

1. For basics of the Quantum Monte Carlo method see this volume, and, especially, its precursor (from which references to the original articles may be taken):
   M. Suzuki editor, *Quantum Monte Carlo methods in equilibrium and nonequilibrium systems*, Taniguchi symposium, Springer series in Solid State Physics ; 74 (1987).
2. P. C. Hohenberg and B. I. Halperin, Rev. Mod. Phys. **49** (1977) 435.
3. R. H. Swendsen and J. S. Wang, Phys. Rev. Lett. **58** (1987) 86.
4. R. C. Brower and P. Tamayo, Phys. Rev. Lett. **62** (1989) 1087;
   U. Wolff, Phys. Rev. Lett. **62** (1989) 361, Nucl. Phys. **B322** (1989) 759, and Phys. Lett. **228B** (1989) 379.
5. For reviews, see e.g.:
   U. Wolff, in *Lattice '89*, Capri 1989, N. Cabbibo et al., editors, Nucl. Phys. B (Proc. Suppl.) **17** (1990) 93;
   A. D. Sokal, in *Lattice '90*, Tallahassee 1990, U. M. Heller et al., editors, Nucl. Phys. B (Proc. Suppl.) **20** (1991) 55.
6. D. Kandel and E. Domany, Phys. Rev. **B43** (1991) 8539.
7. H. G. Evertz, M. Hasenbusch, M. Marcu, K. Pinn and S. Solomon, Phys. Lett. **254B** (1991) 185, and in *Workshop on Fermion Algorithms*, Jülich 1991, H. J. Herrmann and F. Karsch editors, Int. J. Mod. Phys. **C3** (1992) 235.
8. R. Ben-Av, D. Kandel, E. Katznelson, P. Lauwers and S. Solomon, J. Stat. Phys. **58** (1990) 125.
9. D. Kandel, R. Ben-Av and E. Domany, Phys. Rev. Lett. **65** (1990) 941.
10. M. Hasenbusch, G. Lana, M. Marcu and K. Pinn, *Cluster algorithm for a solid-on-solid model with constraints*, Phys. Rev. **B46** (1992) 10472.
11. H.G. Evertz, G. Lana and M. Marcu, Phys. Rev. Lett. **70** (1993) 875.
12. H.G. Evertz and M. Marcu, in *Lattice 92*, Amsterdam 1992, ed. J. Smit et al., Nucl. Phys. B (Proc. Suppl.) **30** (1993) 277.
13. E. H. Lieb, Phys. Rev. Lett. **18** (1967) 1046;
    E. H. Lieb and F. Y. Wu, *Two-dimensional Ferroelectric Models*, in *Phase Transitions and Critical Phenomena* Vol. **1**, C. Domb and M. S. Green, editors, (Academic, 1972) p. 331.
14. R. J. Baxter, *Exactly Solved Models in Statistical Mechanics* (Academic, 1989).
15. U.J. Wiese and H.P. Ying, Bern preprint, bulletin board cond-mat/9212006.
16. M. A. Novotny and H. G. Evertz, in *Quantum Monte Carlo Methods in Condensed Matter Physics*, ed. M. Suzuki (World Scientific 1993).

QUANTUM MANYBODY SPIN SYSTEMS IN RANDOM FIELDS AND ANISOTROPIES

P. Reed
School of Computing and Information Systems
University of Sunderland
Sunderland, SR2 7EE
U.K.

ABSTRACT

The spin-1 antiferromagnetic and ferromagnetic chain in the presence of two-fold random anisotropy have been investigated by Trotter-Suzuki decomposition and transfer matrix methods. Results for the energy and the effect of randomness on the Haldane gap are presented for the antiferromagnetic chain. For the ferromagnetic chain results for the magnetization and susceptibility and given which indicate the irrelevance of the anisotropy.

The spin-½ chain frustrated by random bonds and fields has also been investigated. Results for the entropy and magnetization are given.

INTRODUCTION

The theory of interacting quantum spin systems is a challenging problem in physics still producing new and often surprising results. Exact results are rare and mostly restricted to one dimension [1], with some complete solutions, theorems and near theorems having been found. Amongst these are the Bethe ansatz groundstate solution of the spin-½ antiferromagnet [2], which can be extended to other "integrable" systems, and Katsura [3] solution for the thermodynamics of the spin-½ XY chain. More recently results for the groundstate and first excited state of the spin-1 antiferromagnetic chain have been produced which now have the status of a near theorems [3]. Outside the few exact results found further insights have been provided by spin-wave theory which has given a good theoretical framework for understanding quantum spin systems [4]. In the presence of randomness such as might be expected to be present in real systems the problem becomes even more complex. Here there is a lack of exact results and spin wave theory fails due to the lack of translational invariance. In the weak randomness limit other approaches have been adopted such as the coherent-anisotropy-field approximation of Huber [5] used to investigate the random anisotropy field spin-1 ferromagnetic chain. Random field models have been investigated in the continuum limit in a field theoretical framework for the ferromagnetic spin-½ chain [6,7], and the random bond spin-½ antiferromagnetic chain has also been investigated using renormalisation group methods [8].

In the theory of classical non-quantum systems one of the most powerful tools used to unlock their critical behaviour has been simulation. Indeed it might be argued that the spin-glass problem was "solved" by this approach [9]. Simulation is now available for the study of quantum systems with randomness [10]. However, few such simulation have been completed and most of the numerical work has been directed to diagonalization of small two dimensional clusters frustrated by random bonds [11]. One of the reasons for this may be the difficulty with the troublesome "sign" problem which bedevils quantum Monte Carlo simulation of random frustrated systems. A strategy for circumventing this problem is to use the transfer matrix method of Binder and Morgenstern [12]. This method is totally general and can be used for any sort of decomposition including large cell decomposition [13]. This is the method adopted in two recent simulations [14,15], and a brief account of this approach is given in the appendix.

The effect of randomness on both spin-1 and spin-½ Hamiltonians has been investigated using transfer matrix methods.

METHOD

The approach is the same for all Hamiltonians discussed and uses the decomposition method of Suzuki [10] to transform the one dimensional quantum system into a two dimensional classical system. The classical analogue system is composed of a square lattice with site variables that can take 2s+1 values, where s is the quantum number of the original spins.

Each of the Hamiltonians considered has the same form consisting of two particle nearest neighbour terms $\{H_{i, i+1}\}$ and single site terms $\{H_i\}$. So the Hamiltonian is

$$H = \sum_{i=1}^{n-1} H_{i,i+1} + \sum_{i=1}^{n} H_i \qquad (1)$$

where the length of the chain is n. Both $\{H_{i, i+1}\}$ and $\{H_i\}$ may contain random potentials. Equation 1 is rewritten

$$H = \left\{ \sum_{\substack{i \\ even}} H_{i,i+1} + H_i \right\} + \left\{ \sum_{\substack{i \\ odd}} H_{i,i+1} + H_i \right\}$$

$$= H_e + H_0. \qquad (2)$$

The following decomposition [10] is now used

$$\left(e^{-\beta(H_i + H_0)/m} \right)^m = \left(e^{\frac{-\beta H_e}{m}} e^{\frac{-\beta H_0}{m}} \right)^m + 0\left(\frac{1}{m}\right)$$

and the partition function Z and free energy calculated as

$$Z = Trace < \left(e^{\frac{-\beta H_e}{m}} e^{\frac{-\beta H_0}{m}} \right)^m > + 0 \left(\frac{1}{m^2} \right)$$

(3)

and

$$F = \frac{-1}{\beta} \log_e Z.$$

Note that the error term in Z is quadratic in $\frac{1}{m}$. This is because the decomposition used here satisfies the conditions of the evenness theorem [16] even though H_e and H_0 may only be Hermitian. The form of the error term is in fact a polynomial in $\left(\frac{1}{Tm} \right)^2$ when T is temperature. This known form can be used to extrapolate to the limit m→∞ where decomposition is exact. However the presence of T in the denominator restricts the validity of the results to $\frac{1}{Tm} < 1$. Finally complete sets are inserted between the exponential operators in equation 3. This transforms the system into a 2m x n lattice, periodic in the m direction and with (2s+1)-valued classical variable at each site. The details of this procedure can be found in many places including [17].

The lattice is shown in the appendix in figure A1. The shaded squares indicate four site interactions. The weight of these interactions is calculated from

$$< s_i s_{i+1} | e^{-\beta \left(H_{i;i+1} + \frac{H_i}{2} + \frac{H_{i+1}}{2} \right)} | s'_{i+1} \ s'_i >.$$

(4)

This expression is easily evaluated numerically by matrix multiplication using a representation of $H_{i;i+1} + \frac{H_i}{2} + \frac{H_{i+1}}{2}$ as the direct product of spin matrices. Since the Hamiltonian contains random terms the interactions within the shaded squares will differ from column to column, though they are the same in each column.

Z is calculated using the method outlined in the appendix. Thermodynamic functions are found by numerical differentiation using finite differences.

SPIN-1 MODEL IN RANDOM ANISOTROPY FIELD

The Hamiltonian for the spin-1 linear chain in random 2-fold anisotropy fields is taken to be

$$H = -\left\{ J\sum_{i=1}^{n-1} \underline{S}_i \underline{S}_{i+1} + D\sum_{i=1}^{n} (\underline{N}_i \underline{S}_i)^2 + h\sum_{i=1}^{n} S_i^z \right\} \qquad (5)$$

Here $\underline{S}_i$ are spin-1 operators, D is the strength of the anisotropy field, $\underline{N}_i$ are unit vectors randomly oriented and h is an external magnetic field. In the above notation J > 0 corresponds to ferromagnetic and J < 0 to antiferromagnetic coupling. Of these J > 0 is perhaps the most physically relevant corresponding as it does to a model [18] describing compounds of transition metals and rare earths. In such materials the model proposes that each magnetic ion has a local easy axis arising from interactions with crystal fields. The antiferromagnetic case (J < 0) is also interesting because of the result of Haldane [3] that the D = h = 0 model has a gap between the groundstate and first excited state. This gap is also predicted to survive non-random 2-fold anisotropy [19]. This anisotropy can take the form of a mixture of easy plane and easy axis anisotropy. It is therefore not clear what effect randomness may have particularly since thermal fluctuations in fact increase this gap [20]. This question is investigated together with a discussion of the energy.

Antiferromagnetic Coupling (J < 0)

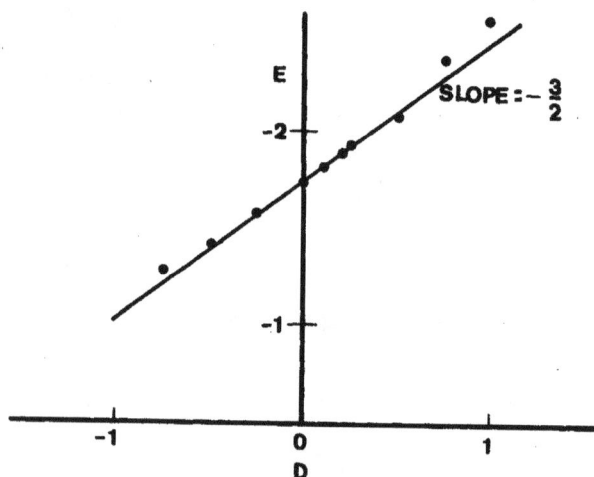

Figure 1. Antiferromagnetic groundstate energy E against anisotropy field strength D, both in units of |J|, for spin-1 Hamiltonian (equation 5). Linear extrapolation.

The method outlined in the previous section has been used to collect data for a single chain of length n = 252 spins. Thus only one realisation of the anisotropy field was used; however, the length is sufficiently long to produce thermodynamically typical results. The calculation was done for m = 4,5,6. These values of the so called "Trotter" dimension though modest are sufficient for extrapolation purposes. All results quoted in this subsection involve extrapolation to m = ∞ using linear fitting and the values for m = 5 and 6 or quadratic fitting to the data for m = 4,5 and 6. The extrapolation used is indicated in the figure captions.

Figure 1 shows the energy for h = 0 and a range of D values. The results have been projected to T = 0. The data for $|D| \leq 0.5|J|$ is seen to fit to high accuracy the form

$$E(D) = E(0) - \frac{2}{3} D, \tag{6}$$

where E(D) is the energy at anisotropy strength D. This result can be confirmed by first order perturbation theory. This would predict a shift in groundstate energy per site of

$$-\frac{D}{n} < 0 | \sum_i (N_i \cdot S_i)^2 | 0 >. \tag{7}$$

Where $<0|$ is the groundstate which is a singlet and hence rotationally invariant. It is thus easy to see that

$$< 0 | S_i^\alpha S_i^\beta | 0 > = 0 \text{ for } \alpha \neq \beta.$$

So 7 becomes

$$\frac{-D}{n} < 0 | \sum_i \left[(N_i^x S_i^x)^2 + (N_i^y S_i^y)^2 + (N_i^z S_i^z)^2 \right] | 0 >,$$

and again because of the rotational invariance of $<0|$ on average we have

$$< 0 | \sum_i (N_i^x S_i^x)^2 | 0 > = < 0 | \sum_i (N_i^y S_i^y)^2 | 0 > = < 0 | \sum_i (N_i^z S_i^z)^2 | 0 >$$

$$= \frac{1}{3} s(s+1) \sum_i (N_i^z)^2 = \frac{2}{9} n.$$

So the correction to the groundstate energy is just $\frac{-2}{3} D$ in agreement with the data and equation 6. The reduction in groundstate energy for D > 0 is no indication of how the gap is effected. To analyse this the susceptibility is investigated using the approach of Kubo [21]. The existence of a mass gap above the perturbed groundstate, which if it is assumed still has the property $<\phi|(\sum_i s_i^x)^2|\phi> = 0$ should result in the uniform susceptibility in the z direction decaying to zero as temperature is reduced. This behaviour is similar to that which occurs for the one dimensional Ising antiferromagnet. Thus conversely a finite intercept on the χ-axis is taken as a signature of a zero gap. Results are shown in figures 2a and 2b.

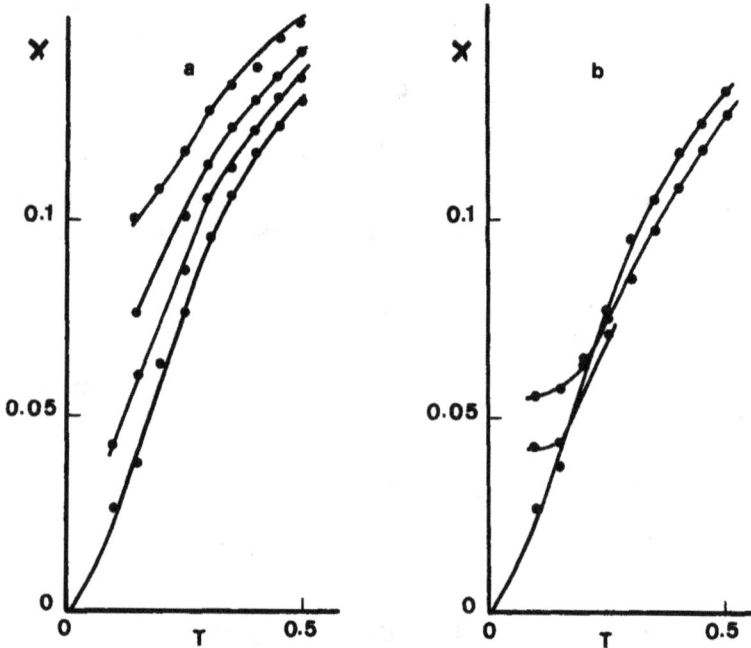

Figure 2. Susceptibility of antiferromagnetic spin-1 chain (equation 5) against temperature T in units of | J| (a) shows plots for D = -0.75| J|, -0.5| J|, -0.25| J|, 0 from top to bottom.   (b) shows plots for D = | J|, 0.75| J|, 0 from top to bottom. Linear extrapolation.

For D < 0 it is clear that the susceptibility terminates on a finite intercept on the χ axis. A similar conclusion follows for D > 0, however the effect is less obvious. The data in figures 2a and 2b has been obtained for large values of | D| and does not exclude the possibility of a small region near D ≈ 0 where the gap survives. To investigate this data has been collected for D = 0.1| J| and 0.2| J| the results fitted to the form suggested by Kubo.

$$\chi \sim \frac{e^{\frac{-g}{T}}}{T^{\frac{1}{2}}} \tag{8}$$

where g is the gap. Plots of $\log(T^{\frac{1}{2}}\chi)$ against $\frac{1}{T}$ in figure 3 show the expected linear behaviour indicating $g \neq 0$ for D = 0. However the results for D = 0.2| J| appear to develop curvature and this would indicate that g = 0. A similar plot is found for D = 0.1| J|. It might be that equation 8 is not the appropriate fitting form for $D \neq 0$ so the result g = 0 is not without uncertainty.

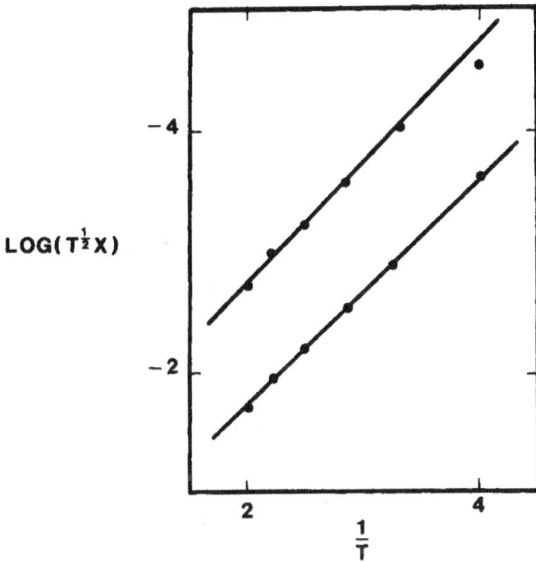

Figure 3. Log-linear plot of $T^{\frac{1}{2}}\chi$ against $\frac{1}{T}$, in units of | J|, for antiferromagnetic spin-1 chain (equation 5). The upper line is for D = 0.2| J| and the lower for D = 0. Quadratic extrapolation.

Ferromagnetic Coupling (J > 0)

For ferromagnetic coupling and D > 0 Hamiltonian 1 is equivalent to the model proposed in [18] to describe compounds of transition metals and rare earths. This model was originally cast in terms of quantum spin operators, however, it has been investigated almost exclusively in its classical form where the quantum spins are replaced by classical spins of the same symmetry. In its classical manifestation the model has been endowed with different understandings of its critical properties at different times, and so the situation might be thought to be confused.

In its classical form the model has been investigated in different limits and for different dimensions of spin and understanding has been complicated by the tendency to treat the results from these special cases as though they are interchangeable. A brief outline of the genesis of the model is given in [15]. The salient points would appear to be as follows:-

In the limit D→∞ the model becomes Ising-like since the spins are forced to lie along the easy axis. The model becomes equivalent to the Ising Hamiltonian

$$H = -\sum_{<ij>} \underline{N_i}\underline{N_j}\delta_i\delta_j \quad \text{with } \delta_i = \pm 1.$$

This model is just a spin-glass with a special distribution of couplings. Numerical work in two [22] and three [23] dimensions does indeed confirm that this model is a spin glass and hence ferromagnetism is absent. This result is taken as support for the belief that ferromagnetism is absent for this model for all dimensions less than four. However, it might be expected that if any values of D are to be special then one of them would be D→∞.

The conclusion that ferromagnetism was absent in less than four dimensions rested on two widely believed results. The first coming from renormalisation group analysis indicated that the effect of the random anisotropy is to reduce the critical properties to those of the pure system in two dimensions less [24]. Unfortunately a similar result for the random field model is now known to fail in the three dimensional Ising model. Further Fisher [25] has cast doubt on the renormalisation group approach by showing there is no perturbative fixed point in 4 + ε dimensions.

The second result is a very appealing argument similar to the domain wall energy balance argument of Imry and Ma [26]. This essentially equates anisotropy field energy gain to Bloch wall energy loss resulting from misalignment to predict that the former is dominant in less than four dimensions and so ferromagnetism is destroyed. Simulation has not given support to this [27,28] where evidence suggests the survival of ferromagnetism for planar XY spins. The implication that if the argument fails for such a system it probably fails more generally.

Quantum fluctuations have previously not been incorporated into investigations of this model. It must thus be thought uncertain as to whether ferromagnetism is destroyed by random anisotropy in less than four dimensions. A study, even in one dimension which includes such effects might give an indication of what happens in higher dimensions.

Results for the magnetisation of the spin-1 Hamiltonian for D = J are shown in figure 4 for temperatures between 0.8J and 0.3J. It is seen that as system size n is increased there is no tendency for the magnetisation to diminish. This behaviour is entirely consistent with a first order transition to a groundstate with non-zero magnetisation at T=0. Thus the random anisotropy has not succeeded in destroying the ferromagnetic groundstate even in one dimension and this result should have some implications for what might happen in higher dimensions. To understand figure 4 better it is probably worth pointing out that although the magnetisation is formally zero at non-zero temperatures (the system is after all one dimensional) the effect of the finite differences used to calculate the magnetisation is to smear the first order transition over a non-zero temperature range. On the other hand if there was no first order transition there would be nothing to smear and the results for the magnetisation would be zero.

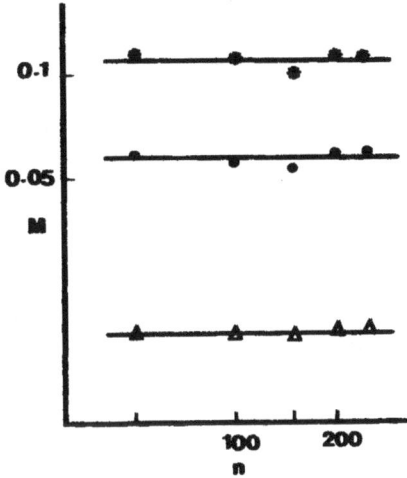

Figure 4. Log-Log plot of magnetisation M against system size n for spin-1 ferromagnetic chain (equation 5) with D = J. △ is for T = 0.7J. • is for T = 0.4J. * is for T = 0.3J. Quadratic extrapolation.

One possible flaw in the above reasoning is that D may only become relevant at much lower temperatures and longer length scales than used to calculate the data in figure 4. This remains to be investigated.

The susceptibility has also been calculated and fitted to the form T^γ. In the temperature range considered all the values of γ lay in the range 1.978 and 2.187. These values are entirely consistent with $\gamma=2$ which is the spin wave result.

SPIN-½ MODEL IN RANDOM FIELDS AND WITH RANDOM COUPLINGS

Two-fold random anisotropy studied in the previous section does not couple to spin-½ operators due to their commutation relationships. The Hamiltonian trivially reverting to the pure system. Random fields do however couple. Two related models have been investigated [14], the simulation uses only the value m=9. This restricts the validity of the simulation to T > 0.1J.

Model I: Random Bonds

The random bond non-isotropic spin-½ chain in a non-zero uniform magnetic field has the Hamiltonian

$$H = -\sum_{<ij>} J_{ij}\left(S_i^z S_j^z + \Delta\left(S_i^x S_j^x + S_i^y S_j^y\right)\right) - F\sum_i S_i^z \qquad (9)$$

where $\left\{S_i^x, S_i^y, S_i^z\right\}$ are Pauli spin-½ matrices and $\{J_{ij}\}$ are random variables taken to the ±J with equal probability, F is a uniform magnetic field, the summation is over nearest neighbours and Δ is the anisotropy parameter.

All results were obtained from a single long chain containing 1000 spins. The partition function has been obtained for a single external field strength of $F = J$ and two values of the anisotropy parameter. These were $\Delta=1$ (isotropic) and $\Delta=0.2$ (easy axis).

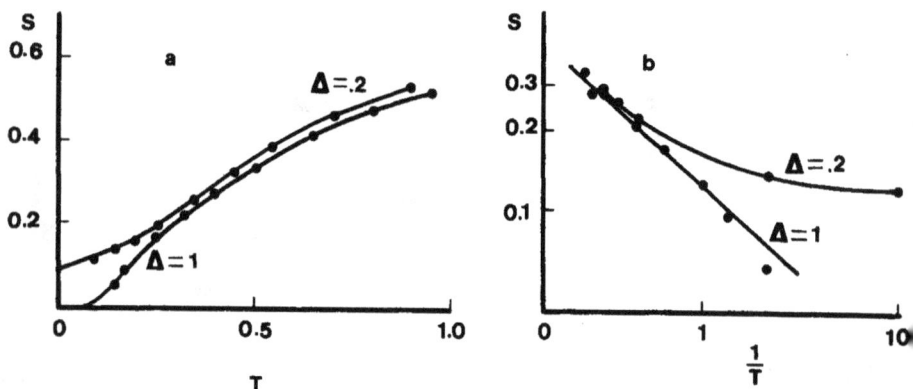

Figure 5. (a) Entropy s, in units of Boltzmann constant, against temperature in units of J for spin-½ chain (equation 9) and $\Delta=1$ and $\Delta=0.2$. (b) Log-linear plot of data in (a).

Results for the entropy are shown in figure 5a and 5b. The implication from these results is that for $\Delta=1$ there is zero entropy at T=0 while there is finite entropy for $\Delta=0.2$. The approach of the $\Delta=1$ curve to zero has been investigated. It is not possible to fit the results to a power low approach of the form T^x. Figure 5b shows the data on a log-linear plot. For $\Delta=1$ the results are clearly consistent with an exponential approach to zero of the form

$$s \sim e^{-(c/T)}.$$

However the data at the lowest temperature is seen to pull away from linearity indicating the possibility of crossover to even more rapid approach to zero. The result for $\Delta=0.2$ are also shown on figure 5b confirming the approach to a non-zero value already indicated in figure 5a. This result should be compared with that of Puma and Fernandez [39] for the random field Ising model. Their result for the T=0 entropy for that model was 0.125k compared with 0.1k for the present case, k being the Boltzmann constant. Thus quantum effects seems to have suppressed but not eliminated T=0 entropy.

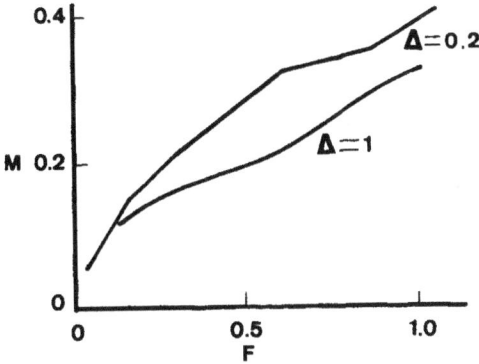

Figure 6. Magnetization M against field strength F in units of J for the spin-½ chain (equation 9) for T=0.1J and Δ=0 and 0.2.

Figure 6 shown plots of the magnetisation for varying applied field F. For $\Delta=0.2$ distinct discontinuities of slope are apparent, particularly if comparison is made with $\Delta=1$ case which is also shown.

To try to interpret this result consideration can be given to the analogous result for the one dimensional random field Ising model. Here exact calculation at T=0 shows that the magnetisation is a devil's staircase [30,32]. Even though the results here are not given at T=0 but at T=0.1J, the lowest temperature accessible, the indications are of similar structure to the magnetisation for the quantum case. That is the magnetisation is not everywhere an analytic function of external field. It is at present a matter of speculation as to whether the non-analyticities are of the same nature in both the quantum and Ising cases.

Model II: Random Fields

The isotropic spin-½ ferromagnetic chain in both a uniform and random magnetic field has the Hamiltonian

$$H = -J\sum_{\langle ij \rangle} \left(S_i^x S_j^x + S_i^y S_j^y + S_i^z S_j^z \right) - F\sum_i S_i^z - \sum_i h_i S_i^z \qquad (10)$$

Here $\{h_i\}$ are external random fields. Two cases are considered:

case (a): $\{h_i\}$ distributed continuously according to a Gaussian distribution with a mean 0 and standard deviation J.

case (b): $\{h_i\}$ distributed discretely taking $\pm J$ with equal probability.

A single chain containing 250 spins has been studied. Results for the entropy for the case external field F=J and continuous random fields (case (a)) and discrete random fields (case (b)) are shown in figure 7. The results for discrete random fields indicate clearly a non-zero entropy at T=0. For continuous Gaussian fields the implication is less clear but zero entropy is indicated.

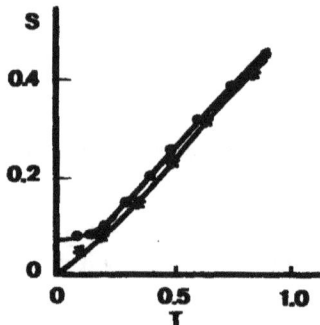

Figure 7. Entropy s, in units of Boltzmann constant, against temperature T, in units of J, for spin-½ chain (equation 10). • is for discrete random fields (case a). * is for Gaussian fields (case b).

Results have also been obtained for other values of the uniform external field. The data here is rather weak and it will be necessary to access even lower temperatures before any definite prediction can be made. The tentative conclusion so far which is put forward only as a speculation is that for the Gaussian field the entropy is always zero for all F≤J. However for the discrete case there may be a crossover to zero entropy at lower field strengths.

APPENDIX

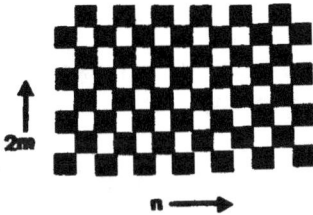

Figure A1. 2m x n classical lattice resulting from decomposition using equation 3. The shaded squares denote 4-site interactions.

In this short appendix an outline of the transfer matrix method will be given [12]. This method is quite general and can be used for checkerboard or large cell decompositions, however only the former will be discussed. The method can also be implemented on fast vector processors or massively parallel machines with high efficiency.

Figure A1 shows the end result of insertion of complete sets of states between the exponential operators of equation 3. At each site of the lattice there is a 2s+1 classical variable and the shaded squares denote the four spin interactions. The lattice is periodic in the m direction. Denote the site variables at location ij by s_{ij}, and the weights of the shaded squares in the jth-column by

$$E\left(s_{i,j},\ s_{i+1,j},\ s_{i,j+1},\ s_{i+1,j+1}\right). \tag{A1}$$

Where the value of E is calculated using equation 4. Firstly the vector of weights is set up for the first column of spins. Call this vector $Z_1\left(\{s_{i,1}\}\right)$. In the first column of spins we have the condition

$$s_{2i,1} = s_{2i+1,1}; \quad i=1,...m \tag{A2}$$

with $s_{2m+1,1} = s_{1,1}$, which just corresponds to putting

$$Z_1\left(\{s_{i,1}\}\right) = 1$$

for all configurations where A2 is satisfied and zero otherwise. We now replace $s_{1,1}$ and $s_{2,1}$ by $s_{1,2}$ and $s_{2,2}$ by taking the trace over $s_{1,1}$ and $s_{2,1}$. The new vector of weights is calculated using

$$Z_2\left(\{s_{i,1}\}_2\right) = \sum_{\substack{s_{1,1} \\ s_{2,1}}} E\left(s_{1,1}, s_{2,1}, s_{1,2}, s_{2,2}\right) Z_1\left(\{s_{i,1}\}\right) \tag{A3}$$

where $\{s_{i,1}\}_2$ has the first two elements of $\{s_{i,1}\}$ replaced by the first two elements of $\{s_{i,2}\}$. The process is now iterated replacing all subsequent pairs of spins in row 1 by the corresponding spins in row 2 using the general expression

$$Z_k\left(\{s_{i,1}\}_{2k}\right) = \sum_{\substack{s_{2k-1,1} \\ s_{2k,1}}} E\left(s_{2k-1,1}, s_{2k,1}, s_{2k-1,2}, s_{2k,2}\right) Z_{k-1}\left(\{s_{i,1}\}_{2(k-1)}\right) \tag{A4}$$

When k=m the process is complete and all spins in the first row have been traced over. The third row and subsequent rows are added in a similar way until the entire lattice has been summed over. If n is even the partition function is calculated by summing the elements of the final vector which correspond to condition A2. If n is odd then the summation is over those elements which satisfy

$$s_{2j+1,n} = s_{2j+2,n} \quad j=0, \dots, m-1.$$

References

1.   For a review see D.C. Mattis, The Many-Body Problem (World Scientific Singapore 1993) p671.

2.   H. Bethe, Z Physik, **71**, (1931) 205 (and also see ref 1).

3.   I. Affleck, Review, J.Phys. C1, (1989) 3047. I. Affleck, T. Kenedy, E. Lieb and H. Tasaki, Phys. Rev. Letts, **59**, (1987) 799 and Comm. Math. Phys **115**, (1988) 477.

4.   M. Wortis, Phys. Rev. **132**, (1963) 85; also see C.P. Enz., A Course in Many-Body Theory applied to Solid State Physics (World Scientific, Singapore 1992), p.271.

5.   D.L. Huber, Phys. Rev., **B37**, (1987) 3497.

6.   C.A. Doty and D.S. Fisher, Phys. Rev., **B45**, (1992) 2167.

7.   D.S. Fisher, Phys. Rev. Letts. 63, (1992) 534.

8.   S-K Ma, C. Dasgupta and C-K Hu, Phys. Rev. Letts. **43**, (1979) 1434.

C. Dasgupta and S-K Ma, Phys. Rev., **B22**, (1979) 1305.
J.E. Hirch, Phy. Rev., **B22**, (1980) 5355.

9. K. Binder and A.P. Young, Rev. Mod. Phys. **58**, (1986) 801.

10. M. Suzuki, Prog. Theor. Phys. **56**, (1976) 1454 (also see M. Suzuki in this volume and references therein).

11. S. Bacci, E. Gagliano and E. Dagatto, Phys. Rev. **B44**, (1991) 285.
E. Dagatto and A. Moreo, Phys. Rev. Letts, **19**, (1989) 2148.
E. Dagatto and A. Moreo, Phys. Rev., **B39**, (1989) 4744.

12. I. Morgenstern and K. Binder, Phys. Rev., **B22** (1980) 288.

13. P. Reed, preprint.

14. P. Reed, J.Phys.A, **25**, (1992) 5861.

15. P. Reed, J.Phys.A. (Letters), to appear (1993).

16. M. Suzuki, Phys. Rev., **B31**, (1985) 2957.

17. M. Barma and B. Shastry, Phys. Rev., **B18**, (1978) 3351.
J.J. Cullen and D.P. Landau, Phys. Rev., **B27**, (1983) 297.

18. R. Harris, M. Plischke and M. Zuckermann, Phys. Rev. Letts, **31**, (1973) 160.

19. O. Golinelli, Th Jolicoeur and R. Lacuze, J.Phy.C., to appear (1993).

20. P. Gaveau, J.P. Boucher, L.P. Regnault, T. Goto and J.P. Renard, J.Appl.Phys. **69**, (1991) 5956.

21. K. Kubo, Phys. Rev., **B46**, (1992) 866.

22. A.J. Bray and M.A. Moore, J.Phys., **C18**, (1985) L139.

23. C. Jayaprakash and S. Kirkpatric, Phys. Rev., **B21**, (1979) 4072.
A. Chakrobarti, Phys. Rev., **B36**, (1987) 5747.

24. R.A. Pelcovits, Phys. Rev., **B19**, (1979) 465.

25. D.S. Fisher, Phys. Rev., **B21**, (1985) 7233.

26.   Y. Imry and S-K. Ma, Phys. Rev. Letts, **35**, (1975) 1399.

27.   P. Reed, J.Phy.A., **24**, (1991) L117.

28.   P. Fish, Phy. Rev. Letts, **66**, (1991) 204 and Phys. Rev., **B46**, (1992) 242.

29.   M. Puma and J.F. Fernandez, Phys. Rev., **B18**, (1978) 1391.

30.   B. Derrida, J. Vannimenus and Y. Pomeau, J.Phys., **C11**, (1978) 4749.

31.   T.M. Nieuwenhuizen and J.M. Luck, J.Phys.A., **19**, (1986) 1207.

32.   R. Bruinsma and G. Aeppli, Phys. Rev. Letts, **50**, (1983) 1494.

INHOMOGENEITY EFFECTS IN QUANTUM SPIN SYSTEMS

S. Miyashita, J. Behre[†] and S. Yamamoto[‡]
Graduate School of Human and Environmental Studies,
Kyoto University, Kyoto 606, Japan
[†]Institut für Theoretische Physik, Universität Hannover, Hannover, Germany
[‡]Department of Physics, College of General Education,
Osaka University, Toyonaka, Osaka 560, Japan

Abstract

Inhomogeneity effects in quantum spin systems are studied by quantum Monte Carlo methods. In particular, the ground-state and thermodynamic properties of diluted $S = 1/2$ Heisenberg antiferromagnets on the square lattice and $S = 1$ antiferromagnetic Heisenberg chains with open boundaries are investigated. Methodological problems which are inherent in inhomogeneous systems are are also described.

1 Introduction

Quantum Monte Carlo (QMC) methods bsaed on the Suzuki-Trotter decomposition [1] have been extensively applied to quantum spin systems and proved to be very powerful. Compared with the diagonalization method, QMC methods have a great advantage in treating large systems. In two-dimensional quantum spin systems, the existence of the long range order in the ground state has been confirmed [2-4]. Furthermore, the nature of the symmetry breaking of the order with respect to the symmetry of the interaction has been investigated [5]. The quantum effect leads to various interesting phenomena not only these homogeneous systems but also in inhomogeneous systems.

Firstly we investigate the correlation function of the order parameter in the diluted Heisenberg antiferromagnet on the square lattice [6-8]. At short disance, the antiferromagnetic correlations are enhanced by the dilution. The enhancement of the correlation shows a systematic spatial distribution according to the positions of non-magnetic sites. The range of the enhancement, however, is finite, and the correlations are reduced at long distance. Thus the long range order is reduced by the dilution. The concentration dependence of the long range order is investigated, and turns out to be qualitatively different from that in the corresponding classical systems. In the classical system, the long range order as a function of the concentration, namely the concetration dependence of the size of the percolated cluster, is convex cueve but in the present model it is a concave shape. Furthermore, the data may suggest that even the percolation threshold is different from the classical value ($\delta_C = 1 - p_C \simeq 0.41$ for the site percolation on the square lattice). The model has no long range order at finite temperatures because of the infrared divergence of the fluctuation, and thus the temperature dependence of the correlation length ξ of the order parameter is investigated.

Secondly we investigate properties of the $S = 1$ antiferromagnetic Heisenberg chain. It has been shown that the quantum effect has an essential influence on the ground state of the one-dimensional $S = 1$ Heisenberg antiferromagnet, which leads to the Haldane problem [9,10]. After the Haldane's conjecture, the ground-state properties of the $S = 1$ antiferromagnetic Heisenberg chain have been studied in detail. In particular, the existence of a gap between the ground state and the first excited state as well as the exponential decay of the spin correlation function in the ground state have been confirmed by many numerical investigations [11-13]. An analytically solvable model which exhibits typical properties of the Haldane system has been introduced by Affleck, Kennedy, Lieb and Tasaki (AKLT) [14]. The nature of the ground state has been clarified using the AKLT model. The ground state is not a simple paramagnetic phase, but exhibits a hidden order. In this system, inhomogeneity causes interesting effects. The ground state of open chains has fourfold degeneracy, and at the edges of the chain, magnetic moments with $S = 1/2$ appear although the original spin of the system is 1. These properties are found to be valid in the present model if the chain is long enough, which was first pointed out by the diagonalization method [15] and later confirmed for much longer chains by a Monte Carlo method [16]. Here QMC study of the model on finite open chains is presented. The distribution of the local magnetizations is examined and the appearance of the $S = 1/2$ moments is demonstrated. The properties at finite temperatures have also been reported recently [17]. It has been found that the gap has no essential influence on the behavior of the specific heat. But the appearance of the $S = 1/2$ moments at the edges and the fourfold degeneracy result in a Curie-type divergence of the susceptibility. It has been pointed out that there exists an intrinsic order parameter in the Haldane phase [18]. This order is called the 'hidden order', and represents antiferromagnetic alignments of ± 1 spins, remaining after omitting all the sites with spin projection 0. The temperature dependence of the correlation length for the long range hidden order is investigated. The mechanism of a collapse of the long range order at finite temperatures is clarified, as we observe a qualitative difference between quantum mechanical and thermal fluctuations. The former survives even in the ground state, while the latter vanishes exponentially as the temperature decreases.

2   Diluted $S = 1/2$ Heisenberg antiferromagnets on the square lattice

Properties of the Heisenberg antiferromagnet $S = 1/2$ diluted with non–magnetic impurities on the square lattice are investigated by quantum Monte Carlo simulations. We show data of the correlation functions and thermodynamic properties for system sizes up to $L \times L = 16 \times 16$, temperature down to $T = 0.1J$, and impurity concentrations up to $\delta = 37.5\%$.

2.1 Quantum Monte Carlo method on the diluted lattice

The Hamiltonian of the model is

$$\mathcal{H} = 2J \sum_{<i,j>} \varepsilon_i \varepsilon_j \mathbf{S}_i \cdot \mathbf{S}_j, \tag{2.1}$$

where $\varepsilon_i = 1$ or 0. The positions with $\varepsilon_i = 0$ denote the diluted sites and are hereafter called 'impurities'. They are chosen randomly from the lattice points. The concentration of the impurities is defined by $\delta = 1 - \sum_i \varepsilon_i / L^2$. The checker board decomposition [19] is used in the present simulation. Namely, the Hamiltonian is divided into two parts within each of which all terms are commutable:

$$\mathcal{H} = \mathcal{H}_1 + \mathcal{H}_2,$$

$$\mathcal{H}_1 = 2J \sum_{i,j} \mathbf{S}_{2i-1,2j-1} \cdot \mathbf{S}_{2i,2j-1} + \mathbf{S}_{2i,2j-1} \cdot \mathbf{S}_{2i,2j} + \mathbf{S}_{2i,2j} \cdot \mathbf{S}_{2i-1,2j} + \mathbf{S}_{2i-1,2j} \cdot \mathbf{S}_{2i-1,2j-1},$$

$$\mathcal{H}_2 = 2J \sum_{i,j} \mathbf{S}_{2i,2j} \cdot \mathbf{S}_{2i+1,2j} + \mathbf{S}_{2i+1,2j} \cdot \mathbf{S}_{2i+1,2j+1} + \mathbf{S}_{2i+1,2j+1} \cdot \mathbf{S}_{2i,2j-1} + \mathbf{S}_{2i,2j+1} \cdot \mathbf{S}_{2i,2j}.$$
$$\tag{2.2}$$

With the standard construction of the $(d+1)$-dimensional Ising model $\{\sigma_i\}$, we have a three-dimensional lattice with eight-spin interactions (Fig. 1).

Here the Ising spin equlas ± 1 for the states $|\pm\rangle$ $(S^z|\pm\rangle = \pm\frac{1}{2}|\pm\rangle)$, and 0 for the impurity sites. Each interacting cube corresponds to the local Boltzmann factor, namely

$$\rho(\sigma_1, \sigma_2, \sigma_3, \sigma_4, \sigma_1', \sigma_2', \sigma_3', \sigma_4') = \langle \sigma_1, \sigma_2, \sigma_3, \sigma_4 | \exp(\frac{-\beta \mathcal{H}_{(i)}}{n}) | \sigma_1', \sigma_2', \sigma_3', \sigma_4' \rangle. \tag{2.3}$$

In the present case the conservation law of the magnetization holds for each cube such that $\sigma_1 + \sigma_2 + \sigma_3 + \sigma_4 = \sigma_1' + \sigma_2' + \sigma_3' + \sigma_4'$. In the computer program, we assign each cube (eight spins) a number from 1 to 3^7. In the simulation we provide a flip of spins around an impurity (or impurity cluster), which is called the 'loop flip', in order to avoid non-ergodic confinements of the configuration. If we use only local flips and straight global flips, as in the simulation of pure systems, then configurations in the present model are easily frozen at low temperatures, because the world line cannot pass through an impurity site, and leads to erroneous results [6,7]. Examples of the local flips are shown as the dotted lines in Fig. 2.

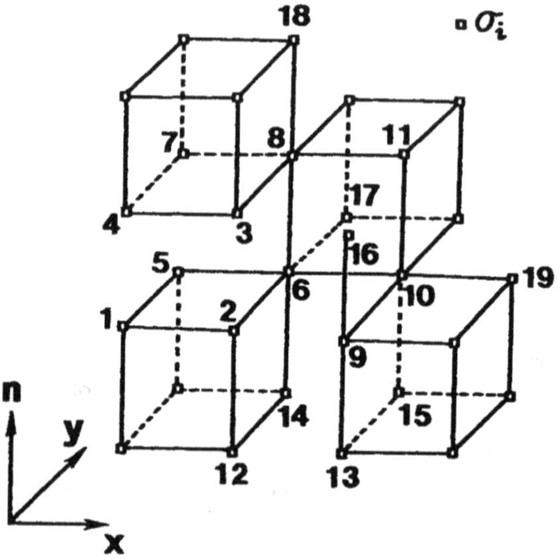

Figure 1: The $(d+1)$-dimensional Ising lattice with the eight-spin interactions. This is the figure caption.

2.2   Short-range correlations

For small lattices, it has been found by the spin-wave calculation and also by the diagonalization method that the spin correlation functions near the impurities are enhanced [20,21]. In the present model the Hamiltonian of the system consists of the nearest-neighbor interactions and the average energy per bond should be enhanced in diluted systems. But the further-neighbor correlations are also found to be enhanced. Thus the range of the enhancements has been investigated, but it is found to be rather short. Besides the range of the enhancement, we have also found that the enhancement appears in a systematic way according to the configuration of impurities. An example of the spatial distribution of the enhancements is given in Fig. 2.

2.3   Long-range order

Although the short range correlations are enhanced, the correlation functions are reduced at long distances. The square of the staggered magnetization $< N_z^2 >$ per site, namely

$$\frac{\langle N_z^2 \rangle}{L^4} = \frac{1}{L^4} \langle (\sum_{i \in A} \sigma_i - \sum_{i \in B} \sigma_i)^2 \rangle, \tag{2.4}$$

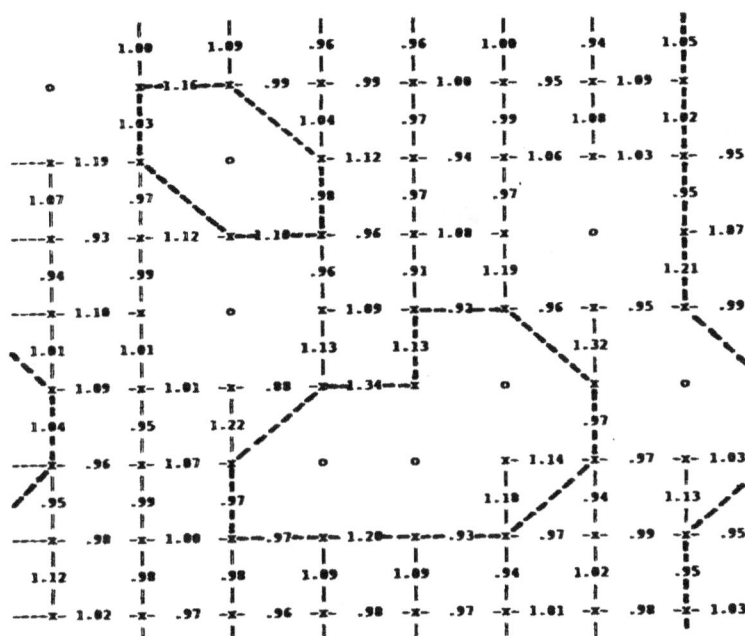

Figure 2: Nearest neighbor spin correlations functions at $T = 0.2J$ normalized by the pure value. Examples of the loop flip and nonstraight Marcu flip are shown as dotted lines.

represents the long range order of the system. This quantity is the sum of the staggered correlations between all the pairs of sites. If the long range order s exists one expects, the following form of correlation:

$$\langle \mathbf{S}_i \cdot \mathbf{S}_j \rangle \sim s^2 + \frac{a}{|i-j|} + \cdots, \qquad (2.5)$$

and thus also the following size dependence

$$\frac{\langle N_z^2 \rangle}{L^4} \sim s^2 + \frac{A}{L} + \cdots, \qquad (2.6)$$

where a and A are positive constants. In Fig. 3 the size dependence of $\langle N_z^2 \rangle / L^2$ is shown for several concentrations. The extrapolated values give s^2 for each δ. The concentration dependence is presented in Fig. 4. Here we find that the extrapolated values are arranged into a concave shape.

This behavior is very different that in from the corresponding classical systems, where the percolated cluster increases as $(p - p_C)^\beta$. The value of β is believed to be 5/36 [22], and thus s^2 has the concentration dependence $(\delta_C - \delta)^{2\beta}$, which has a convex shape. Furthermore, the critical concentration seems to be smaller than the classical

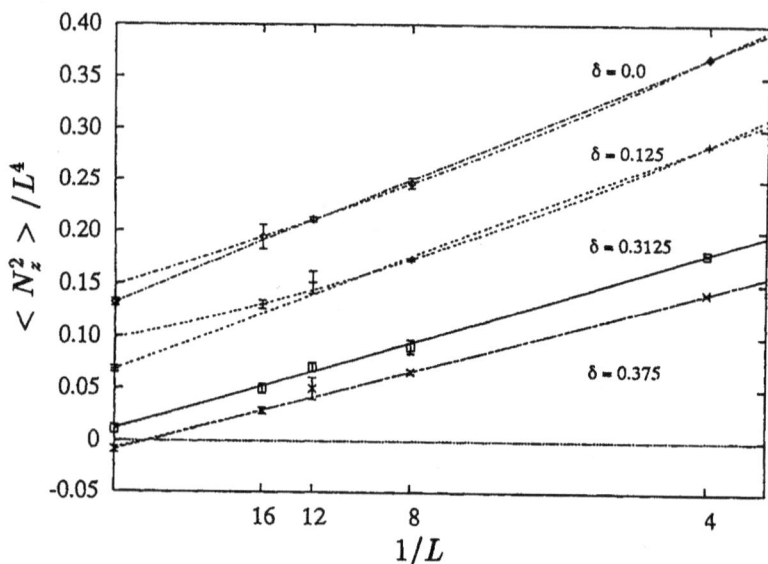

Figure 3: Size dependence of $\langle \mathcal{N}_z^2 \rangle / L^4$ for various concentrations of impurities. From Ref. [8].

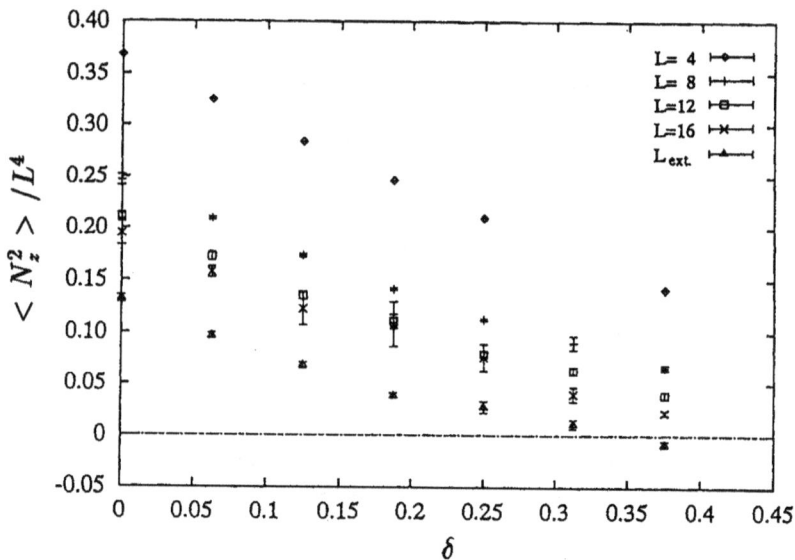

Figure 4: Concentration dependence of $\langle \mathcal{N}_z^2 \rangle / L^4$ and the extrapolated values ($\triangle$). From Ref. [8].

value 0.41 ($= 1-p_C$), although the error in estimation is very large. If $\delta_C < 0.41$, then the system in the pregion, $\delta_C < \delta < 0.41$, is in a quantum disordered phase, which is qualitatively suggested by studies of the non-linear sigma model [23,24]. The critical value is naively estimated as $\delta_C \sim 0.345$. The extrapolated values are negative for $\delta > \delta_C$. This is simply attributed to the fact that the extrapolation form Eq. (2.6) is no more valid. If $s = 0$, then automatically $A = 0$ and we have to use the dependence

$$\frac{\langle \mathcal{N}_z^2 \rangle}{L^4} \propto \frac{B}{L^2} + \cdots . \tag{2.7}$$

2.4 Temperature dependence of the correlation length

As we have seen in the previous section, the long range order exists for concentrations $\delta < \delta_C$. As far as the long range order exists, the description of the non-linear sigma model is correct and thus we expect that the correlation length diverges as

$$\xi(T, \delta) = C(\delta) \exp(\frac{2\pi \rho_S(\delta)}{T}), \tag{2.8}$$

where C is the prefactor and ρ_S is the spin stiffness, which depends on δ in the present case [8,25]. The correlation length is estimated from the spin correlation function as $\langle \sigma_i^z \sigma_j^z \rangle = A r^{-\lambda} e^{r/\xi}$ with $r = |i - j|$. Here we put $\lambda = 1/2$.

In Fig. 5[b] the spin correlation functions for various concentrations are given for $T = 0.1J$. The concentration dependence of ρ_S is given in Fig. 6.

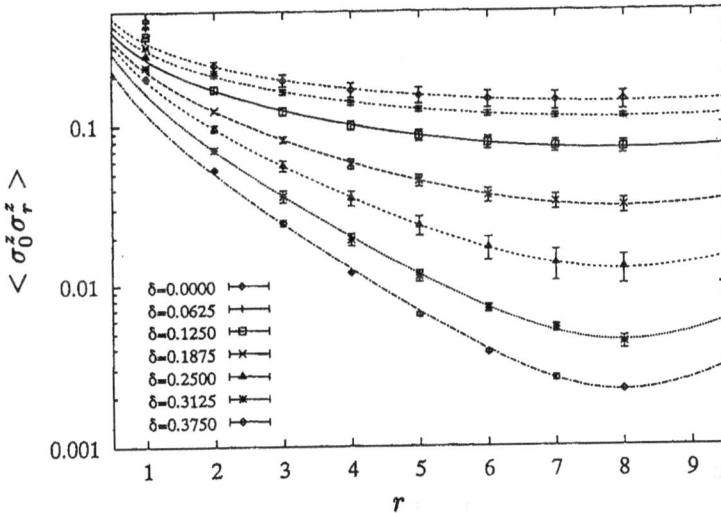

Figure 5: Spin correlation function $\langle \sigma_0^z \sigma_r^z \rangle$ at $T = 0.1J$ for various concentrations. From Ref. [8].

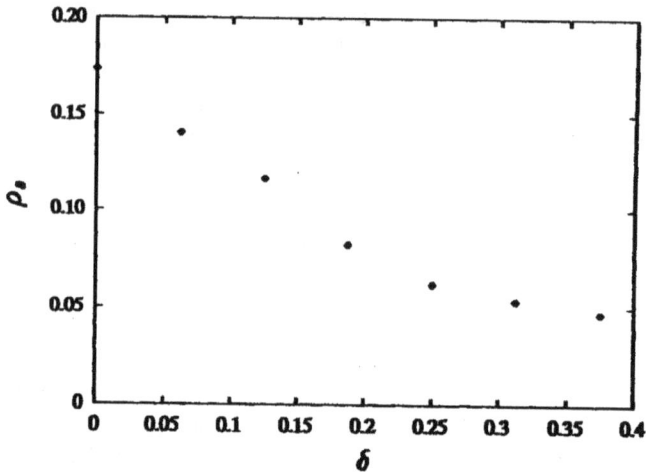

Figure 6: Concentration dependence of the spin stiffness constant. From Ref. [8].

3 $S = 1$ antiferromagnetic Heisenberg chains with open boundaries

Properties of $S = 1$ antiferromagnetic Heisenberg chains at temperatures down to $T = 0.05J$ are investigated by quantum Monte Carlo simulations with the checkerboard decomposition. The Hamiltonian of the model is

$$\mathcal{H} = J \sum_i \mathbf{S}_i \cdot \mathbf{S}_{i+1} . \tag{3.1}$$

3.1   Magnetization profiles

In open chains with an odd number of spins, the total spin in the ground state is 1 and the magnetization M_z equals ± 1 or 0. In a state with $M_z = \pm 1$, the expectation value of each spin $\langle S_i^z \rangle$ is nonzero. In short chains, $\langle S_i^z \rangle$'s make a staggered alignment over the whole chain, but in long chains the alignment appears only at the boundaries and it decays exponentially [16,26,27]. The data of the local moments $\{\langle S_i^z \rangle\}_{i=1,L}$ for $L = 65$ and $L = 97$ are shown in Fig. 7.

The local magnetization decays exponentially with the correlation length $\xi \sim 6$, which is the same as in the bulk [13]. Summation of the local moments at a boundary gives 0.5, which is nothing but the above-mentioned edge moment with $S = 1/2$. The appearance of the moments with $S = 1/2$ has been observed experimentally in diluted chains [28].

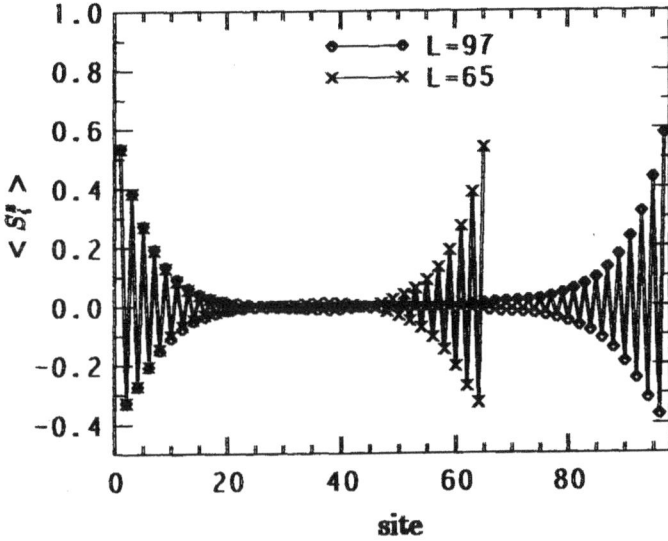

Figure 7: The magnetization profile in the open-odd chain. From Ref. [11].

3.2 *Thermodynamic properties*

The data of the specific heat and the magnetization are given as properties at finite temperatures [17]. Data of the specific heat for a free chain with an odd number of spins (open odd-chain) are plotted in Fig. 8.

The extrapolated values agree with those obtained by other methods [29]. The data are obtained directly from a quantity corresponding to

$$C = \frac{1}{k_B T^2} \frac{\partial^2}{\partial \beta^2} \ln Z. \qquad (3.2)$$

The data are consistent with the temperature derivative of the energy. The data for a free chain with an even number of spins (open even-chain) and for periodic chains give similar shapes. Here it should be noted that the shape in Fig. 8 seems to be of the Shottkey type, but the peak is located at T_P which is about twice of the gap $\Delta E (\sim 0.4J)$, namely $T_P \sim 0.8J \sim 2\Delta E$. On the other hand, the Shottkey specific heat has a peak at the temperature $T_P \simeq 0.42 \times$ (the energy gap between the two levels). This difference is important when the value of the coupling constant J is estimated from the observed specific heat.

In Fig. 9, the temperature dependence of the magnetic susceptibility per spin, χ^N / N, where

$$\chi^N = \langle (\sum_i^N S_i^z)^2 \rangle - \langle (\sum_i^N S_i^z) \rangle^2, \qquad (3.3)$$

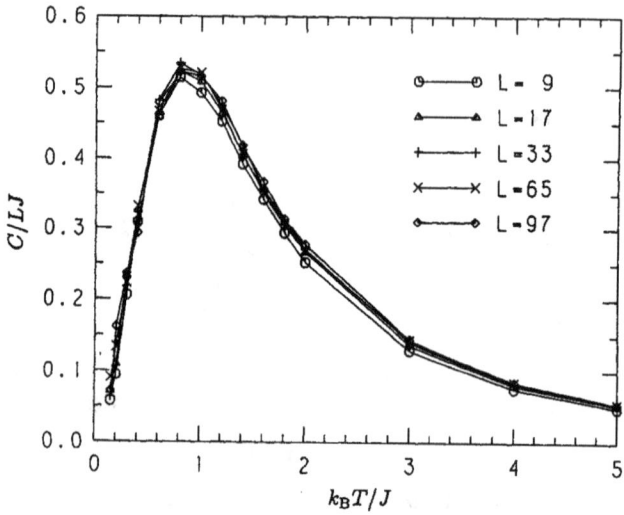

Figure 8: The specific heat of the open-odd chains. From Ref. [12].

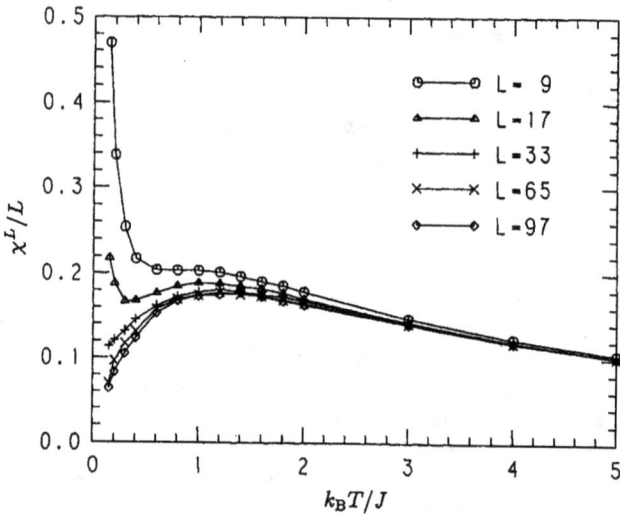

Figure 9: The magnetic susceptibility for the open chains. From Ref. [12].

of the open odd-chains is shown. Here a Curie-type divergence is observed. It is well known for free antiferromagnetic chains with an odd number of spins that a Curie-type divergence takes place because of the existence of an unpaired spin. But in the present system, the divergence is observed even in open even-chains, although the susceptibility becomes zero at $T = 0$ in periodic chains, as expected. The divergence in the present model is not simply due to the unpaired spin but it is due to the fourfold degeneracy of the ground state, where states with the total spin $S = 1$, namely those with $M_z \neq 0$, are included. Thus the divergence in open even-chains is one of the inherent phenomena of the present model. More precisely, the edge moments with $S = 1/2$ are responsible for the divergence. The diverging part is extracted by subtracting the bulk susceptibility estimated from the periodic chains, namely

$$\chi_{\text{Curie}} = \chi_{\text{open}}^{2L+1} - \chi_{\text{periodic}}^{2L}. \tag{3.4}$$

Analyzing this quantity, we estimated the amplitude of the divergence as

$$\chi_{\text{Curie}} \simeq \frac{0.5}{k_{\text{B}}T}. \tag{3.5}$$

The value 0.5 is consistent with $2 \times S(S+1)/3$ for $S = 1/2$. The Curie-type divergence has also been observed experimentally [30]. Using the present result, we can estimate the number of free chains in the sample.

3.3 The hidden order

In this section the correlation function of the hidden order is investigated.

A snapshot of the spin configuration of the ground state is shown in Fig. 10, where the abscissa represents the chain direction and the ordinate represents the Trotter direction, namely the imaginary time. Here, we find a well-developed hidden ordering, namely antiferromagnetic alignments of ±1 spins remaining after omitting all the sites with spin projection 0. We also find some anti-domains created by quantum fluctuation (the enclosed region in Fig. 10).

The nonlocal string order parameter has been introduced [18] as

$$O_{\text{string}}^z = -S_1 \exp[i\pi \sum_{j=2}^{L-1} S_j^z]S_L^z. \tag{3.6}$$

Using a nonzero sequence of spins $\{\tilde{S}_i\}_{i=1,\tilde{L}}$, where $\tilde{L}$ is the number of nonzero spins, we detect here the hidden order. In the present model without field, the spin projections ±1 and 0 appear with equal probabilities and thus we expect $\tilde{L} \simeq \frac{2}{3}L$, although this quantity fluctuates. The hidden order parameter is defined here as

$$O_{\text{LR}} = \frac{1}{\tilde{L}} \sum_{i=1}^{\tilde{L}} (-1)^i \tilde{S}_i^z. \tag{3.7}$$

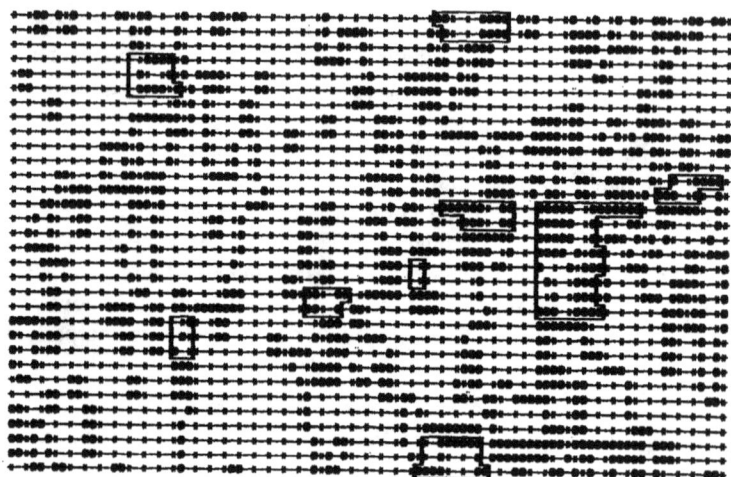

Figure 10: A snapshot of QMC for S=1 antiferromagnetic chain ($L = 97$, $n = 16$, and $T = 0.05$). Anti-domains are surrounded by lines. From Ref. [12].

At finite temperatures the hidden order is of a short range. To detect the correlation, we have observed $\langle O^2_{LR} \rangle$, from which the correlation length ξ_D is estimated as

$$\langle O^2_{LR} \rangle \propto \frac{1}{\bar{L}} \int_0^{\bar{L}} e^{-\tau/\xi_D} d\tau \simeq \frac{\xi_D}{\bar{L}}. \qquad (3.8)$$

The temperature dependence of $< O^2_{LR} >$ and its $\bar{L}$-dependence are shown in Fig. 11. The temperature dependence of ξ_D is presented in Fig. 12, which shows an intrinsic ordering process of the present model [17,31]. In the ground state the hidden order is of long range although it is disturbed by quantum fluctuation. Because the present system is of one dimension, if there is thermal fluctuation at all, then the long range order has to vanish. Thus the quantum and thermal fluctuations should have qualitatively very different characteristics.

In Fig. 13, a snapshot of the spin configuration at a finite temperature is shown. Here we find non-temporal domain walls (shown as arrows) are found besides the bubble-like temporal fluctuation representing the quantum fluctuation. The non-temporal domain walls represent the thermal fluctuation, which destroys the hidden long range order at finite temperatures.

The authors would like to thank Dr. Dorota Lipowska for her critical reading of the manuscript.

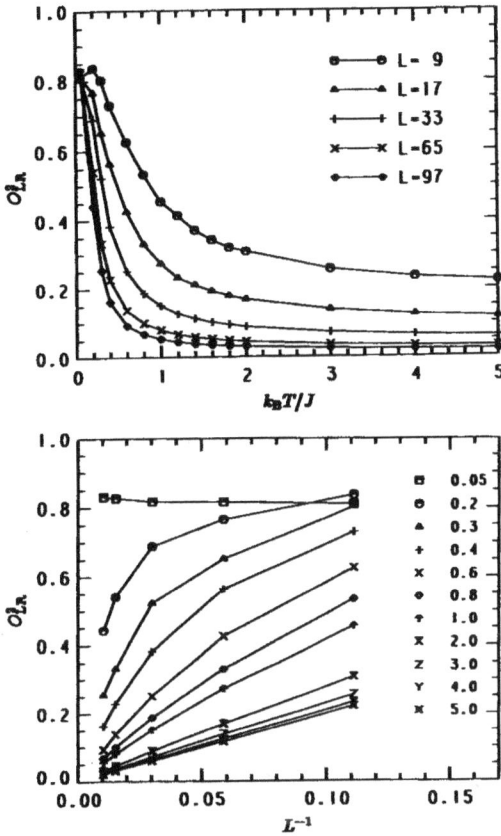

Figure 11: The temperature and size dependences of $\langle O_{LR}^2 \rangle / L$. From Ref. [12].

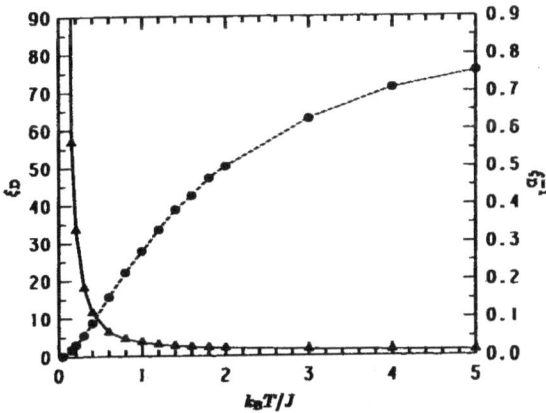

Figure 12: The temperature dependence of ξ_D. From Ref. [12].

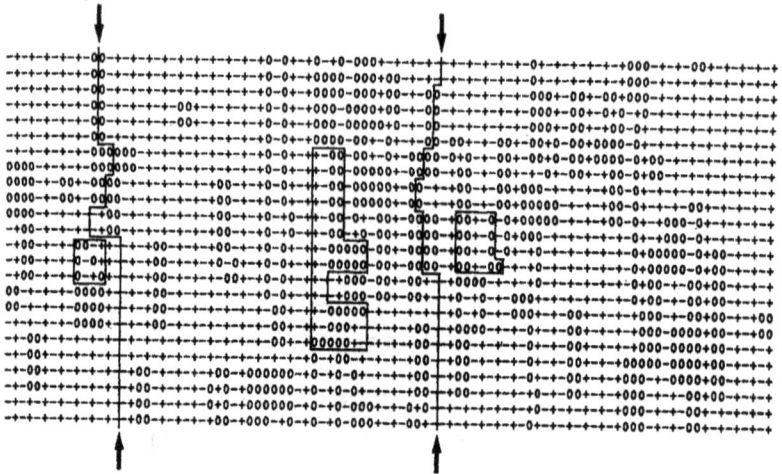

Figure 13: A snapshot of QMC for S=1 antiferromagnetic chain ($L = 97$, $n = 12$, and $T = 0.2$). Domain walls are shown as lines. From Ref. [12].

References

1. M. Suzuki, *Commun. Math. Phys.* **51**(1976)183 and 1454. M. Suzuki, *Quantum Monte Carlo Method in Equilibrium and Nonequilibrium Systems*, ed. M. Suzuki (Springer, Berlin 1987) p2.

2. J. D. Reger and A. P. Young, *Phys. Rev.* **B37**(1988)5978.

3. S. Miyashita, *J. Phys. Soc. Jpn.* **57**(1988)1934. H. Q. Ding and M. S. Makivić, *Phys. Rev. Lett.* **64**(1990)1449. T. Barnes, *Int. J. Mod. Phys.* C **2**(1991)659.

4. Manousakis E., *Rev. Mod. Phys.* **63**(1991)1.

5. S. Miyashita, *Quantum Simulations of Condensed Matter Phenomena*, ed. J. D. Doll and J. E. Gubernatis (World Scientific, Singapore, 1989) p.229. M. Kikuchi, Y. Okabe and S. Miyashita, *J. Phys. Soc. Jpn.* **59**(1990)492.

6. J. Behre and S. Miyashita, *J. Phys. A: Math. Gen.* **25**(1992)4745.

7. S. Miyashita and J. Behre, *Computational Approaches in Condensed Matter Physics*, Springer Proceedings in Physics **70**, ed. by S. Miyashita, M. Imada and H. Takayama, (Springer, Berlin, 1992) p. 145

8. J. Behre and S. Miyashita, unpublished.

9. F. D. M. Haldane, *Phys. Rev. Lett.* **50**(1983)1153 and *Phys. Lett.* **93A** (1983) 464.

10. I. Affleck, *J. Phys. Condens. Matt.* **1**(1989)3047.

11. R. Botet and R. Julien, *Phys. Rev.* **B27**(1983)613.

12. K. Kubo and S. Takada, *J. Phys. Soc. Jpn.* **55**(1986)438.

13. H. Betsuyaku, *Phys. Rev.* **B34** (1986)8125.
14. R. M. Nightingale and H. W. J. Blöte, *Phys. Rev.* **B33**(1986)6545.
15. M. Takahashi, *Phys. Rev. Lett.* **62**(1989)2313.
16. M. Takahashi, *Phys. Rev.* **B38**(1988)5188.
17. K. Nomura, *Phys. Rev.* **B40**(1989)2421.
18. S. Liang, *Phys. Rev. Lett.* **64**(1990)1597.
19. I. Affleck, T. Kennedy, E. H. Lieb and H. Tasaki , *Phys. Rev. Lett.* **59**(1987)799 and *Commun. Math. Phys.* **115**(1988)477.
20. T. Kennedy, *J. Phys. Condens.* **2**(1990)5737.
21. S. Miyashita and S. Yamamoto, *Phys. Rev.* **B47**(1993).
22. S. Miyashita and S. Yamamoto, *J. Phys. Soc. Jpn.* **62**(1993)1459 and
23. S. Yamamoto and S. Miyashita, *Phys. Rev.* **B48**(1993).
24. G. Gomes-Santos, *Phys. Rev. Lett.* **63**(1990)790.
25. M. den Nijs and K. Rommelse, *Phys. Rev.* **B40**(1989)4709.
26. H.-J. Mikeska, *Europhys. Lett.* **19**(1992)39.
27. T. Kennedy and H. Tasaki, *Phys. Rev.* **B45**(1992)304 and *Commun. Math. Phys.* **147**(1992)431.
28. E. Loh, Jr., D. J. Scalapino and P. M. Grant, *Phys. Rev.* **B31**(1985)4712.
29. N. Nagaosa, N. Y. Hatsugai and M. Imada, *J. Phys. Soc. Jpn.* **58**(1989)978.
30. N. Bulut, D. Hone, D. J. Scalapino and E. Y. Loh, *Phys. Rev. Lett.* **62**(1989) 2192 and K. J. B. Lee and P. Schlottmann, *Phys. Rev.* **B42**(1990)4426.
31. J. Alder, Y. Meir, A. Aharony and A. B. Harris , *Phys. Rev.* **B41**(1990)9183.
32. J. W. Essam, *Phase Transitions and Critical Phenomena* vol. 2, Ed. C. Domb and M. S. Green (Academic, 1972) p. 197.
33. S. Chakravarty, B. I. Halperin and D. R. Nelson, *Phys. Rev. B* **39**(1989)2344.
34. Yanagisawa T.,*Phys. Rev. Lett.* **68**(1992)1026.
35. E. Manousakis and R. Salvador, *Phys. Rev. Lett.* **60**(1988)840 and *Phys. Rev. B* **45**(1992)7570.
36. M. Kaburagi, I. Harada and T. Tonegawa, *J. Phys. Soc. Jpn.* **62**(1993).
37. S. R. White, *Phys. Rev. Lett.* **69**(1992)2863.
38. M. Hagiwara, K. Katsumata, I. Affleck, B. I. Halperin and J. R. Renard, *Phys. Rev. Lett.* **65**(1990)3181.
39. S. H. Glarum, S. Geshwind, K. M. Lee, M. L. Kaplan and J. Michel , *Phys. Rev. Lett.* **67**(1991)1614.
40. H. W. J. Blöte, *Physica* **79B**(1975)427.
41. H. Betsuyaku and T. Yokota, *Prog. Theor. Phys.* **75**(1986) 427.
42. O. Avenel et. al, *Phys. Rev.* **B46**(1992)8655.
43. K. Kubo, *Phys. Rev.* **B42**(1992)866.

THE QUANTUM TRANSFER MATRIX AND ITS APPLICATION TO QUANTUM SPIN CHAINS

Kenn Kubo

Institute of Physics, University of Tsukuba,
Ibaraki 305, Japan
e-mail: kkubo@ph.tsukuba.ac.jp

Abstract

The quantum transfer matrix method based on the checkerboard decomposition and its application to one-dimensional spin systems are briefly reviewed. It is demonstrated that the extrapolation in the quantum limit is quite effective so that the properties at fairly low temperatures can be described by using approximants with a modest number of Trotter slicings.

1 Introduction

Since the proposal of the numerical approach to the quantum many-body systems based on the generalized Trotter formula by Suzuki[1], various methods have been developed and utilized according to this approach. Among them the quantum transfer-matrix (QTM) method based on the generalized Trotter formula is a useful tool for investigating one-dimensional quantum systems. It was first applied to the one dimensional spin systems by Betsuyaku in 1984[2]. Since then it has been applied quite successfuly not only to spin systems[3] but also to fermion systems[4] by many investigators. The method is very simple and its numerical part is suited to parallel computing. The finite-temperature properties of the one-dimensional quantum systems can be obtained by this method accurately except for very low temperatures, where smallness of the Trotter slicings prevents an accurate extrapolation. In some fortunate cases, however, the critical properties of the phase transition at $T = 0$ can be analyzed by this method.

In this paper we review the method based on the checkerboard decomposition and the results of its application to the $S = 1/2$ and the $S = 1$ XXZ chains done by Takada and the present author. We also discuss the $S = 1/2$ frustrated spin chain called the Δ chain. Though a detailed explanation of the QTM method was given by Betsuyaku previously[5], we review the method in the next section for the selfcontainedness of the description. The result for the $S=1/2$ XXZ model was also reviewed previously[6].

2   Quantum Transfer Matrix

Let us consider a one-dimensional quantum system with the Hamiltonian

$$H = \sum_{i=1}^{N} H_i, \tag{1}$$

where H_i's are local Hamiltonians and satisfy the condition $[H_i, H_j] = 0$ for $|i-j| \geq 2$ ($N = $ even). The Hilbert space of the state vectors of the system is spanned by the basis vectors $\{|\mathbf{q}\rangle = \prod_{i=1}^{N+1} |q_i\rangle\}$ where $|q_i\rangle$ is an eigenstate of the local operator Q_i, i.e., $Q_i |q_i\rangle = q_i |q_i\rangle$, and $\mathbf{q}$ denotes the $(N+1)$-dimensional vector $(q_1, q_2, ..., q_{N+1})$. The dimensionality of Q_i is denoted as d. We assume that

$$\langle \mathbf{q} | H_i | \mathbf{q}' \rangle = \langle q_i, q_{i+1} | H_i | q_i', q_{i+1}' \rangle \prod_{j \neq i, i+1} \delta_{q_j, q_j'}. \tag{2}$$

For example, spin operators $S_i^\alpha (\alpha = x, y$ or $z)$ can be taken as Q_i in the case of usual spin Hamiltonians with nearest-neighbor interactions. Since it is generally difficult to obtain the matrix elements of $\rho = e^{-\beta H}$, we approximate the density matrix as

$$\rho \simeq \rho_M = [e^{-\Delta \tau H_o} e^{-\Delta \tau H_e}]^M, \tag{3}$$

where $\Delta \tau = \beta/M$, $H_o = \sum_{i=\text{odd}} H_i$ and $H_e = \sum_{i=\text{even}} H_i$. Then the approximate partition function is expressed as

$$Z_M = \text{Tr} \, \rho_M = \sum_{\{q_{i,j}\}} \prod_{i=1}^{N} \prod_{j=1}^{M} W_{i,2j-[i]}, \tag{4}$$

where $W_{i,j} = \langle q_{i,j+1}, q_{i+1,j+1} | e^{-\Delta \tau H_i} | q_{i,j}, q_{i+1,j} \rangle$, $[i] = \text{Mod.}(i,2)$ and $\{q_{i,2M+1}\} = \{q_{i,1}\}$. This quantity is regarded as the partition function of a classical lattice system in two dimensions. The statistical weight of a configuration of four lattice points surrounding the plaquette with the lattice point (i,j) as its left down corner is given by $W_{i,j}$. The system is depicted as a checkerboard lattice as shown in Fig. 1. The periodic boundary condition is imposed in the Trotter (imaginary time) space. As the boundary condition in the real space, we adopt the open one for the sake of simplicity. Of course, the thermodynamic properties do not depend on the boundary condition. Let us define transfer marices $\{T_i\}$ as follows

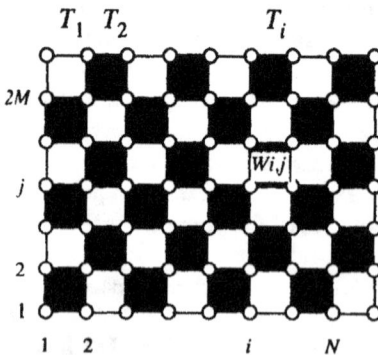

FIG. 1. The checkerboard lattice representing the approximate classical systems.

$$\ll \mathbf{q}_i \mid T_i \mid \mathbf{q}_{i+1} \gg \equiv \prod_{j=1}^{M} W_{i,2j-[i]}. \tag{5}$$

We denote a vector in the d^{2M}-dimensional space where the transfer matrices operate as $\mid u \gg$ and its adjoint as $\ll u \mid$. Here $\mid \mathbf{q}_i \gg$ is the basis vector corresponding to a $2M$-dimensional vector $\mathbf{q}_i = (q_{i,1}, q_{i,2}, ..., q_{i,2M})$. We assume that $\langle q, q' \mid H_i \mid p, p' \rangle$ is independent of i. Then $T_i = T_o$ for an odd i, and $T_i = T_e$ for an even i. Thus Z_M is rewritten as

$$Z_M = \ll \psi_L \mid (T_o T_e)^{N/2} \mid \psi_R \gg, \tag{6}$$

where ψ_L and ψ_R are the vectors corresponding to boundary conditions. For free boundaries, for example, they are defined as $\ll \psi_L \mid \mathbf{q} \gg \equiv \prod_{j=1}^{M} \delta_{q_{2j}, q_{2j+1}}$ and $\ll \mathbf{q} \mid \psi_R \gg \equiv \prod_{j=1}^{M} \delta_{q_{2j-1}, q_{2j}}$. Let us consider a local operator A_i which satisfies $[A_i, Q_k] = 0$ for $k > i+1$ and $k < i$. We approximate the thermal average of A_i with

$$\langle A_i \rangle_M = (2Z_M)^{-1} \mathrm{Tr} \left[\rho_M (A_i + e^{\Delta \tau H_e} A_i e^{-\Delta \tau H_e}) \right], \tag{7}$$

which is written down with the use of the transfer matrix as

$$\langle A_i \rangle_M = (2Z_M)^{-1} \ll \psi_L \mid \prod_{j=1}^{i-1} T_j \; (\widetilde{A_i^2} + \widetilde{A_i^1}) \prod_{j=i+1}^{N} T_j \mid \psi_R \gg, \tag{8}$$

where

$$\ll \mathbf{q}_i \mid \widetilde{A_i^{2l}} \mid \mathbf{q}_{i+1} \gg$$
$$= \begin{cases} \langle q_{i,2l}, q_{i+1,2l} \mid A_i e^{-\Delta \tau H_i} \mid q_{i,2l-1}, q_{i+1,2l-1} \rangle \prod_{j \neq l} W_{i,2j-1} & \text{for } i = \text{odd} \\ \langle q_{i,2l+1}, q_{i+1,2l+1} \mid e^{-\Delta \tau H_i} A_i \mid q_{i,2l}, q_{i+1,2l} \rangle \prod_{j \neq l} W_{i,2j} & \text{for } i = \text{even} \end{cases}$$

and

$$\ll \mathbf{q}_i \mid \widetilde{A_i^{2l-1}} \mid \mathbf{q}_{i+1} \gg$$
$$= \begin{cases} \langle q_{i,2l}, q_{i+1,2l} \mid e^{-\Delta \tau H_i} A_i \mid q_{i,2l-1}, q_{i+1,2l-1} \rangle \prod_{j \neq l} W_{i,2j-1} & \text{for } i = \text{odd} \\ \langle q_{i,2l-1}, q_{i+1,2l-1} \mid A_i e^{-\Delta \tau H_i} \mid q_{i,2l-2}, q_{i+1,2l-2} \rangle \prod_{j \neq l} W_{i,2j-2} & \text{for } i = \text{even} \end{cases}$$

The correlation function of two operators at i and j sites $(i < j)$ is approximated as

$$\langle A_i B_j \rangle_M = (2Z_M)^{-1} \sum_{l=1,2} \ll \psi_L \mid \prod_{k=1}^{i-1} T_k \; \widetilde{A_i^l} \prod_{k=i+1}^{j-1} T_k \; \widetilde{B_j^l} \prod_{k=j+1}^{N} T_k \mid \psi_R \gg. \tag{9}$$

The Duhamel two-point function (canonical correlation function) $\langle A_i; B_j \rangle$ defined by

$$g_{A_i B_j} \equiv \beta^{-1} \int_0^\beta d\tau \langle A_i(\tau) B_j \rangle, \tag{10}$$

where $A_i(\tau) = e^{\tau H} A_i e^{-\tau H}$, is necessary to obtain susceptibilities and is approximated as

$$\langle A_i; B_j \rangle_M = (4M Z_M)^{-1} \sum_{l=1}^{2M} \sum_{m=1,2} \ll \psi_L \mid \prod_{k=1}^{i-1} T_k \; \widetilde{A}_i^l \prod_{k=i+1}^{j-1} T_k \; \widetilde{B}_j^{\overline{m}} \prod_{k=j+1}^{N} T_k \mid \psi_R \gg .$$

(11)

In some cases an operator $P_i (i = 1, ..., N)$ such that $[P_i, H_j] = 0$ for $i \neq j$ appears in H_i in addition to Q_i and Q_{i+1}. For example, if we take the Hamiltonian of the cluster of $m+1$ spins, $(S_i \; ; \; i = mi - m + 1, ..., mi + 1)$, as H_i in a spin chain with nearest-neighbor interactions, then $m - 1$ operators $S_{mi-m+2}^\alpha, ..., S_{mi}^\alpha$ are considered as P_i. In such a case we use the representation with basis vectors $\{|q\rangle \otimes |p\rangle\}$, where $|p\rangle$ is the eigenstate of P_i's. Then (5) is modified as

$$\ll q_i \mid T_i \mid q_{i+1} \gg = \sum_{P_i} \prod_{j=1}^{M} (q_{i,j+1}, p_{i,j+1}, q_{i+1,j+1} \mid e^{-\Delta \tau H_i} \mid q_{i,j}, p_{i,j}, q_{i+1,j}) \, \delta_{p_{i,j+1}, p_{i,j+2}},$$

(12)

where $\tilde{j} = 2j - [i]$. Thus we have to sum over the eigenstates of P_i in a correlated way. This increases computational time appreciably.

One can show that the above approximants $\langle A_i \rangle_M$, $\langle A_i B_j \rangle_M$ and $\langle A_i; B_j \rangle_M$ are even functions of M if $[A_i, B_j] = 0$ and approach the exact values as $M \to \infty$[1,7,8]. If $[A_i, B_{i+1}] \neq 0$, we have to adopt slightly complicated formulas for $\langle A_i B_{i+1} \rangle_M$ and $\langle A_i; B_{i+1} \rangle_M$ to certify the even property of the approximants. In practical cases, we often encounter the situations where all the matrix elements of $e^{-\Delta \tau H_i}$, A_i and B_j are real and $[A_i, B_j] = 0$. Then the above formulas are simplified since the term involving $\widetilde{B}_i^1$ and that with $\widetilde{B}_i^2$ give the same contributions. For example, the spin correlation functions $\langle S_i^\alpha S_j^\alpha \rangle$ and $\langle S_i^\alpha; S_j^\alpha \rangle$ in the spin systems are expressed in a very simple way by employing the representation which diagonalizes $\{S_i^\alpha\}$.

Let us denote the eigenvalues of $T_M = T_e T_o$ as $\{\lambda_m; m = 0, 1, ...\}(|\lambda_0| > |\lambda_1| > ...)$ and the corresponding right and left eigenvectors as $|u_m \gg$ and $\ll v_m|$ $(\ll v_m | u_{m'} \gg = \delta_{mm'})$. Note that $\ll v_m | \neq \ll u_m |$ since T_M is not generally a normal matrix. We consider also another transfer matrix $\widetilde{T}_M = T_e T_o$, which has the same eigenvalues and $|\widetilde{u_m} \gg = \lambda_m^{-1/2} T_e |u_m \gg$ and $\ll \widetilde{v_m} | = \lambda_m^{-1/2} T_o \ll v_m |$ as eigenvectors. In many cases T_o and T_e are real and symmetric. Then $\ll \widetilde{v_m} | = \ll u_m |$ and $|\widetilde{u_m} \gg = |v_m \gg$.

In the thermodynamic limit $(N \to \infty)$, the approximants are expressed as

$$F_M = -k_B T \ln \lambda_0,$$

(13)

$$\langle A_i \rangle_M = 2^{-1} \lambda_0^{-1/2} \ll w | (\widetilde{A}_i^1 + \widetilde{A}_i^2) | w' \gg,$$

(14)

and

$$\langle A_i B_j \rangle_M = 2^{-1} \lambda_0^{-(j-i+1)/2} \sum_{l=1,2} \ll w | \widetilde{A}_i^l \prod_{k=i+1}^{j-1} T_k \; \widetilde{B}_j^l | w'' \gg,$$

(15)

where F_M is the approximant for the free energy, $\ll w |$, $|w' \gg$ and $|w'' \gg$ are the appropriate eigenvectors of the maximum eigenvalue, and $\langle A_i; B_j \rangle_M$ is expressed in the same way.

We can calculate the above approximants straightforwardly by using computers. The main part of the computations is successive multiplications of the small matrices $W_{i,j}$ by d^{2M}-dimensional vectors. We have to store several such vectors in the memory of a computer. As M increases, the necessary memory size increases quite rapidly. For example, a double-precision vector for $M = 7$ needs about 38Mbytes for $S = 1$ ($d = 3$) spin systems. We can often reduce the dimensionality of the vectors utilizing the conservation laws inherent in the transfer matrices. For the above example, the memory size of a vector can be reduced to about 5Mbytes in the S^z representation for the XXZ model, since we can restrict the vector to the subspace without staggered magnetization. Nevertheless, the restricted memory size forces us to terminate our calculations at rather small M compared to that used in the quantum Monte Carlo (QMC) simulations. This is the most serious problem in utilizing the QTM method to analyze low temperature properties where the approximation (3) leads to large errors. We are, however, able to estimate the thermal quantities at fairly low temperatures by extrapolating finite-M results to the infinite M through a simple procedure.

3 Extrapolation procedure

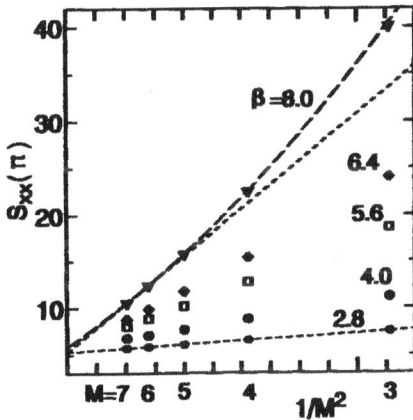

FIG. 2. M-dependence of the apppproximants of $S(\pi)$ in the $S = 1$ AFH model. Dotted and dashed lines represent the linear and the quadratic fits to the highest M's.

Let C_M denote any of the approximants given by (6), (8), (9) and (11) or any quantity derived from them. As C_M approaches the exact value C as an even function of M, we assume an asymptotic expansion in terms of M^{-2} as

$$C_M = C + \sum_{l=1}^{\infty} c_l M^{-2l}. \quad (16)$$

We show as an example the M-dependence of the approximants of the spin structure factor at $k = \pi$ in the $S = 1$ antiferromagnetic Heisenberg (AFH) model.

When we have data of K approximants with $M = M_1, M_2, ..., M_K$, we can estimate C and $c_i (1 \leq i \leq K - 1)$ by fitting the data with the series (16). In contrast to the case of the QMC simulations, which are necessarily accompanied by statistical errors, in the QTM method one can compute approximants almost exactly. The error involved in the estimated C comes from the truncation of the series (16) at the Kth term and it increases with decreasing temperature. We show in Table 1 the estimates of $S(\pi)$ from different sets of the approximants. We see in the table that different extrapolations agree quite well even at the lowest temperature. We can roughly evaluate the uncertainty of the estimates from the scattering of the

extrapolated values. Though the scattering increases with decreasing temperature, one can see that the extrapolation is quite effective and $S(\pi)$ can be estimated with an accuracy of a few percent even at the temperature which is $1/8$ of the coupling constant.

Table 1 Approximants and estimates extrapolated from $(M_1,...,M_K)$ approximants of $S(\pi)$ in the $S = 1$ AFH model.

β	$M = 3$	$M = 4$	$M = 5$	$M = 6$	$M = 7$	$(4,5,6)$	$(5,6,7)$	$(4,5,6,7)$
2.8	7.55955	6.61801	6.17484	5.93180	5.78437	5.37340	5.37263	5.37225
4.0	11.2066	8.84754	7.76145	7.17263	6.81779	5.83743	5.83954	5.83521
5.6	18.6382	12.7812	10.2324	8.89448	8.10478	5.97580	5.98475	5.98909
6.4	24.0538	15.4058	11.7785	9.91349	8.82816	5.95073	5.97658	5.98911
8.0	40.0851	22.5455	15.7698	12.4461	10.5735	5.82675	5.90407	5.94156

4 $S = 1/2$ XXZ model

In this and in the following section we show the results obtained by applying the QTM method to the XXZ model. The model is described by the Hamiltonian

$$H = \sum_{i=1}^{N}(S_i^x S_{i+1}^x + S_i^y S_{i+1}^y + \Delta S_i^z S_{i+1}^z), \qquad (17)$$

where Δ is the anisotropy parameter of the exchange interactions. The AFH, the XY and the ferromagnetic Heisenberg models are contained in this model and correspond to $\Delta = 1$, 0 and -1, respectively. The $S = 1/2$ model was solved exactly by Bethe's method and the ground-state property is nowadays well known. The $S = 1$ model has been investigated quite intensively in the last decade to test Haldane's prediction, which we will dicuss in the next section.

In this section we consider the $S = 1/2$ model[8]. This model is the simplest non-trivial quantum manybody system and therefore might be the best object to demonstrate the effectiveness of the QTM method. The ground state of the model can be classified into three phases; the ferromagnetic Ising (FI) phase for $\Delta < -1$, the XY phase for $-1 < \Delta < 1$, and the antiferromagnetic Ising (Néel) phase for $1 < \Delta$. In the XY phase the ground state is critical, i.e. the spin correlation functions exhibit the power-law decay with respect to the distance between spins, and there exists no energy gap between the ground state and the excited states. In the other two phases the ferromagnetic or the antiferromagnetic long-range order exists in the ground state. The deviation of the correlation function from the limiting value decays as an exponential function of the distance, and there exists an excitation energy gap. The AFH model at the point $\Delta = 1$ has the same properties as in the XY phase.

In the following we concentrate on the critical properties in the XY phase. We investigate the correlation functions between S_i^α's, namely

$$f_\alpha(r) \equiv \langle S_i^\alpha S_{i+r}^\alpha \rangle \tag{18}$$

and

$$g_\alpha(r) \equiv \langle S_i^\alpha ; S_{i+r}^\alpha \rangle. \tag{19}$$

These correlations are expressed in this case simply as

$$f_\alpha(r) = Z_M^{-1} \sum_{s,s'} \ll \psi_L \mid \prod_{k=1}^{i-1} T_k \mid s \gg s_1 \ll s \mid \prod_{k=i+1}^{j-1} T_k \mid s' \gg s_1' \ll s' \mid \prod_{k=j+1}^{N} T_k \mid \psi_R \gg \tag{20}$$

and

$$g_\alpha(r) = (2M Z_M)^{-1} \sum_{s,s'} \ll \psi_L \mid \prod_{k=1}^{i-1} T_k \mid s \gg s_1 \ll s \mid \prod_{k=i+1}^{j-1} T_k \mid s' \gg$$
$$\times (\sum_{l=1}^{2M} s_l') \ll s' \mid \prod_{k=j+1}^{N} T_k \mid \psi_R \gg \tag{21}$$

if we employ the transfer matrices in the S^α-representation. We assume the scaling form as

$$f_\alpha(r), g_\alpha(r) \sim (-1)^r A\phi(r/\xi)e^{-r/\xi}, \tag{22}$$

where $\phi(x)$ is normalized as $\lim_{x\to\infty}\phi(x) = 1$. At low temperatures $A \sim T^\sigma$ and $\xi \sim T^{-\nu}$. The staggered susceptibility is given by

$$\chi_\alpha(\pi) = \sum_{r=-\infty}^{\infty} (-1)^r g^\alpha(r) \tag{23}$$

and is proportional to $T^{-\gamma}$ at low temperatures. We have index relations

$$\sigma = \nu\eta \quad \text{and} \quad \gamma = \nu(1-\eta)+1, \tag{24}$$

where η is the correlation index at $T = 0$. To determine the above critical indices, we calculated the approximants of the correlation functions employing the QTM. The data obtained for a fixed M were fitted very nicely to (22), from which ξ_M and A_M were estimated. Then we extrapolated ξ_M, A_M and χ_M to $M = \infty$ using (16). We estimated the critical indices from

$$\gamma = \frac{d\ln\chi}{d\ln\beta}, \quad \nu = \frac{d\ln\xi}{d\ln\beta} \quad \text{and} \quad \eta = \frac{d\ln A}{d\ln\xi} \tag{25}$$

in the limit $\beta \to \infty$. The differentiation was done numerically. In practice we have
to estimate the critical indices at some finite temperature, as the extrapolation fails
at very low temperatures. We show in Fig. 3 the temperature dependence of several
approximants and extrapolated values of $\frac{d\ln\xi}{d\ln\beta}$ as functions of β. We see in Fig. 3
that at higher temperatures the convergence of the extrapolation is very good but the
extrapolated values still reveal rather strong temperature dependence. At $\beta = 4 \sim 5$
they show a stationary behavior with respect to β. At lower temperatures, i.e. $\beta \geq 6$,
the extrapolated values are scattered and also start to show strong temperature

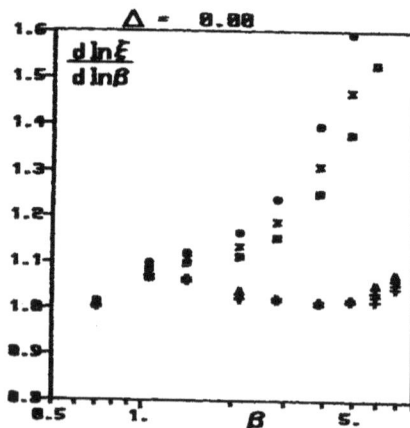

FIG. 3 dlogξ/dlogβ vs. β for Δ=0. Circles,
crosses and squares represent approximants with
M=7, 8 and 9, respectively. Triangles, asterisks
and pluses represent estimates from the sets of
approximants (6,7,8), (7,8,9), and (6,7,8,9), respectively.

dependence. This behavior should
be attributed to the failure of the
extrapolation. We have to estimate
the critical index at medium temperatures, where the extrapolated
data show good convergence and at
the same time exhibit a stationary
behavior with respect to β. This is
therefore, a rather subtle procedure.
In this case we estimate ν in the
region $3 < \beta < 5$, which leads to
$\nu = 1.014 \pm 0.008$. The value agrees
quite well with the exact $\nu = 1$. We
have estimated ν, γ and η in the
same way for $-0.5 \leq \Delta \leq 0.75$ employing approximants up to $M = 9$.
For the estimation of η, the behavior of data was not stationary
enough and so we have used data at
the lowest temperatures, where extrapolations still show a good convergence. We have estimated η also
by employing the index relation as $\eta_* = 1 - (\gamma - 1)/\nu$. We show the obtained indices
in Table 2. We see that η and η_* agree fairly well.

Table 2 Critical indices of S=1/2 XXZ model. η(LP) denotes the value given by (26).

Δ	ν	γ	η	η_*	η(LP)
-0.5	1.00 ±0.01	1.69 ±0.02	0.30 ±0.02	0.31 ±0.03	0.3333
-0.25	1.01 ±0.01	1.603±0.005	0.39 ±0.01	0.40 ±0.01	0.4196
0.0	1.014±0.008	1.516±0.001	0.50 ±0.02	0.491±0.005	0.5
0.25	0.99 ±0.01	1.43 ±0.01	0.55 ±0.01	0.57 ±0.01	0.5804
0.5	1.003±0.005	1.36 ±0.01	0.64 ±0.01	0.65 ±0.01	0.6667
0.75	0.990±0.005	1.25 ±0.02	0.746±0.005	0.75 ±0.02	0.7699

Our result is in a good agreement with the exact value of η derived by Luther and Peschel as[9]

$$\eta = 1/2 + \pi^{-1}\arcsin\Delta. \qquad (26)$$

For larger $|\Delta|$ we did not estimate critical indices as data did not show a stationary behavior at all in the region where the good convergence was obtained.

We have shown above that we can determine even the critical indices, which are defined in the low-temperature limit, by making use of a rather modest number ($M \leq 9$) of Trotter slicings. This was brought about by the success of the extrapolation which utilizes higher-order terms in M^{-2}, and was made possible by the precise determination of the approximants by the transfer-matrix calculation.

5 $S = 1$ XXZ model

Haldane predicted in 1983 that the AFH model with integer spins has a quite different ground state from that of the $S = 1/2$ model[10]. According to his theory, the spin correlations decay with a finite correlation length and a finite energy gap exists between the ground and the excited states in the integer-spin systems. This type of the ground state is realized in the XXZ model with integer spins in some region of Δ surrounding $\Delta = 1$. Therefore, in the $S =$ integer model, the four-ground state phases are realized according to the value of Δ; the FI ($\Delta < -1$), the XY ($-1 < \Delta < \Delta_1$) , the Haldane ($\Delta_1 < \Delta < \Delta_2$) and the Néel ($\Delta_2 < \Delta$) phases.

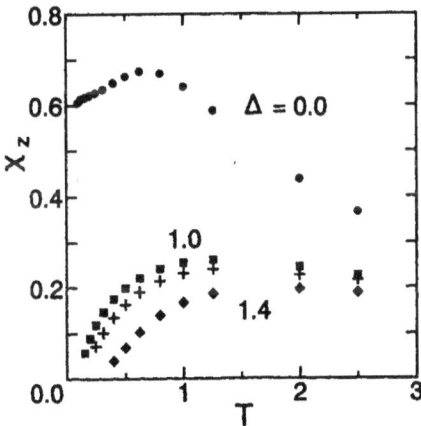

FIG. 4. The susceptibility of the $S=1$ XXZ model. Pluses represent the data for $\Delta=1.1$.

Since Haldane's prediction the problem has been intensively investigated by many workers both analytically and numerically[11]. The numerical study has been concentrated on the $S = 1$ model. The existence of the energy gap was confirmed by scaling analyses of finite-cluster diagonalizations[12], the QTM method in the Trotter direction[13] and also by the QMC simulations[14,15]. The finiteness of the correlation length was confirmed by QTM method utilizing approximants with M up to 6[16]. Afterwards it was confirmed by the QMC methods[17,18].

In the following we show the results obtained by using the transfer matrix with M up to 7[19]. We examine four values of Δ as representatives of the XY ($\Delta = 0$), the Haldane ($\Delta = 1$ and 1.1) and the Néel ($\Delta = 1.4$) phases. The extrapolation based on

(16) was employed and the averages of the extrapolated values for $M = (5,6,7)$ and $(4,5,6,7)$ are shown in the figures presented here as the estimates at $M = \infty$, and the differences betweeen them are considered as the estimation errors. These errors are usually too small to be noticed in the figures shown here. In Fig. 4 we show the uni-

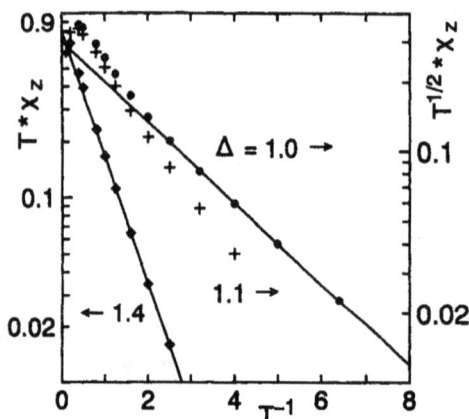

FIG. 5 The logarithmic plot of $\chi_z T^\epsilon$ vs. T.

form susceptibility χ_z in the z direction as a function of T. We observe that χ_z rapidly decreases as $T \to 0$ for $\Delta = 1, 1.1$ and 1.4. This result indicates that there exists an excitation energy gap for these Δ's. The energy gap is estimated from the temperature dependence. At low temperatures χ_z can be fitted to the expression

$$\chi_z \sim D T^{-\epsilon} e^{-E_0/T} \qquad (27)$$

with $\epsilon = 1/2$ for $\Delta = 1$ and 1.1, and $\epsilon = 1$ for $\Delta = 1.4$, as shown in Fig. 5. The energy gap E_0 is estimated to be 0.40 ± 0.01 for $\Delta = 1$, which is in a good agreement with the results obtained by the QMC methods. We tabulate E_0 for other values of Δ in Table 3. In Fig. 6 we show the temperature dependence of the correlation length $\xi_x(T)$ of $g_x(r)$. We can see that $\xi_x(T)$ approaches a finite value as $T \to 0$ for $\Delta = 1, 1.1$ and 1.4. At higher temperatures $\xi_x(T)^{-1}$ decreases linearly with T. It should be exponentially dependent on T^{-1} at low temperatures if an energy gap exists. We fitted the data at low temperatures to $\xi_x(T) = \xi_x(0) + C T^{-1} e^{-E_0/t}$ with $E_0 = 0.40$ as we have not enough number of data at low temperatures to estimate both E_0 and $\xi_x(0)$ directly. We have obtained

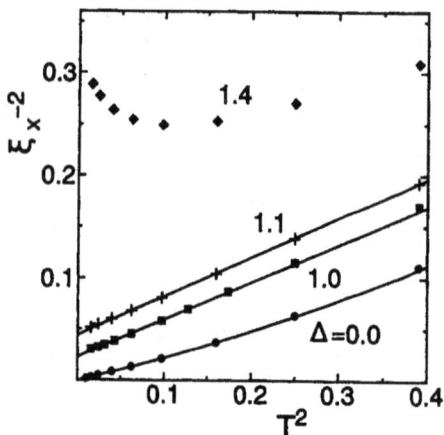

FIG. 6. The correlation length of $g_x(r)$ for $S=1$ XXZ model. The curves are fittings, which are linear ($\Delta=1, 1.1$) and quadratic ($\Delta=0$) with respect to T^2.

$\xi_x(0) = 6.1$, which is comparable with $5.5[17]$ and $6.25[18]$ obtained by the QMC methods. If we simply extrapolate $\xi_x(T)^{-2}$ with a quadratic function of T^2, we obtain $\xi_x(0) = 6.48 \pm 0.04$ which may be considered as an upperbound on $\xi_x(0)$. It was argued by den Nijs and Rommelse[20] on the basis of the exact ground state of the Affleck-Kennedy-Lieb-Tasaki model[21] that hidden order parameters exist in the Haldane-phase ground state. The order is described by the string correlation function

$$h_\alpha(r) = -\langle S_i^\alpha \exp(i\pi \sum_{k=i+1}^{i+r-1} S_k^\alpha) S_{i+r}^\alpha \rangle. \tag{28}$$

In the Haldane-phase ground state, $h_\alpha(r)$ approaches a finite value as $r \to \infty$ for all α, while in the Néel state a finite value is reached only for $\alpha = z$. To apply the QTM method, we have to replace T_k lying between s_1 ans s_1' in (20) with the string transfer matrix $\hat{T}_k$ given by

$$\ll t \mid \hat{T}_k \mid t' \gg = \exp(i\pi t_1) \ll t \mid T_k \mid t' \gg. \tag{29}$$

We show the correlation lengths ξ_{sz} and ξ_{sx} of $h_z(r)$ and $h_x(r)$ as functions of temperature in Figs. 7 and 8. We see that ξ_{sz} apparently diverges as $T \to 0$ for all of $\Delta=1$, 1.1 and 1.4 but ξ_{sx} does so only for $\Delta=1$ and 1.1. The result is consistent with the fact that

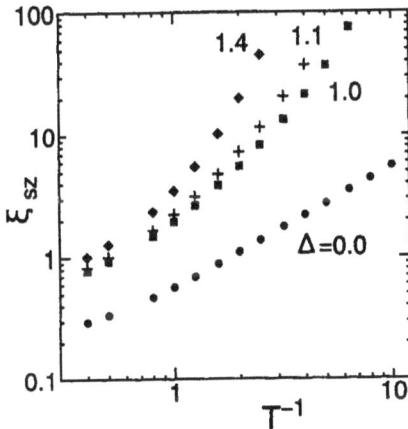

FIG. 7. The correlation length of $h_z(r)$ for $S=1$ XXZ model.

$\Delta=1$ and 1.1 belong to the Haldane phase and $\Delta=1.4$ to the Néel phase. From the temperature dependence of ξ_{sz} and ξ_{sx}, we can estimate the energy gap E_0, and the results are tabulated in Table 3. It is interesting that for $\Delta=1.1$ different values of E_0 are obtained from ξ_{sz} and ξ_{sx}. We believe that these two energy gaps correspond to different types of excitations. We have also obtained the correlation length ξ_z of $f_z(r)$, which diverges only for $\Delta=1.4$ as $T \to 0$. The energy gap estimated from ξ_z is also tabulated in Table 3.

The results for $\Delta=0$ indicate that the ground state is critical. We have estimated the critical indices as $\nu = 1.02 \pm 0.01$ and $\eta = 0.22 \pm 0.01$. The latter result is consistent with that for $\Delta_1 > 0$ since η is expected to be 1/4 at $\Delta = \Delta_1$. This point is rather subtle, however, since the present estimate may include larger errors than given above, as can be seen from the estimate of ν (exact ν should equal 1). Indeed, the evaluation of Δ_1 is very difficult since the phase transition at $\Delta = \Delta_1$ is of the Kosterlitz-Thouless type. So far different investigations have given scattered

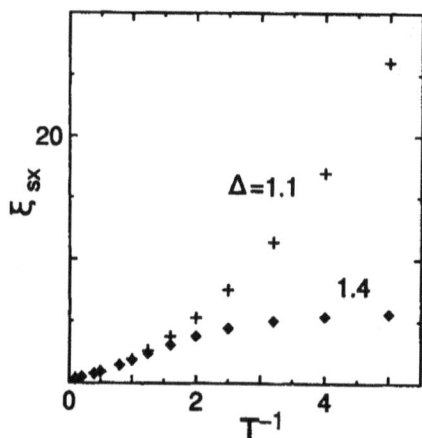

FIG. 8. The correlation length of $h_x(r)$ for $S=1$ XXZ model.

Table 3 Excitation energy gap of $S=1$ XXZ model estimated from various quantities.

Δ	1.0	1.1	1.4
E_0 from χ_z	0.40	0.57	1.56
E_0 from ξ_{sx}	0.33	0.48	1.41
E_0 from ξ_{sz}	0.33	0.21	
E_0 from ξ_z			0.56

estimates of Δ_1, ranging from 0 to 0.6[22]. On the other hand, Δ_2 was determined rather precisely as $\Delta_2 \simeq 1.19$[23]. Above we have clarified a qualitative difference between finite-temperature properties of the systems belonging to different ground-state phases. Furthermore, the obtained energy gap and $\xi_x(0)$ are in good agreement with numerical $T=0$ measurements. It is interesting that the present results show that two different excitations influence spin correlations at finite temperatures. One of them has an energy gap decreasing with Δ for $1 < \Delta < \Delta_2$, and the other has an energy gap increasig monotonously for $\Delta_1 < \Delta$. These excitations seem to correspond to the longitudinal and the transverse magnons discussed previously by Betsuyaku[13]. To further clarify the nature of these excitations, we need more extensive and more precise data at low temperatures.

Recently several quasi-one-dimensional materials with $S=1$ Ni ions such as $Ni(C_2H_8N_2)_2NO_2ClO_4$ (NENP) were found to reveal a characteristic behavior of the Haldane-phase ground state[24]. The QTM calculation done by Delica et al. explained very nicely the experimental value of the susceptibility by taking into account both the single-site anisotropy and the anisotropy in the g-factor[25]. The Haldane problem in the $S=2$ system was treated also by the QTM method[26].

6   The Δ chain

In this section we consider a fully-frustrated spin chain called the Δ chain. By "a fully-frustrated system" we mean a system whose classical (mean-field) ground state exhibits an infinite continuous degeneracies due to frustration. In such a system, construction of the linear-spin-wave theory based on one of the ordered classical ground states leads to at least one spin-wave mode, with its frequency vanishing for all wavevectors. The classical ground state may be considered disordered. It is an interesting problem to investigate the properties of the quantum-mechanical fully-frustrated systems.

A typical example of such a systems is the AFH model on the kagomé lattice. The recent intensive study of this system has been inspired by the experiments on the ^{3}He layer adsorbed on graphite[27] and also on the compound $SrCr_8Ga_4O_{19}$[28]. Whether an ordered ground state is realized or not in the $S = 1/2$ AFH model on the kagomé lattice is still an open question[29].

The full frustration is not restricted to two dimensions but is realized also in one and three dimensions. In one dimension we have a very simple example called Δ chain, which is described by the Hamiltonian (1) with

$$H_i = S_{2i-1} S_{2i} + S_{2i} S_{2i+1} + S_{2i-1} S_{2i+1}. \tag{30}$$

The classical ground state of H_i is realized by the configuration of the three spins making an angle $2\pi/3$ with each other, as shown in Fig. 9. One can easily see that three successive spins S_{2i}, S_{2i+1} and S_{2i+2} can be rotated simultaneously without raising the energy. The zero-temperature specific heat is $3/4k_B$ per spin and the linear-spin-wave theory leads to two dispersionless zero-energy modes. Monti and Sütő showed rigorously that the $S = 1/2$ system has dimer ground states and the excited states have a finite energy gap[30].

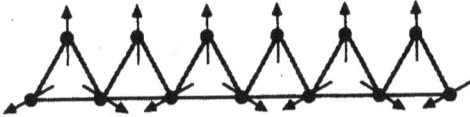

FIG. 9. A classical ground state of the Δ chain.

The ground states for the perodic boundary conditions are twofold degenerate and identical to those of the Majumdar-Ghosh (MG) model, which is frustrated but not "fully"[31]. The characteristic feature of the full frustration appears in the excitation spectrum and the low-temperature thermodynamics. Low-lying excited states of finite systems with the periodic boundary conditions were obtained by a numerical diagonalization of the Hamiltonian. The lowest excited state for each wavevector k is a triplet ($S = 1$) state, and its excitation energy δE_1 is shown in Fig. 10. The k-dependence of δE_1 for finite systems is very weak and it decreases with increasing N. Extrapolating δE_1 from $N = 6$, 8 and 10 to ∞, we obtained $\delta E_1(N = \infty) = 0.219$ at $k = \pi$, and 0.218 at $k = 0$. We also calculated the second-lowest excited state ($S = 0$), which also appears to converge to a dispersionless mode with energy identical to the lowest excitation. These features of the excitation spectrum are quite different from those of the MG model, which is the peculiarity of the fully frustrated system. These peculiar features of the excitation spectrum should influence the specific heat. We have calculated the specific heat by the QTM method with $M = 4 \sim 8$. We adopted the formula (12) taking S_{2i-1}^z as Q_i and S_{2i}^z as P_i. We calculated λ_0 and $|u_0 \gg$ by the power method and obtained the free energy F_M from (13), and the energy E_M from (14) taking H_i as A_i. The entropy was obtained from F_M and E_M through $S_M = (E_M - F_M)/k_B T$, and we calculated the specific heat C_M by differentiating E_M or S_M numerically. Both results agree very well except for very low temperature. The results are depicted in Fig. 11 as functions of the temperature. We have not extrapolated the finite-M results to infinite M as the data at low temperatures seem to vary still too drastically

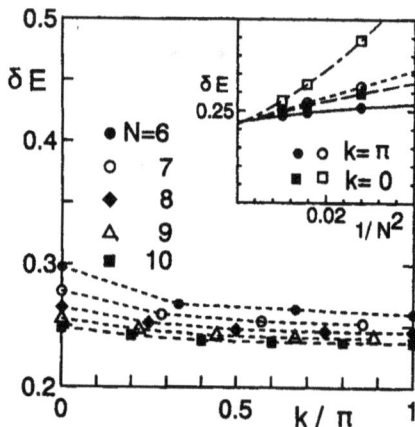

FIG. 10.   The excitation spectrum of finite $S=1/2$ Δ chain. The inset depicts the size dependence at $k=\pi$ and $k=0$.

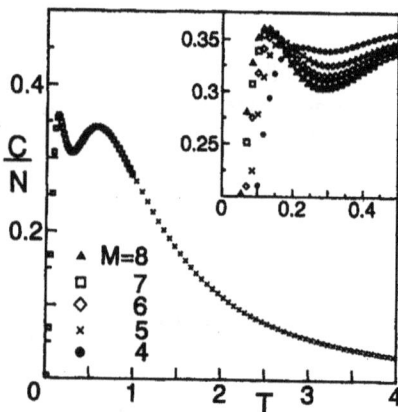

FIG. 11.   The specific heat of the $S=1/2$ Δ chain. The inset shows the M-dependence of the low-temperature peak.

when M increases to apply a simple extrapolation procedure. The values of C_M's as functions of the temperature, however, clearly exhibit a two-peak structure. The high temperature broad peak at $T \simeq 0.58$ is a common feature of one dimensional antiferromagnets. The rather sharp peak at a lower temperature is the peculiarity of this system. This peak grows and shifts to the lower temperature with increasing M, which excludes the possibility that this peak is a finite-M effect. The peak position is at $T \simeq 0.12$ for $M = 8$, and the peak height is $0.18k_B$ per spin. The entropy contributed by this peak is roughly estimated as $\sim$20 percent of the total entropy, which indicates that $2^{0.4N}$ states accumulate at $\delta E \sim 0.2$. This cotribution to the entropy is comparable to the reduction of the zero-temperature specific heat, i.e. $1/4$ in the classical model.

Above we have shown that the full frustration leads to a Schottky-type low-temperature peak in the specific heat of the Δ chain. One might expect that this property may be realized in systems in higher dimensions if their ground states have no magnetic order and their excitations have finite energy gaps. No such system is known so far in higher dimensions.

7 Summary

We showed the results for several one-dimensional spin systems obtained by the QTM method. It was exhibited that a simple extrapolation is very powerful and leads to quite accurate results down to fairly low temperatures by using a modest number of Trotter slicings. For the $S=1/2$ XXZ model, even the correlation index in the ground state was precisely determined for $\Delta \simeq 0.5$. For the $S=1$ model the obtained ground-state correlation length and excitation energy gap are in good agreement with the results by other methods. Our results demonstrate that there are cases where the QTM method may be used even for studying the ground-state properties. Generally speaking, however, it is difficult to apply the QTM method at very low temperatures as only small number of Trotter slicings is available. A possible way to remedy this failure partially is to take a large-size cluster as a local Hamiltonian H_i. This method was employed by several authors[32]. Finally, we want to stress that the QTM method is potentially very useful to analyze the experimental results of quasi-one-dimensional systems as was displayed in Ref. 25. One can calculate almost any observable quantity easily and accurately by this method if the temperature is not too low (which is usually the case in the experiments). Physical parameters of the studied materials such as exchange constants, anisotropy parameters and so on, may be extracted by comparing the theoretical results with the experimental ones without resorting to approximate theories.

8 Acknowledgements

The author wishes to express his gratitude to Professor S. Takada for helpful discussions and constant encouragement. He is also indebted to Dr. D. Lipowska for her valuable advices to improve the manuscript. This work was partially supported by a Grant-in-Aid for Scientific Research on Priority Areas by the Ministry of Education, Science and Culture.

9 References

1. M. Suzuki, Commun. Math. Phys. **51** (1976) 183; Phys. Rev. **B31** (1985) 2957.
2. H. Betsuyaku, Phys. Rev. Lett. **53** (1984) 629.
3. See for earlier works, Refs. 5 and 6 and also references therein.
4. For example, T. Yokota and H. Betsuyaku, Prog. Theor. Phys. **75** (1986) 46; N. Nagaosa, J. Phys. Soc. Jpn. **56** (1987) 2450; M. Imada, J. Phys. Soc. Jpn. **59** (1990) 4121.
5. H. Betsuyaku, in *Quantum Monte Carlo Methods*, ed. M. Suzuki (Springer, Berlin, 1987) p. 50.

6. S. Takada, in *Quantum Monte Carlo Methods*, ed. M. Suzuki (Springer, Berlin, 1987) p. 86.
7. M. Suzuki, Phys. Lett. **113A** (1985) 299; J. Stat. Phys. **43** (1986) 883.
8. T. Takada and K. Kubo, J. Phys. Soc. Jpn. **55** (1986) 1671.
9. A. Luther and I. Peschel, Phys. Rev. **139** (1974) 2911; N. M. Bogoliubov, A. G. Izergin and V. E. Korepin, Nucl. Phys. **B275**[FS17] (1986) 687.
10. F. D. M. Haldane, Phys. Lett. **93A** (1983) 464; Phys. Rev. Lett. **50** (1983) 1153.
11. For a review, see I. Affleck, J. Phys. C **1** (1989) 3047.
12. R. Botet and R. Jullien, Phys. Rev. **27** (1983) 613 ; ibid **B28**(1983) 3914 ; U. Glaus and T. Schneider, Phys. Rev. **B30** (1984) 273.
13. H. Betsuyaku, Phys. Rev. **B36** (1987) 799.
14. M. P. Nightingale and H. W. J. Blöte, Phys. Rev. **B33** (1986) 659.
15. M. Takahashi, Phys. Rev. Lett. **62** (1989) 2313.
16. K. Kubo and S. Takada, J. Phys. Soc. Jpn. **55** (1986) 438.
17. M. Takahashi, Phys. Rev. **B38** (1988) 5188.
18. K. Nomura, Phys. Rev. **B40** (1989) 2421.
19. K. Kubo, Phys. Rev. **B46** (1992) 866.
20. M. den Nijs and K. Rommelse, Phys. Rev. **B49** (1989) 4709.
21. I. Affleck, T. Kennedy, E. H. Lieb and H. Tasaki, Phys. Rev. Lett. **59** (1987) 799 ; Commun. Math. Phys. **115** (1988) 477.
22. For example, H. J. Shulz and T. Ziman, Phys. Rev. **B33** (1986) 6546 ; T. Sakai and M. Takahashi, J. Phys. Soc. Jpn. **59** (1990) 2688 ; see also Refs. 13 and 16.
23. See, for example, T. Sakai and M. Takahashi in Ref. 22.
24. For example, J. P. Renard, M. Verdaguer, L. P. Regnault, W. A. C. Erkelelens, J. Rossat-Mignod and W.G. Stirling, Europhys. Lett. **3** (1987) 945 ; J. P. Renard, M. Verdaguer, L. P. Regnault, W. A. C. Erkelelens, J. Rossat-Mignod, J. Ribas, W.G. Stirling and C. Vettier, J. Appl. Phys. **63** (1988) 3538 ; K. Katsumata, H. Hori, T. Takeuchi, M. Date, A. Yamagishi and J. P. Renard, Phys. Rev. Lett. **63** (1989) 86.
25. T. Delica, K. Kopinga, H. Leschke and K. K. Mon, Europhys. Lett. **15** (1991) 55.
26. N. Hatano and M. Suzuki, J. Phys. Soc. Jpn. **62** (1993) 1346.
27. D. S. Greywall, Phys. Rev. **B41** (1990) 1842 and references therein.
28. C. Broholm, G. Aeppli, G. P. Espinosa and A. S. Cooper, Phys. Rev. Lett. **65** (1990) 3173 and references therein.
29. For example, C. Zeng and V. Elser, Phys. Rev. **B42** (1990) 8436 ; A. B. Harris, C. Kallin and A. J. Berlinsky, Phys. Rev. **B45** (1992) 2899 ; S. Sachdev, Phys. Rev. **B45** (1992) 12377 ; R. R. P. Singh and P. A. Huse, Phys. Rev. Lett. **68** (1992) 1766 ; A. Chubukov Phys. Rev. Lett. **69** (1992) 832.

30. F. Monti and A. Sütö, Phys. Lett. **69A** (1992) 197 ; Helv. Phys. Acta. **65** (1992) 560.
31. C. K. Majumdar and D. Ghosh, J. Math. Phys. **10** (1969) 1388 ; C. K. Majumdar, J. Phys. **C3** (1970) 911.
32. T. Tsuzuki, Prog. Theor. Phys. **73** (1985) 1352 ; *ibid* **75** (1986) 225 ; I. Harada, T. Kimura and T. Tonegawa, J. Phys. Soc. Jpn. **57** (1988) 2799.

TRANSFER MATRICES IN QUANTUM MANY-BODY SYSTEMS

Tohru Koma
Department of Physics, Gakushuin University,
Mejiro, Toshima-ku, Tokyo 171, Japan

Abstract

We review the quantum transfer matrix method based on the path integral idea, and summarize useful related theorems. The eigenvalues of the transfer matrix in a quantum many-body system lead to the free energy and the correlation lengths in the thermodynamic limit. For calculating these eigenvalues efficiently, the thermal Bethe ansatz method and the Monte Carlo power method are presented. We treat also transfer matrices in quantum many-body systems at zero temperature.

1 Introduction

The transfer matrix method [1, 2] has been widely used to study thermal properties of classical many-body systems. The method reduces the calculations of thermal quantities to eigenvalue problems of transfer matrices. In particular, for a translation invariant system with finite range interactions, the maximum eigenvalue of the transfer matrix leads to the free energy in the thermodynamic limit, and the second and third eigenvalues lead to the interfacial tension or the correlation lengths.

In 1941, Kramers and Wannier [1] used the transfer matrix method to determine the critical temperature of the square lattice Ising model assuming the existence of a single phase transition. Later, in 1944, Onsager [3] obtained the free energy of the square lattice Ising model by diagonalization of the transfer matrix, and confirmed the existence of the phase transition. This approach was further developed by Baxter to the commuting transfer matrix method and to the corner transfer matrix method for the so-called exactly solvable models [4].

For translation invariant one-dimensional classical lattice systems with finite range interactions, the application of the Perron-Frobenius theorem to the transfer matrices yields immediately the result that the free energies in the thermodynamic limit are analytic at finite temperature with respect to their interactions, so that there is no phase transition [5, 6]. The same result for quantum spin systems with finite range interactions on a one-dimensional lattice was obtained by Araki [7]. This approach can be extended to classical continuum systems and also to classical systems with long-range interactions [6, 8].

Fröhlich, Simon and Spencer [9] used the transfer matrix technique to develop the infrared bound method for establishing the existence of phase transitions for classical lattice systems. In particular, their method works well also for the problem of the continuous symmetry breaking, as in the classical isotropic Heisenberg model.

Generally speaking, the transfer matrices for quantum systems have much more complicated forms [7, 10, 11] than those for classical systems. This situation makes it very difficult to treat the eigenvalue problems of the transfer matrices of quantum systems.

Instead of treating directly the transfer matrix of a quantum system, Betsuyaku and Suzuki proposed the approximate transfer matrix method [12, 13, 14] in which one first transforms the partition function of the quantum system into that of a classical system employing the path integral idea [15, 16, 17, 18], and then applies the transfer matrix method to the classical system so obtained. This approach makes transfer matrices of quantum systems much more tractable than previously, in spite of the fact that one must make an extrapolation to recover the original partition function of a quantum system [12, 19, 20, 21]. In fact, this "quantum transfer matrix" method has been applied to various quantum systems [20, 22, 23, 24, 25, 26, 27, 28, 29, 30, 31].

In the present review, we mainly concentrate on the transfer matrices derived from the path integral representations of the partition functions of quantum systems.

The present review is organized as follows. In Section 2, we explain the basic approximation scheme for thermal quantities of quantum systems, and state some theorems [14, 20, 23, 24] useful for the following arguments. In Section 3, as a typical example, the spin-1/2 XXZ Heisenberg chain [15, 16, 32] is treated, and the explicit expression of the transfer matrix in the "spatial" direction [20] is written down. In Section 4, we apply the Bethe ansatz method to the transfer matrix so obtained to reduce the eigenvalue problem of the transfer matrix to a much more tractable problem of solving the so-called Bethe ansatz equations [20]. In Section 5, we explain the Monte Carlo power method introduced by Koma [33]. In Section 6, which is independent of the rest of the review, we treat transfer matrices at zero temperature for particular quantum systems [34, 35].

2   Approximation Scheme

In this section, we introduce some approximations [23] for the density matrix of a general quantum system, and state some useful theorems [14, 20, 23].

We consider a translation invariant quantum system on $L_1 \times L_2 \times \cdots \times L_d$ parallelepiped lattice Λ with $N = L_1 \cdots L_d$ sites. (We identify Λ as a subset of the d-dimensional simple hypercubic lattice $\mathbf{Z}^d$.) Suppose that the interactions are represented by finite dimensional matrices and have finite range.

We wish to apply the path integral idea so as to treat the quantum system as a kind of a classical system. For this purpose, we shall introduce some useful approximations of the density matrix.

The simplest (and the most popular) approximation of the partition function is the so-called Trotter decomposition (or, more accurately, the Lie decomposition)

$$Z_{N,M} = \text{Tr} \left[e^{-\beta \hat{H}_N^{(1)}/M} e^{-\beta \hat{H}_N^{(2)}/M} \right]^M \tag{2.1}$$

for the partition function with the Hamiltonian $\hat{H}_N = \hat{H}_N^{(1)} + \hat{H}_N^{(2)}$, where Tr denotes the trace of a matrix, and β is the inverse temperature. The free energy f in the thermodynamic limit is given by

$$f = -\frac{1}{\beta} \lim_{\Lambda \uparrow \mathbf{Z}^d} \lim_{M \uparrow \infty} \frac{1}{N} \log Z_{N,M}. \tag{2.2}$$

Theorem 1: [20, 23] *Suppose that two matrices $\hat{H}_N^{(1)}$ and $\hat{H}_N^{(2)}$ are written as*

$$\hat{H}_N^{(j)} = \sum_{x \in \Lambda^{(j)}} \hat{h}_x^{(j)} \tag{2.3}$$

($j = 1, 2$) in terms of the local Hamiltonians $\hat{h}_x^{(j)}$ ($x \in \Lambda^{(j)}; j = 1, 2$) which satisfy $[\hat{h}_x^{(j)}, \hat{h}_y^{(j)}] = 0$ for any sites $x \neq y$ in $\Lambda^{(j)}$, where $\Lambda^{(j)}$ ($j = 1, 2$) are sublattices of the lattice Λ. Then the order of limits in (2.2) can be interchanged as

$$f = -\frac{1}{\beta} \lim_{M \uparrow \infty} \lim_{\Lambda \uparrow \mathbf{Z}^d} \frac{1}{N} \log Z_{N,M}. \tag{2.4}$$

Proof: This theorem can be proved by using an inequality related to that of Peierls's. See [20, 23] for the details. ∎

A similar theorem was first proved in 1985 by Suzuki [14], who showed the interchangeability of the order of the limits in (2.2) under the assumption that the interactions of the corresponding classical system obtained due to the path integral idea are bounded and have finite range.

As a generalization of (2.1), Suzuki and Inoue [14, 23] introduced the approximate partition function

$$Z_{N,M}^{(p)} := \text{Tr} \left(\hat{\rho}_{N,M}^{(p)} \right)^M \tag{2.5}$$

with the symmetrized decomposition

$$\hat{\rho}_{N,M}^{(p)} := \left(\hat{Q}_{N,M}^{(p)} \right)^\dagger \hat{Q}_{N,M}^{(p)} \tag{2.6}$$

with

$$\hat{Q}_{N,M}^{(p)} := e^{-\beta \hat{H}_N^{(p)}/2M} e^{-\beta \hat{H}_N^{(p-1)}/2M} \dots e^{-\beta \hat{H}_N^{(1)}/2M} \tag{2.7}$$

for the Hamiltonian

$$\hat{H}_N = \sum_{j=1}^{p} \hat{H}_N^{(j)}. \tag{2.8}$$

The free energy f in the thermodynamic limit is given by

$$f = \lim_{\Lambda \uparrow \mathbf{Z}^d} \lim_{M \uparrow \infty} f_{N,M}^{(p)} \tag{2.9}$$

with the approximate free energy

$$f_{N,M}^{(p)} := -\frac{1}{\beta N} \log Z_{N,M}^{(p)}. \tag{2.10}$$

For the approximate free energy $f_{N,M}^{(p)}$, Suzuki and Inoue proved the following theorem which is an extension of Theorem 1.

Theorem 2: [23] *Suppose that the matrices $\hat{H}_N^{(j)}$ ($j = 1, 2, \cdots, p$) are written as*

$$\hat{H}_N^{(j)} = \sum_{x \in \Lambda^{(j)}} \hat{h}_x^{(j)} \tag{2.11}$$

($j = 1, 2, \cdots, p$) in terms of the local Hamiltonians $\hat{h}_x^{(j)}$ ($x \in \Lambda^{(j)}; j = 1, 2, \cdots, p$) which satisfy $[\hat{h}_x^{(j)}, \hat{h}_y^{(j)}] = 0$ for any sites $x \neq y$ in $\Lambda^{(j)}$, where $\Lambda^{(j)}$ ($j = 1, 2, \cdots, p$) are sublattices of the lattice Λ. Then the order of limits in (2.9) can be interchanged as

$$f = \lim_{M \uparrow \infty} \lim_{\Lambda \uparrow \mathbf{Z}^d} f_{N,M}^{(p)}. \tag{2.12}$$

Proof: See [23] for the details. ∎

Let us follow the idea introduced by Suzuki in [15]. Using the path integral idea, we can transform the partition function $Z_{N,M}^{(p)}$ into the partition function of a $(d+1)$-dimensional classical spin system. When the interactions of the classical system so obtained are of finite range, the partition function $Z_{N,M}^{(p)}$ can be rewritten as

$$Z_{N,M}^{(p)} = \text{Tr}\hat{W}^L, \tag{2.13}$$

where $\hat{W}$ is the transfer matrix in the first "spatial" direction. Here we have taken $L_1 = rL$ ($L = 1, 2, \cdots$) with a positive integer r which is determined by the decomposition of (2.6). The representation (2.13) combined with (2.12) gives

$$
\begin{aligned}
f &= -\frac{1}{\beta} \lim_{M \uparrow \infty} \lim_{L_d \uparrow \infty} \cdots \lim_{L_2 \uparrow \infty} \frac{1}{L_d \cdots L_2} \lim_{L \uparrow \infty} \frac{1}{rL} \log \left[\sum_j (\lambda_j)^L \right] \\
&= -\frac{1}{\beta r} \lim_{M \uparrow \infty} \lim_{L_d \uparrow \infty} \cdots \lim_{L_2 \uparrow \infty} \frac{1}{L_d \cdots L_2} \log \lambda^{\max},
\end{aligned}
\tag{2.14}
$$

where λ_j are the eigenvalues of the transfer matrix $\hat{W}$. A remarkable feature of the formula (2.14) is that the free energy f in the thermodynamic limit is determined only by the maximum eigenvalue $\lambda^{\max}$ of the transfer matrix $\hat{W}$. This fact makes the calculation of the free energy surprisingly easy, as will be demonstrated in the following sections.

Next, let us treat the thermal average of observables in a similar way. Consider an observable $\hat{O}$ which is a hermitian matrix, and introduce the approximate thermal average

$$\langle\hat{O}\rangle_{N,M}^{(p)} := \frac{1}{N}\frac{\mathrm{Tr}\left[\left(\hat{Q}_{N,M}^{(p)}\right)^{\dagger}\hat{O}\hat{Q}_{N,M}^{(p)}\left(\hat{\rho}_{N,M}^{(p)}\right)^{M-1}\right]}{\mathrm{Tr}\left(\hat{\rho}_{N,M}^{(p)}\right)^{M}}. \tag{2.15}$$

The thermal average of $\hat{O}$ in the thermodynamic limit is given by

$$\langle\hat{O}\rangle := \lim_{N\uparrow\infty}\lim_{M\uparrow\infty}\langle\hat{O}\rangle_{N,M}^{(p)}. \tag{2.16}$$

Note that the approximate thermal average (2.15) can be written as

$$\langle\hat{O}\rangle_{N,M}^{(p)} = -\frac{\partial}{\partial B}f_{N,M}^{(p)}(B)\Big|_{B=0} \tag{2.17}$$

where, as the approximate free energy $f_{N,M}^{(p)}(B)$ in the external field, we have taken

$$f_{N,M}^{(p)}(B) = -\frac{1}{\beta N}\log Z_{N,M}^{(p)}(B) \tag{2.18}$$

with

$$Z_{N,M}^{(p)}(B) := \mathrm{Tr}\left[\left(\hat{Q}_{N,M}^{(p)}\right)^{\dagger}e^{\beta B\hat{O}/M}\hat{Q}_{N,M}^{(p)}\right]^{M}. \tag{2.19}$$

In order to show the interchangeability of the order of the limits in (2.16), we state the following lemma.

Lemma 3: *The approximate free energy $f_{N,M}^{(p)}(B)$ (2.18) is concave with respect to the external field B, i.e.,*

$$\frac{\partial^{2}f_{N,M}^{(p)}(B)}{\partial B^{2}} \leq 0. \tag{2.20}$$

Proof: It is not hard to show that

$$\left(\frac{\partial^{2}}{\partial B^{2}}Z_{N,M}^{(p)}(B)\right)Z_{N,M}^{(p)}(B) \geq \left(\frac{\partial}{\partial B}Z_{N,M}^{(p)}(B)\right)^{2} \tag{2.21}$$

by using the Schwarz inequality $\left(\mathrm{Tr}\hat{A}^{\dagger}\hat{A}\right)\left(\mathrm{Tr}\hat{B}^{\dagger}\hat{B}\right) \geq \left|\mathrm{Tr}\hat{A}\hat{B}\right|^{2}$, where $\hat{A}, \hat{B}$ are matrices. The statement of the lemma follows from the inequality (2.21). ∎

The concavity of $f_{N,M}^{(p)}(B)$ leads to the following theorem.

Theorem 4: *If the free energy*

$$f(B) = \lim_{N \uparrow \infty} \lim_{M \uparrow \infty} f_{N,M}^{(p)}(B) \tag{2.22}$$

in the thermodynamic limit is differential at $B = 0$ with respect to the external field B, then we have

$$\langle \hat{O} \rangle = - \left. \frac{\partial f(B)}{\partial B} \right|_{B=0} = - \lim_{M \uparrow \infty} \lim_{N \uparrow \infty} \left. \frac{\partial f_{N,M}^{(p)}(B)}{\partial B} \right|_{B=0}. \tag{2.23}$$

In other words, if there is no discontinuity of the thermal average $\langle \hat{O} \rangle$ at the point $B = 0$, then the order of limits in (2.16) can be interchanged as

$$\langle \hat{O} \rangle = \lim_{M \uparrow \infty} \lim_{N \uparrow \infty} \langle \hat{O} \rangle_{N,M}^{(p)} \tag{2.24}$$

due to (2.17).

Proof: It is not hard to verify the statement of the theorem along the line of Griffiths [36]. ∎

3   Application to the Spin-1/2 XXZ Heisenberg Chain

As a typical example of the formulation in Section 2, we consider the spin-1/2 XXZ Heisenberg chain, and obtain the explicit expression of the transfer matrix [20] in the "spatial" direction by constructing the path integral representation of the partition function [32].

The Hamiltonian of the spin-1/2 XXZ Heisenberg chain is given by

$$\hat{H}_N = \sum_{j=1}^{N} \hat{h}_j, \tag{3.1}$$

where the number of sites $N = 2n$ is even and

$$\hat{h}_j := \left(\frac{1}{2} - 2\hat{\mathbf{s}}_j \cdot \hat{\mathbf{s}}_{j+1} + 2\Delta \hat{s}_j^z \hat{s}_{j+1}^z \right) - B \left(\hat{s}_j^z + \hat{s}_{j+1}^z \right) \tag{3.2}$$

with the periodic boundary condition $\hat{s}_{N+1}^\alpha = \hat{s}_1^\alpha$ ($\alpha = x, y, z$). Here $\hat{s}_j^\alpha$ are the Pauli spin matrices at the jth lattice site, while Δ and B are the anisotropy parameter and the external field, respectively. We take

$$\hat{H}_N^{(1)} = \sum_{\ell=1}^{n} \hat{h}_{2\ell-1} \quad \text{and} \quad \hat{H}_N^{(2)} = \sum_{\ell=1}^{n} \hat{h}_{2\ell}. \tag{3.3}$$

Then the path integral representation of the partition function (2.1) can be rewritten as

$$Z_{N,M} = \sum_{\omega^1, \cdots, \omega^{2M}} W^{(1)}(\omega^1, \omega^2) W^{(2)}(\omega^2, \omega^3) \cdots W^{(1)}(\omega^{2M-1}, \omega^{2M}) W^{(2)}(\omega^{2M}, \omega^1) \quad (3.4)$$

where, for $j = 1, 2$, we set

$$W^{(j)}(\omega, \omega') := \left(\omega \left| \exp\left[-\beta \hat{H}_N^{(j)}/M\right] \right| \omega' \right). \quad (3.5)$$

The basis states are given by $|\omega) := |s_1\rangle \otimes |s_2\rangle \otimes \cdots \otimes |s_N\rangle$ and

$$|s) := \begin{cases} \begin{pmatrix} 1 \\ 0 \end{pmatrix} & \left(s = \frac{1}{2}\right) \\ \begin{pmatrix} 0 \\ 1 \end{pmatrix} & \left(s = -\frac{1}{2}\right) \end{cases} \quad (3.6)$$

The matrix elements (3.5) are easy to write down because of the locality of $\hat{h}_j$ (3.2). We then get

$$W^{(1)}(\omega, \omega') = \prod_{\ell=1}^{n} \left(s_{2\ell-1}, s_{2\ell} \left| \exp[-\beta \hat{h}_{2\ell-1}/M] \right| s'_{2\ell-1}, s'_{2\ell} \right) \quad (3.7)$$

and

$$W^{(2)}(\omega, \omega') = \prod_{\ell=1}^{n} \left(s_{2\ell}, s_{2\ell+1} \left| \exp[-\beta \hat{h}_{2\ell}/M] \right| s'_{2\ell}, s'_{2\ell+1} \right). \quad (3.8)$$

To obtain the expression of the transfer matrix $\hat{W}_M$ in the "spatial" direction, we define the local transfer matrix $\hat{V}_k$ at "time" k by

$$\left(s_j^k, s_{j+1}^k \left| \hat{V}_k \right| s_{j+1}^{k+1}, s_{j+1}^{k+1} \right) = \left(-s_j^k, s_{j+1}^k \left| \exp[-\beta \hat{h}_j/M] \right| s_j^{k+1}, -s_{j+1}^{k+1} \right). \quad (3.9)$$

Here, we note that, since there is the local conservation law in the "temporal" direction

$$- s_j^k + s_{j+1}^k = s_j^{k+1} - s_{j+1}^{k+1}, \quad (3.10)$$

which is the standard consequence of the symmetry of the local Hamiltonian (3.2), we have also the local conservation law in the "spatial" direction [20]

$$s_j^k + s_j^{k+1} = s_{j+1}^k + s_{j+1}^{k+1}. \quad (3.11)$$

As an expression of the transfer matrix $\hat{V}_k$, we use

$$\hat{V}_k := \exp[-\beta B(\hat{\tau}_k^z - \hat{\tau}_{k+1}^z)/2M] \hat{V}_k^{(0)} \exp[\beta B(\hat{\tau}_k^z - \hat{\tau}_{k+1}^z)/2M] \quad (3.12)$$

and

$$\hat{V}_k^{(0)} := \frac{1}{2} D + 2D^{-1}(\hat{\tau}_k^x \hat{\tau}_{k+1}^x + \hat{\tau}_k^y \hat{\tau}_{k+1}^y) + 2D(1 - 2C)\hat{\tau}_k^z \hat{\tau}_{k+1}^z \quad (3.13)$$

with $D = \exp[\beta\Delta/2M]$ and $C = \exp[-\beta/M]\cosh(\beta/M)$ where $\hat{\tau}_k^\alpha$ ($\alpha = x, y, z$) are the Pauli spin matrices at time k.

Using (3.4),(3.7),(3.8) and (3.9), we can rewrite the approximate partition function as

$$Z_{N,M} = \text{Tr}\left[\hat{R}_M^{(1)}\hat{R}_M^{(2)}\right]^n \tag{3.14}$$

in terms of the transfer matrices

$$\hat{R}_M^{(1)} := \prod_{k=1}^M \hat{V}_{2k-1} \quad \text{and} \quad \hat{R}_M^{(2)} := \prod_{k=1}^M \hat{V}_{2k} \tag{3.15}$$

in the "spatial" direction with the periodic boundary condition $\hat{\tau}_{2M+1}^\alpha = \hat{\tau}_1^\alpha$ ($\alpha = x, y, z$).

Similarly, we can write down the representation of the approximate correlation function in terms of the matrices of (3.15) as

$$\langle\!\langle \hat{o}_i \hat{o}_j' \rangle\!\rangle_{N,M} := \frac{\text{Tr}\left[\hat{o}_i \hat{o}_j' \left(e^{-\beta\hat{H}_N^{(1)}/M} e^{-\beta\hat{H}_N^{(2)}/M}\right)^M\right]}{\text{Tr}\left(e^{-\beta\hat{H}_N^{(1)}/M} e^{-\beta\hat{H}_N^{(2)}/M}\right)^M}$$

$$= \frac{\text{Tr}\left[\tilde{o}_1 \left(\hat{R}_M^{(1)}\hat{R}_M^{(2)}\right)^m \tilde{o}_1' \left(\hat{R}_M^{(1)}\hat{R}_M^{(2)}\right)^{n-m}\right]}{\text{Tr}\left(\hat{R}_M^{(1)}\hat{R}_M^{(2)}\right)^n} \tag{3.16}$$

for the two local observables $\hat{o}_i$ and $\hat{o}_j'$ at the sites i and j, respectively, where we take $|i - j| = 2m$ ($m = 1, 2, \cdots$) for simplicity, and $\tilde{o}_1$ and $\tilde{o}_1'$ at time 1 are the corresponding operators in the "spatial" direction.

The approximate correlation function (3.16) is slightly different from $\langle \hat{o}_i \hat{o}_j' \rangle_{N,M}^{(2)}$ of (2.15). But the difference can be controlled as

$$\left|\langle\!\langle \hat{o}_i \hat{o}_j' \rangle\!\rangle_{N,M} - \langle \hat{o}_i \hat{o}_j' \rangle_{N,M}^{(2)}\right| \le \frac{\text{const.}}{M}, \tag{3.17}$$

where the constant on the right-hand side is independent of the chain length N. The bound (3.17) implies that the interchangeablity of Theorem 4 holds also for $\langle\!\langle \hat{o}_i \hat{o}_j' \rangle\!\rangle_{N,M}$.

The transfer matrices $\hat{R}_M^{(1)}$ and $\hat{R}_M^{(2)}$ have the following two remarkable properties.

First, we have $[\hat{R}_M^{(j)}, (\hat{T}_M)^2] = 0$ ($j = 1, 2$), where the operator $\hat{T}_M$ shifts any periodic array of the spin states by one lattice unit backward in the "temporal" direction. Owing to this property, we have immediately

$$\hat{W}_M = \hat{R}_M^{(1)}\hat{R}_M^{(2)} = (\hat{U}_M)^2 (\hat{T}_M)^2 \tag{3.18}$$

with $\hat{U}_M := (\hat{T}_M)^{-1}\hat{R}_M^{(2)}$. This implies that the eigenvalue problem of $\hat{W}_M$ can be reduced to that of $\hat{U}_M$.

Secondly, from (3.12), (3.13) and (3.15), we have $[\hat{R}_M^{(j)}, \hat{\tau}_{\text{tot}}^z] = 0$ $(j = 1, 2)$, where

$$\hat{\tau}_{\text{tot}}^z := \sum_{k=1}^{2M} \hat{\tau}_k^z. \tag{3.19}$$

This implies that the matrix $\hat{U}_M$ can be written in the block diagonal form with the direct sum of matrices as

$$\hat{U}_M = \bigoplus_{k=0}^{2M} \hat{U}_{M;k}, \tag{3.20}$$

where the subscript k denotes the restriction to the k-down-spin subspace. This property makes the calculation of the free energy and the estimates of the correlation lengths very easy [20] in the following way.

Theorem 5: *The maximum eigenvalue $\lambda_M^{\max}$ of the transfer matrix $\hat{U}_M$ is equal to the maximum eigenvalue $\lambda_{M;M}^{\max}$ of the matrix $\hat{U}_{M;M}$.*

Proof: See [20] for the details. ∎

Therefore, in order to obtain the free energy f in the thermodynamic limit of the present model using the formula (2.14), it is sufficient to obtain only the maximum eigenvalue $\lambda_{M;M}^{\max}$ of the transfer matrix $\hat{U}_{M;M}$.

It is believed that the correlation length ξ corresponding to a certain correlation function is given by

$$\xi = \lim_{M \uparrow \infty} \left[\log\left(\lambda_M^{\max}/\lambda_M'\right)\right]^{-1}, \tag{3.21}$$

where λ_M' is an appropriate eigenvalue of $\hat{U}_M$. However, such relations are proved only for special models, such as the XY chain [23, 24].

As for the present XXZ model with $\Delta \geq 1$ and $B = 0$, we are able to prove the bounds [37]

$$\left|\langle \hat{s}_i^+ \hat{s}_j^- \rangle\right| \leq \text{const.} \times \exp[-|i - j|/\tilde{\xi}_{\text{xy}}] \tag{3.22}$$

and

$$\left|\langle \hat{s}_i^z \hat{s}_j^z \rangle\right| \leq \text{const.} \times \exp[-|i - j|/\tilde{\xi}_z] \tag{3.23}$$

for the correlations in the thermodynamic limit, where $\hat{s}_j^{\pm} := \hat{s}_j^x \pm i\hat{s}_j^y$,

$$\tilde{\xi}_{\text{xy}} := \lim_{M \uparrow \infty} \left[\log\left(\lambda_M^{\max}/\lambda_{M;M-1}^{\max}\right)\right]^{-1} \quad \text{and} \quad \tilde{\xi}_z := \lim_{M \uparrow \infty} \left[\log\left(\lambda_M^{\max}/\lambda_{M;M}^{\text{second}}\right)\right]^{-1}. \tag{3.24}$$

Here, $\lambda_{M;M-1}^{\max}$ and $\lambda_{M;M}^{\text{second}}$ are the maximum eigenvalue of $\hat{U}_{M;M-1}$ and the second largest eigenvalue of $\hat{U}_{M;M}$, respectively. To prove the bounds (3.22) and (3.23), we make use of Theorem 4, the representation (3.16), the analyticity of the free energy f [7] and the positivity of the matrix $\hat{R}_M^{(2)}$. (Without $\Delta \geq 1$ and $B = 0$, $\hat{R}_M^{(2)}$ is not positive.) The last condition $\hat{R}_M^{(2)} \geq 0$ is probably redundant, but we are not able to remove it for some technical reasons.

4   Thermal Bethe Ansatz

In this section, we review the thermal Bethe ansatz method [20, 25, 26, 27, 28, 29, 30].

As is well known, the energy eigenvalue problem of the spin-1/2 XXZ Heisenberg model (3.1) can be treated by the Bethe ansatz method [38]. More generally, the method can be applied to systems with the special property that all the scattering amplitudes for many quasiparticles can be written in terms of the two-body scattering amplitudes [39].

To calculate the free energy from the standard formula $Z = \sum_i e^{-\beta E_i}$, one must get all the energy eigenvalues E_i [40]. However, the completeness of the system of the Bethe eigenstates has not yet been proved. (It was proved only partially [41].)

But if we apply the Bethe ansatz method to the eigenvalue problem of the transfer matrix $\hat{U}_M$ introduced in the previous section, then it is enough to obtain only the maximum eigenvalue, which leads to the free energy in the thermodynamic limit by the formula (2.14). The present approach, which we call the thermal Bethe ansatz method, is thus free from the unproven assumption on the completeness of the system of the Bethe eigenstates.

In order to obtain the maximum eigenvalue $\lambda_{M;k}^{max}$ of the transfer matrix $\hat{U}_{M;k}$, we assume that the eigenvectors of $\hat{U}_{M;k}$ have the Bethe ansatz form [20, 42]

$$\Phi_k := \sum_{1 \leq y_1 < \cdots < y_k \leq 2M} \sum_P A_P F(z_{P1}, y_1) \cdots F(z_{Pk}, y_k) |y_1, \cdots, y_k\rangle \qquad (4.1)$$

with

$$F(z, y) := \begin{cases} a_j z_j^{(y+1)/2} & (y = \text{odd}) \\ z_j^{y/2} & (y = \text{even}) \end{cases} \qquad (4.2)$$

$(j = 1, 2, \cdots, k; k = 0, 1, 2, \cdots, 2M)$, where $|y_1, \cdots, y_k\rangle$ is the state with all the spins up except those k spins at the sites $y_1, \cdots, y_k$; the summations run over all the possible distributions of k down spins, and all the permutations P of $(1, 2, \cdots, k)$. The numbers z_j, a_j and A_P are determined by

$$z_j = \frac{D^{-1}E\zeta_j - D^{-2} + (DC)^2}{\zeta_j(\zeta_j - D^{-1}E^{-1})} \quad , \quad a_j = \frac{\zeta_j - D^{-1}E^{-1}}{DC} \qquad (4.3)$$

and

$$\frac{A_{(P1,\cdots,P(j+1),Pj,\cdots,Pk)}}{A_{(P1,\cdots,Pj,P(j+1),\cdots,Pk)}} = -\frac{\zeta_{Pj}\zeta_{P(j+1)} - 2DSE^{-1}\tilde{\Delta}\zeta_{Pj} + (DSE^{-1})^2}{\zeta_{Pj}\zeta_{P(j+1)} - 2DSE^{-1}\tilde{\Delta}\zeta_{P(j+1)} + (DSE^{-1})^2}, \qquad (4.4)$$

where $S = 1 - C, E = \exp[\beta B/M]$ and $\tilde{\Delta} = D^2 - S^{-1}\sinh(\beta\Delta/M)$. Here $\zeta_1, \cdots, \zeta_k$ are determined by the system of the Bethe ansatz equations

$$\left[\frac{D^{-1}E\zeta_j - D^{-2} + (DC)^2}{\zeta_j(\zeta_j - D^{-1}E^{-1})}\right]^M = (-1)^{k-1} \prod_{\ell=1}^{k} \frac{\zeta_j\zeta_\ell - 2DSE^{-1}\tilde{\Delta}\zeta_\ell + (DSE^{-1})^2}{\zeta_j\zeta_\ell - 2DSE^{-1}\tilde{\Delta}\zeta_j + (DSE^{-1})^2} \qquad (4.5)$$

with $j = 1, 2, \cdots, k$. The eigenvalues of the transfer matrix $\hat{U}_{M;k}$ are given by

$$\lambda_{M;k} = (DS)^{M-k}\zeta_1\zeta_2\cdots\zeta_k. \qquad (4.6)$$

A clear presentation of physics underlying the Bethe ansatz method can be found in Sutherland's lecture note [39]. The basis of the thermal Bethe ansatz calculation is summarized in the following theorem.

Theorem 6: [20] *For any given three parameters, i.e., for the temperature $\beta \geq 0$, the anisotropy parameter $\Delta \in \mathbf{R}$, the external field $B \in \mathbf{R}$, the maximum eigenvalue $\lambda_{M;k}^{\max}$ of the transfer matrix $\hat{U}_{M;k}$ can be obtained from $\lambda_{M;k}$ (4.6) with a solution $(\zeta_1, \cdots, \zeta_k)$ to the Bethe ansatz equations (4.5).*
Proof: See [20] for the details. ∎

Koma [20] solved numerically the Bethe ansatz equation (4.5) in the isotropic ferromagnetic case $\Delta = 0$, and obtained the free energy and the susceptibility. However, for calculating the susceptibility, he assumed that the limit $M \uparrow \infty$ and the thermodynamic limit can be interchanged. The proof has still been left as an open problem. Later, Yamada [25] calculated the correlation length of the spin-spin correlation for the same model using the formula (3.21). The isotropic antiferromagnetic case was treated by Nomura and Yamada [26]. They calculated the free energy and the correlation length of the spin-spin correlation. Suzuki, Akutsu and Wadati showed that the thermal Bethe ansatz works well also for the spin-1/2 XYZ Heisenberg chain [27]. Takahashi [28] calculated the free energies and the correlation lengths of the spin-spin correlations for the cases with the anisotropy and with the magnetic field.

The thermal Bethe ansatz method can be applied also to the Hubbard chain, as was shown by Koma [29]. Following his formalism, Tsunetsugu [30] calculated the free energy and the correlation length of the spin-spin correlation for the half-filled band case.

5 Monte Carlo Power Method

In the previous section, we have treated the spin-1/2 XXZ Heisenberg chain, which has the special property that it is tractable by the Bethe ansatz method.

But when a system has no such property, one must depend on other methods to obtain eigenvalues of a transfer matrix of a system. For example, one can numerically obtain the eigenvector corresponding to the maximum eigenvalue by multiplying a trial vector by the transfer matrix repeatedly. This is the most popular power method. However, it is difficult to apply the power method to a large matrix, in particular, to a transfer matrix whose size increases exponentially with the size of a system, requiring a correspondingly large computer memory.

Recently, a new approach was proposed to circumvent the above difficulty in a power method [33]. In the present section, we review the new approach, which we call the Monte Carlo power method.

Consider a non-negative matrix $\hat{A}$. (In most applications, the matrix $\hat{A}$ is chosen as the transfer matrix of a many-body system). We assume that there exists a positive integer k such that all the matrix elements of $\hat{A}^k$ are strictly positive in an orthonormal basis $\{u(n)\}_n$. To obtain the maximum eigenvalue, we introduce the Rayleigh quotient

$$\lambda^{(L)} := \frac{\left(\psi, \hat{A}^{L+1}\psi\right)}{\left(\psi, \hat{A}^L\psi\right)} \tag{5.1}$$

$(L = 0, 1, 2, \cdots)$, where ψ is a trial vector with non-negative elements in the basis $\{u(n)\}_n$. Due to the Perron-Frobenius theorem, the maximum eigenvalue $\lambda^{\max}$ of $\hat{A}$ can be obtained as

$$\lambda^{\max} = \lim_{L \uparrow \infty} \lambda^{(L)}, \tag{5.2}$$

because all the matrix elements of $\hat{A}^\ell$ are strictly positive for any integer $\ell \geq k$.

To circumvent the above-mentioned difficulty of the power method, we choose

$$\psi = \sum_n u(n) \tag{5.3}$$

for the trial vector. Then the Rayleigh quotient (5.1) can be written as

$$\lambda^{(L)} = \frac{1}{Z^{(L)}} \sum_{n_1,\cdots,n_{L+1}} A_{\mathrm{S}}(n_1) \prod_{j=1}^{L} W(n_j, n_{j+1}) \tag{5.4}$$

with $A_{\mathrm{S}}(n) := \left(\psi, \hat{A}u(n)\right)$, $W(n,m) := \left(u(n), \hat{A}u(m)\right)$ and

$$Z^{(L)} := \sum_{n_1,\cdots,n_{L+1}} \prod_{j=1}^{L} W(n_j, n_{j+1}). \tag{5.5}$$

The outstanding feature of the representation (5.4) is that it has precisely the form of the thermal average of the observable $A_{\mathrm{S}}(n_1)$ with respect to the configuration $\Omega := \{(n_1, \cdots, n_{L+1})\}$ with the Boltzmann weights

$$\prod_{j=1}^{L} W(n_j, n_{j+1}), \tag{5.6}$$

because the definition of the matrix $\hat{A}$ guarantees that these Boltzmann weights are non-negative.

The representation (5.4) may be of interest in its own right, but it is useful mainly as a basis of numerical methods for calculating the maximum eigenvalues of transfer matrices.

Since the maximum eigenvalue of the transfer matrix leads to the free energy of the system in the thermodynamic limit, one can easily calculate the free energy by combining the standard importance-sampling technique introduced by Metropolis,

Rosenbluth, Rosenbluth, Teller and Teller [43] with the representation (5.4). In fact, as is well known, the size of the memory in simulations with the Metropolis algorithm is roughly proportional to the lattice size of the system (see [44], for example). Thereby the difficulty encountered in the power method can be circumvented.

When a transfer matrix has a particular block diagonal form as (3.20) in Section 3, one can calculate some of the correlation lengths of the system assuming the formula (3.21).

Hatano and Suzuki [31] applied this method to the calculation of the spin-spin correlation length of the spin-2 antiferromagnetic Heisenberg chain. Their results strongly support the Haldane conjecture [45] for the spin-2 case.

Let us emphasize that the present power method is different from that of the transfer-matrix Monte Carlo method introduced by Nightingale and Blöte [46]. The latter is a variant of the Green-function Monte Carlo method by Ceperly and Kalos [47], and does not utilize the idea of importance sampling [48].

6   Transfer Matrices at Zero Temperature

We have reviewed the formalism which is efficient for analyzing quantum systems at finite temperature. To obtain information about a quantum system at zero temperature within the same formalism, one must extrapolate quantities at finite temperatures to those at zero temperature. The procedure sometimes yields enormous errors, which makes it difficult to arrive at conclusive results.

In the present section, we review a method which deals directly with a quantum system at zero temperature. The approach was first proposed by Kubo, Kaplan and Borysowicz [34], who treated the spin-1/2 isotropic antiferromagnetic Heisenberg chain, which satifies the so-called Sutherland relation [49] that the Hamiltonian commutes with the transfer matrix of the six-vertex model with certain Boltzmann weights. The application of the Perron-Frobenius theorem to the Hamiltonian and the transfer matrix leads immediately to the result that the eigenvector with the maximum eigenvalue of the transfer matrix can be identified with the ground state of the quantum system, so that a correlation function of the six-vertex model is identical to a certain correlation function of the ground state of the Heisenberg chain.

Generalizing the idea proposed by Kubo-Kaplan-Borysowicz, we assume that the transfer matrix $\hat{W}_N$ of a classical system and the Hamiltonian $\hat{H}_N$ of a system with N sites are defined on the same Hilbert space, and that the eigenvector φ_N satisfies

$$\hat{W}_N \varphi_N = \lambda_N^{\max} \varphi_N \quad \text{and} \quad \hat{H}_N \varphi_N = E_N^{(0)} \varphi_N \qquad (6.1)$$

with non-degenerate eigenvalues $\lambda_N^{\max}$ and $E_N^{(0)}$ which are the maximum eigenvalue of $\hat{W}_N$ and the ground state energy of $\hat{H}_N$, respectively.

Under the assumptions, the thermal average of the two classical spins s_i and s_j

defined as

$$\langle s_i s_j \rangle_{L \times N, \text{P}} := \frac{\text{Tr}\left[\hat{s}_i \hat{s}_j (\hat{W}_N)^L\right]}{\text{Tr}(\hat{W}_N)^L} \tag{6.2}$$

is related to the quantum mechanical correlation for the ground state φ_N by the relation

$$(\varphi_N, \hat{s}_i \hat{s}_j \varphi_N) = \lim_{L \uparrow \infty} \langle s_i s_j \rangle_{L \times N, \text{P}}, \tag{6.3}$$

where the subscript P indicates the use of periodic boundary conditions in the transfer direction of the transfer matrix.

Following the above formalism, Kubo, Kaplan and Borysowicz [34] performed the Monte Carlo simulation up to $N = 40$ for the six-vertex model to calculate the spin-spin correlation of the spin-1/2 isotropic antiferromagnetic Heisenberg chain. They assumed that the spin-spin correlation of the Heisenberg chain $C(r)$ in the thermodynamic limit behaves as

$$C(r) \sim A(-1)^r \frac{[\log(r/r_0)]^\sigma}{r} \tag{6.4}$$

at the large distance $r = |i - j|$, where A, σ and r_0 are constants. The asymptotic form (6.4) is believed to be correct. They obtained the exponent $\sigma = 0.2 \sim 0.3$. However, their result deviates slightly from $\sigma = 0.5$ predicted by the conformal field theory [50].

Recently, Mizukoshi and Koma [35] performed the Monte Carlo simulation up to $N = 80$, modifying the Kubo-Kaplan-Borysowicz method as follows. Instead of the periodic boundary conditions in the transfer direction for the transfer matrix, they used free boundary conditions. Thereby they avoided the difficulty that the acceptance ratio of global flips decreases [34] very rapidly with the increase of the system size. Consequently, the modification is equivalent to choosing

$$\langle s_i s_j \rangle_{L \times N, \text{F}} := \frac{\left(\psi, (\hat{W}_N)^{L/2} \hat{s}_i \hat{s}_j (\hat{W}_N)^{L/2} \psi\right)}{\left(\psi, (\hat{W}_N)^L \psi\right)} \tag{6.5}$$

instead of (6.2) as the approximate correlation function. Here L is even and the subscript F denotes the "free boundary" condition which is determined by the way we choose a trial vector ψ. Clearly, there arises another problem, namely that one cannot measure the observable $\hat{s}_i \hat{s}_j$ near the free boundaries of the system. But this demerit can be outweighed by choosing the trial vector ψ carefully. Mizukoshi and Koma chose ψ so that the original six-vertex model feels free boundary conditions. It turns out that this choice leads to a more rapid convergence than the periodic boundary condition imposed in (6.2). Consequently, the modification enables us to treat much longer chains.

Mizukoshi and Koma assumed the asymptotic form

$$C(r) \sim A \frac{(-1)^r}{r} [\log(r/r_0)]^\sigma + \frac{B}{r^2}, \tag{6.6}$$

where A, B, r_0, σ are constants. The second term is believed to exist as a correction [51]. They obtained $\sigma = 0.57(3)$, where the numeral in the parenthesis is the statistical error in the last digit. The result appears to be consistent with the prediction of the conformal field theory [50].

We can make the following two remarks.

- This zero-temperature Monte Carlo method can be applied also to the Hubbard chain, which satisfies a similar Sutherland relation [52], namely that the Hamiltonian commutes with the transfer matrix of the double layer six-vertex model.

- It is not clear whether the present method applies effectively to a general quantum system. We must first find a "simple" classical system which satisfies the above requirements, but even this way might not be easy in general.

Acknowledgements

I would like to thank Professor H. Ezawa for continual encouragement. I also thank Professor H. Tasaki for critical reading of the manuscript, and Dr. D. Lipowska and Mr. N. Mizukoshi for many useful comments on the manuscript.

References

1. H. A. Kramers and G. H. Wannier, *Phys. Rev.* **60** (1941) 252, 263.
2. E. W. Montroll, *J. Chem. Phys.* **9** (1941) 706; **10** (1942) 61.
   E. N. Lassettre and J. P. Howe, *J. Chem. Phys.* **9** (1941) 747.
   R. Kubo, *Busseiron Kenkyu* **1** (1943) 1.
   G. F. Newell and E. W. Montroll, *Rev. Mod. Phys.* **25** (1953) 353.
3. L. Onsager, *Phys. Rev.* **65** (1944) 117.
4. R. J. Baxter, *Exactly Solved Models in Statistical Mechanics* (Academic, New York, 1989).
5. M. E. Baur and L. H. Nosanow, *J. Chem. Phys.* **37** (1962) 153.
6. D. Ruelle, *Statistical Mechanics: Rigorous Results* (W. A. Benjamin, New York, 1969) p. 134.
7. H. Araki, *Commun. Math. Phys.* **14** (1969) 120.
8. L. Van Hove *Physica* **16** (1950) 137.
   D. Ruelle, *Commun. Math. Phys.* **9** (1968) 267.
   G. Gallavotti, S. Miracle-Sole and D. Ruelle, *Phys. Lett.* **26A** (1968) 350.
   G. Gallavotti and S. Miracle-Sole, *J. Math. Phys.* **11** (1970) 147.
   G. Gallavotti and F. T. Lin, *Arch. Ration. Mech. Anal.* **37** (1970) 181.
9. J. Fröhlich, B. Simon and T. Spencer, *Commun. Math. Phys.* **50** (1976) 79.
10. G. G. Emch, in *Phase Transitions and Critical Phenomena*, Vol. 1, C. Domb and M. S. Green, eds. (Academic Press, New York, 1972).

11. D. J. Klein and T. L. Welsher, *J. Stat. Phys.* **24** (1981) 555.
12. H. Betsuyaku, *Phys. Rev. Lett.* **53** (1984) 629.
13. M. Suzuki, *Phys. Lett.* **111A** (1985) 440.
    M. Suzuki and H. Betsuyaku, *Phys. Rev.* **B34** (1986) 1829.
14. M. Suzuki, *Phys. Rev.* **B31** (1985) 2957.
15. M. Suzuki, *Commun. Math. Phys.* **51** (1976) 183; *Prog. Theor. Phys.* **56** (1976) 1454.
16. H. De Raedt and A. Lagendijk, *Phys. Rep.* **127** (1985) 233.
17. M. Suzuki, *J. Stat. Phys.* **43** (1986) 883.
18. M. Suzuki, ed., *Quantum Monte Carlo Methods in Equilibrium and Nonequilibrium* (Springer, Berlin, 1986).
19. M. Suzuki, *Phys. Lett.* **113A** (1985) 299.
20. T. Koma, *Prog. Theor. Phys.* **78** (1987) 1213; **81** (1989), 783.
21. M. Inoue and M. Suzuki, *Prog. Theor. Phys.* **79** (1988) 30.
22. H. Betsuyaku, *Prog. Theor. Phys.* **73** (1985) 319; **75** (1986) 774; **75** (1986) 808.
    T. Tsuzuki, *Prog. Theor. Phys.* **73** (1985)1352; **75** (1986) 225.
    T. Yokota and H. Betsuyaku, *Prog. Theor. Phys.* **75** (1986) 46.
    K. Kubo and S. Takada, *J. Phys. Soc. Jpn.* **55** (1986) 438.
    S. Takada and K. Kubo, *J. Phys. Soc. Jpn.* **55** (1986) 1671.
    M. Imada and M. Takahashi, *J. Phys. Soc. Jpn.* **55** (1986) 3354.
    T. Delica and H. Leschke, *Physica A* **168** (1990) 736.
    N. Hatano and M. Suzuki, *Prog. Theor. Phys.* **85** (1991) 481.
    T. Delica, K. Kopinga, H. Leschke and K. K. Mon, *Europhys. Lett.* **15** (1991) 55.
    K. Kubo, *Phys. Rev.* **B46** (1992) 866.
    N. Hatano and M. Suzuki, *J. Phys. Soc. Jpn.* **62** (1993) 847.
23. M. Suzuki and M. Inoue, *Prog. Theor. Phys.* **78** (1987), 787.
24. M. Inoue and M. Suzuki, *Prog. Theor. Phys.* **79** (1988), 645.
25. M. Yamada, *J. Phys. Soc. Jpn.* **59** (1990) 848.
26. K. Nomura and M. Yamada, *Phys. Rev.* **B43** (1991) 8217.
27. J. Suzuki, Y. Akutsu and M. Wadati, *J. Phys. Soc. Jpn.* **59** (1990) 2667.
    C. Destri and H. J. de Vega, *Phys. Rev. Lett.* **69** (1992) 2313.
28. M. Takahashi, *Phys. Rev.* **B43** (1991) 5788; Erratum **B44** (1991) 5397; **B44** (1991) 12382.
29. T. Koma, *Prog. Theor. Phys.* **83** (1990) 655.
30. H. Tsunetsugu, *J. Phys. Soc. Jpn.* **60** (1991) 1460.
31. N. Hatano and M. Suzuki, *J. Phys. Soc. Jpn.* **62** (1993) 1346.
32. M. Barma and B. S. Shastry, *Phys. Rev.* **B18** (1978) 3351.
33. T. Koma, *J. Stat. Phys.* **71** (1993) 269.
34. K. Kubo, T. A. Kaplan and J. R. Borysowicz, *Phys. Rev.* **B38** (1988) 11550.
35. N. Mizukoshi and T. Koma, Annual Meeting of the Physical Society of Japan, Sendai, Japan, March 1993.
36. R. B. Griffiths, *J. Math. Phys.* **5** (1964) 1215.

37. T. Koma. Annual Meeting of the Physical Society of Japan, Osaka, Japan, March 1990.
38. H. A. Bethe, *Z. Phys.* **71** (1931) 205.
    C. N. Yang and C. P. Yang, *Phys. Rev.* **150** (1966) 321, 327.
39. B. Sutherland, in *Exactly Solvable Problems in Condensed Matter and Relativistic Field Theory*, B. S. Shastry, S. S. Jha and V. Singh, eds., Lecture notes in Physics. **242** (Springer, Berlin, 1985) p. 1.
40. C. N. Yang and C. P. Yang, *J. Math. Phys.* **10** (1969) 1115.
    M. Gaudin, Phys. Rev. Lett. **26** (1971) 1301.
    M. Takahashi, *Prog. Theor. Phys.* **46** (1971) 401, 1388; **47** (1972) 69.
    M. Takahashi and M. Suzuki, *Prog. Theor. Phys.* **48** (1972) 2187.
    T. C. Dorlas, J. T. Lewis and J. V. Pulé, *Commun. Math. Phys.* **124** (1989) 365.
41. T. Koma and H. Ezawa, *Prog. Theor. Phys.* **78** (1987) 1009.
    T. C. Dorlas, *Commun. Math. Phys.* **154** (1993) 347.
42. R. Z. Bariev, *Teoret. Mat. Fiz.* **49** (1981) 261 [*Theor. Math. Phys.* **49** (1982) 1021].
    T. T. Truong and K. D. Schotte, *Nucl. Phys.* **B220** (1983) 77.
43. N. Metropolis, A. W. Rosenbluth, M. N. Rosenbluth, A. H. Teller and E. Teller, *J. Chem. Phys.* **21** (1953) 1987.
44. K. Binder, ed., *Monte Carlo Methods in Statistical Physics*, 2nd ed. (Springer, Berlin, 1986).
45. F. D. M. Haldane, *Phys. Rev. Lett.* **50** (1983) 1153; *Phys. Lett.* **A93** (1983) 464.
46. M. P. Nightingale and H. W. J. Blöte, *Phys. Rev. Lett.* **60** (1988) 1562.
47. D. M. Ceperly and M. H. Kalos, in *Monte Carlo Methods in Statistical Physics*, 2nd ed., K. Binder, ed. (Springer, Berlin, 1986).
48. J. H. Hetherington, *Phys. Rev.* **A30** (1984) 2713.
49. B. Sutherland, *J. Math. Phys.* **11** (1970) 3183.
50. I. Affleck, D. Gepner, H. J. Schulz and T. Ziman, *J. Phys.* **A22** (1989) 511.
    K. Nomura, Preprint (Tokyo Insitute of Technology, 1992).
51. A. Luther and I. Peschel, *Phys. Rev.* **B12** (1975) 3908.
52. B. S. Shastry. *Phys. Rev Lett.* **56** (1986) 1529, 2453; *J. Stat. Phys.* **50** (1988) 57.
    M. Wadati, E. Olmedilla and Y. Akutsu, *J. Phys. Soc. Jpn.* **56** (1987) 1340.
    E. Olmedilla, M. Wadati and Y. Akutsu, *J. Phys. Soc. Jpn.* **56** (1987) 2298.

MONTE CARLO CALCULATIONS OF ELEMENTARY EXCITATION

Minoru Takahashi

Institute for Solid State Physics, University of Tokyo,
Roppongi, Minato-ku, Tokyo 106 Japan
e-mail: j38a@taka2.issp.u-tokyo.ac.jp

Abstract

We give an efficient Monte Carlo method of calculating the elementary excitation spectrum of quantum systems. The lowest energy with an arbitrary momentum is obtained by the projector Monte Carlo method. This method is applied to the spin-1/2 and spin-1 Heisenberg antiferromagnetic chains of length 32. For the S=1/2 case, the spectrum coincides completely with des Cloiseaux and Pearson's spectrum. For the S=1 case, the spectrum has a gap at the momentum π as was predicted by Haldane. The value of the gap coincides with Nightingale and Blöte's calculation. The spectrum satisfies the variational relation to the structure factor. The properties of $S = 1$ Heisenberg antiferromagnetic chains are completely different from those of $S = 1/2$ Heisenberg antiferromagnetic chains. This method is also applied to the anisotropic $S = 1$ Heisenberg chain. The results are in excellent agreement with those of the diagonalization method and experiment.

1 Introduction

There are two categories of the quantum Monte Carlo method. The first one is for the finite-temperature case. It is sometimes called world-line MC or path-integral MC[1]. The second is for the zero temperature cases[2-6]. It is called projector MC method (PMC) or Green function MC method.

In the first method the low-temperature properties can be derived by using the long imaginary time axis. The checker-board decomposition is based on the exact solution of the two-spin system. One can also use an exact solution of a larger cluster, this is called the large-cluster decomposition. The program becomes complicated, but one can use wider time slices of the imaginary time axis and the acceptance ratio becomes higher. In this world-line Monte Carlo method, the calculation of the correlation function $\rho(l) \equiv < S_i^z S_{i+l}^z >$ is easy. On the other hand, the projector Monte Carlo method is useful for determining the ground state energy. This can be done by using the following formula:

$$E_0 = \lim_{\tau \to \infty} \frac{< j|\mathcal{H}e^{-\tau\mathcal{H}}|i >}{< j|e^{-\tau\mathcal{H}}|i >},\tag{1}$$

if the ground state is not orthogonal to the initial state $|i>$. If the state $|i>$ has the momentum K, the r.h.s. of (1) should converge to the lowest energy in the momentum space K if it is not orthogonal to $|i>$. Thus we consider the following function $B(K,\tau)$:

$$B(K,\tau) \equiv \frac{<\psi|R(-K)\mathcal{H}e^{-\tau\mathcal{H}}R(K)|\varphi>}{<\psi|R(-K)e^{-\tau\mathcal{H}}R(K)|\varphi>}. \qquad (2)$$

Here $|\varphi>$ is the ground-state wave function, and $|\psi>$ is a state with zero momentum. $R(K)$ is an operator which gives the momentum K. Actually the estimation of $B(K,\tau)$ becomes harder as τ becomes large because the statistical error grows larger. This is a kind of the negative sign problem. We apply these Monte Carlo methods to the antiferromagnetic Heisenberg chain (AHC) with spin 1/2 and 1.

For the Heisenberg antiferromagnet in one dimension:

$$\mathcal{H} = J\sum_{i=1}^{N} \mathbf{S}_i \cdot \mathbf{S}_{i+1} + D\sum_{i=1}^{N}(S_i^z)^2, \quad \mathbf{S}_{N+1} \equiv \mathbf{S}_1, \qquad (3)$$

only S=1/2 case is soluble by the Bethe ansatz. The linear spin-wave theory gives the elementary excitation spectrum:

$$\epsilon(K) = 2JS|\sin K|. \qquad (4)$$

On the contrary, it is known that the ground state of this Hamiltonian for $S = 1/2$ has the gapless excitation[7]:

$$\epsilon(K) = \frac{\pi}{2}J|\sin K|. \qquad (5)$$

This means that the gap between the ground state energy and the first excited energy becomes zero in the limit of infinite systems. This fact was rigorously proved by Lieb, Schultz and Mattis[8]. Their proof can be extended to the S =half integer cases[9]. Unfortunately, this proof cannot be extended to the case of integer S. Using the mapping of Hamiltonian (3) to the classical 2-D $O(3)$ non-linear σ model, Haldane[10] predicted that in the integer-S cases the system has a finite energy gap. It is expected that the energy gap decreases as $e^{-\pi S}$. As far as we know, no one has proved rigorously the existence of a gap for the integer S-cases. But we have plenty of numerical evidence for the existence of the gap, especially for the $S = 1$ case. The most direct numerical method for this problem is the exact diagonalization method[11-16]. Initially, the length of the diagonalized chain has often been $N = 12$, but recently the diagonalization of $N = 16$ chain is very popular and $N = 18$ chain is also sometimes used. As for the quantum Monte Carlo calculation, the Hamiltonian (3) does not pose the negative sign problem unlike the antiferromagnet on the triangular lattice or 2-3D fermion problems. Nightingale and Blöte[5] calculated the energy gap for the $S = 1$ case with $N = 32$ spins. They concluded that the energy gap is 0.41J. A rough estimation of the correlation length of this system was done by the author, who used the world-line Monte Carlo method[17]. Nomura[18] performed a more elaborate calculation of the correlation length and obtained $\xi = 6.1$ in units of atomic spacing. The author obtained the lowest-energy eigenvalue for arbitrary momenta of the $N = 32$ chain using the projector Monte Carlo method[19].

From experiments, two kinds of Haldane-gap substances were reported: the first group includes $CsNiCl_3$ and its derivatives $(RbNiCl_3)$[20-24], and the second group comprises $Ni(C_2H_8N_2)_2NO_2ClO_4$ (NENP) and its derivatives (TMNIN, NINO,...)[25-27]. The $CsNiCl_3$ group has an interchain coupling which is large enough to destroy the gap. At the ground state the system is actually gapless because of the three-dimensional order, but it still maintains the character of a Haldane-gap substance. For the NENP group the interchain coupling is sufficiently small, and there is no experimental evidence that the three-dimensional order appears. The anisotropy is not considered to be small. The coupling J is estimated to be 48K and $D/J = 0.2$ for NENP[28].

At $D = 0$ and $Q \neq 0$, the state is a triplet excitation with total spin 1. It has the same energy eigenvalue as $S_z = \pm 1$ states. The result for the isotropic system explained qualitatively the excitation spectrum of NENP, which was observed by the neutron scattering experiments. In the case of $D \neq 0$, however, the triplet states split into $S^z = \pm 1$ doublets and $S^z = 0$ singlets, the splitting is considerable, especially near the zone boundary at $K = \pi$. The Monte Carlo calculation and the numerical diagonalization calculation of the excitation spectrum were therefore performed for the system with anisotropy. The obtained results for the $N = 32$ singlet state agreed quantitatively with the recent neutron scattering experiments[25-27,29]. The diagonalization method for the $N = 14, 16$ and 18 chains was also used in calculating low-lying excitations.

2 Projector Monte Carlo Method

For the calculation of physical quantities in quantum systems, the exact diagonalization method is used in many cases. In this method, however, the size of the system is very restricted. Therefore quantum Monte Carlo methods are used for larger systems. The partition function approach is used widely for finite-temperature properties[1]. On the other hand, the projector Monte Carlo (PMC) method[2-5] is also powerful, especially for the investigation of the ground state. The Green function Monte Carlo method[6] can be regarded as a kind of the PMC method. We assume that all the off-diagonal elements of the Hamiltonian H are zero or negative. Then all the elements of the ground-state wave vector have the same sign. The ground state is represented by a distribution of random walkers. In many cases the Hamiltonian H has a translational symmetry and there is the translation operator T which satisfies:

$$HT = TH, \quad T^N = I, \tag{6}$$

where N is the length of the system. All the eigenvectors of H can be classified by the momentum K:

$$H|l;K> = E_l(K)|l;K>, \quad T|l;K> = e^{iK}|l;K>,$$

$$K = 2\pi k/N, k = 0, 1, 2..., N-1, \tag{7}$$

$$E_1(K) \le E_2(K) \le ...,$$

where $|l, K>$ and $E_l(K)$ are the l-th eigenvector and eigenvalue in K momentum states, respectively. Usually the ground state of H belongs to the subspace $K = 0$. If one can calculate the lowest-energy eigenvalue for a given momentum, it is just the elementary excitation. By the exact diagonalization method, one can calculate the elementary excitation spectrum of small systems. In the PMC method the ground-state energy is given by the following formula:

$$E_1(0) = \lim_{\tau \to \infty} \frac{<\psi_0|He^{-\tau H}|\xi_0>}{<\psi_0|e^{-\tau H}|\xi_0>}, \tag{8}$$

where $|\xi_0>$ and $<\psi_0|$ are some arbitrary vectors. If $|\xi_0>$ and $<\psi_0|$ are states with momentum K, we obtain the lowest-energy eigenvalue with the momentum K. The operator $e^{-\tau H}$ serves as a projector to the lowest-energy states with the momentum K. Tus we are able to obtain the excitation spectrum by the PMC method. One possible choice of $<\psi_0|$ and $|\xi_0>$ with the momentum K is as follows:

$$<\psi_0| = <\psi|R(-K), \quad |\xi_0> = R(K)|0;1>. \tag{9}$$

Here $<\psi|$ is the vector whose elements are all equal to unity and $R(K)$ is the diagonal operator which satisfies:

$$TR(K) = e^{iK}R(K). \tag{10}$$

As $|0;1>$ and $<\psi|$ are zero-momentum states, $|\xi_0>$ and $<\psi_0|$ are K-momentum states. To obtain eq.(8), we should consider the following function:

$$B(K,\tau) = \frac{<\psi|R(-K)He^{-\tau H}R(K)|0;1>}{<\psi|R(-K)e^{-\tau H}R(K)|0;1>}. \tag{11}$$

The function $B(K,\tau)$ should approach the lowest-energy eigenvalue of the momentum K unless $<\psi_0|$ and $|\xi_0>$ are orthogonal to this state. In the PMC method, $|0;1>$ is given as a distribution of walkers. As $R(K)$ is a diagonal operator, the weight of each walker is multiplied by its corresponding value of $R(K)$, after time τ it is multiplied by the new value of $R(-K)$.

Hereafter we restrict ourselves to the antiferromagnetic Heisenberg chain (3) with even N. It commutes with the true translation operator $\tilde{T}(\tilde{T}S_l = S_{l+1})$ and the z-component of the total spin $S = \sum_l S_l^z$. In the PMC calculation of spin systems, a walker is represented by a set of z-component spin values at N sites:

$$(s_1^z, s_2^z, ..., s_N^z), \quad s_l^z = S, S-1, ..., -S. \tag{12}$$

In this representation the off-diagonal elements of H are non-negative. By the unitary transformation $U = \exp(\pi i \sum_{l=\text{even}} S^z)$, eq.(3) and $\tilde{T}$ become as follows:

$$H = U \tilde{H} U^{-1} = \sum_{l=1}^{N} h_l,$$

$$h_l = J(-S_l^x S_{l+1}^x - S_l^y S_{l+1}^y + S_l^z S_{l+1}^z) + (D/2)[(S_l^z)^2 + (S_{l+1}^z)^2], \quad (13a)$$

$$T = U\tilde{T}U^{-1} = \exp[\pi i(NS - S^z)]\tilde{T}. \quad (13b)$$

Then all the off-diagonal elements of H become non-positive and the PMC method is applicable. It should be noted that the true momentum shifts by π from K if $NS^z - S$ is odd. To get the ground-state energy we use eq.(8). At the beginning of calculation, $|\xi_0 >$ is approximated by a set of L random walkers and the weights of the walkers are assumed to be the same. The wave function is approximated by:

$$e^{-\tau H}|\xi_0 >= \sum_{i=1}^{L} w(i,\tau)|i,\tau > . \quad (14)$$

Here $|i,\tau >$ and $w(i,\tau)$ are the i-th walker and its weight at time τ, respectively. The calculation of the ground-state wave function is done by repeating the following two processes.

1) Application of operator $\exp(-\Delta\tau H)$ to the distribution of walkers. Here $\Delta\tau$ is some small time interval. By repeating this operation, we obtain $e^{-\tau H}$. The operator $\exp(-\Delta\tau H)$ is approximated by:

$$[\exp(-\Delta\tau H_1/2r)\exp(-\Delta\tau H_2/r)\exp(-\Delta\tau H_1/2r)]^r, \quad (15a)$$

$$H_1 = \sum_{l=\text{even}} h_l, \quad H_2 = \sum_{l=\text{odd}} h_l. \quad (15b)$$

The matrix $\exp(-\Delta\tau H_1/2r)$ is decomposed into a product of a stochastic matrix and diagonal matrix:

$$\{\exp(-\Delta\tau H_1/2r)\}_{lj} = p_{lj}q_j, \quad \sum_l p_{lj} = 1. \quad (16)$$

The operation of the diagonal matrix is equivalent to multiplying the weight by q_j. The stochastic matrix $\{p_{lj}\}$ makes a jump from j to l with the probability p_{lj}. The spin configuration of a walker is changed by this process. As H is decomposed into $N/2$ pairs, the spin configuration is changed for each spin pair. The operator $\exp(-\Delta\tau H/r)$ is applied in the same way.

2) Reconfiguration of weights.

As time passes, some walkers become very heavy while others become very light. We should create a new set of walkers so that the existence probability is proportional to the weight of the old set:

$$e^{-(\tau+0)H}|\xi_0 >= \sum_{i}^{L} a|i,\tau+0>, \quad a = \sum_{l=1}^{L} w(l,\tau)/L, \quad (17a)$$

$$|i, \tau + 0 >= |j(i), \tau > . \tag{17b}$$

Here $j(i)$ is determined by:

$$h(j-1) \leq i - \alpha < h(j), \quad h(j) \equiv \sum_{i=1}^{j} w(i, \tau)/a, \tag{17c}$$

and α is a random number between 0 and 1. In this way, light walkers are eliminated and heavy walkers are spawned. The total number of walkers remains the same. A walker l after the reconfiguration has as its ancestor the walker $j(l)$ before the reconfiguration. The necessity of this reconfiguration process was pointed out by Hetherington[3]. We sometimes carry out the reconfiguration when the variance of weights becomes large. Repeating the above processes many times, we obtain the approximate ground-state wave function:

$$|0; 1 >= \lim_{\tau' \to \infty} \sum_{i=1}^{L} |i, \tau' > \tilde{w}(i, \tau'),$$

$$\tilde{w}(i, \tau') \equiv w(i, \tau') / \sum_{l} w(l, \tau'), \tag{18}$$

where $\tilde{w}$ is the normalized weight. Taking the long-time average with respect to τ', we find a more accurate wave function. Due to eq.(3), $E_1(0)$ is given by:

$$E_1(0) = \sum_{\tau'} \sum_{i=1}^{L} < \psi | H | i, \tau' > \tilde{w}(i, \tau') / \sum_{\tau'} 1. \tag{19}$$

To calculate eq.(11), we adopt the following operator as $R(K)$:

$$R(K) = \sum_{l=1}^{N} (S_l^z - S) \exp(iKl). \tag{20}$$

Evidently the operator is diagonal and satisfies (10). The i-th walker at time τ_1 has the j-th walker at time τ_2 as its ancestor. We call $j(i; \tau_1, \tau_2)$ the *ancestor function*. It satisfies the following relation:

$$j(i, \tau_1, \tau_3) = j(j(i; \tau_1, \tau_2); \tau_2, \tau_3), \quad \tau_1 > \tau_2 > \tau_3. \tag{21}$$

The ancestor function can be constructed by a successive substitution of i by $j(i; \tau, \tau - \Delta\tau)$. The wave function $e^{\tau H} R(K) |0; 1 >$ is represented as follows:

$$\lim_{\tau' \to \infty} \sum_{i=1}^{L} \tilde{w}(i, \tau') R(j(i; \tau', \tau' - \tau), K, \tau' - \tau)|i, \tau' >, \tag{22}$$

where $R(i, K, \tau') \equiv < i, \tau' | R(K) | i, \tau' >$. Taking the average with respect to τ', we obtain a more accurate wave function. Then $B(K, \tau)$ defined in (11) becomes:

$$\frac{\sum_{\tau'} \sum_{i=1}^{L} < \psi | R(-K) H | i, \tau' > \tilde{w}(i, \tau') R(j(i; \tau', \tau' - \tau), K, \tau' - \tau)}{\sum_{\tau'} \sum_{i=1}^{L} \tilde{w}(i, \tau') R(i, -K, \tau') R(j(i; \tau', \tau' - \tau), K, \tau' - \tau)}. \tag{23}$$

Hence we can calculate $B(K, \tau)$ by a small modification of the ground-state energy calculation. We only need to store the complex value $R(i, K, \tau')$ for each walker i and ancestor function $j(i; \tau', \tau' - \Delta\tau)$. Very old values of these quantities are needless because we do not calculate $B(K, \tau)$ for very large τ.

3 Isotropic Spin-1/2 and Spin-1 Chains

Here we show some results of our calculation for $S^z = 0$.

A) AHC with $S = 1, N = 14$.

For this system the results by the exact diagonalization method are given in ref.13. In Fig.2, $B(K, \tau)$ is plotted for $K = 0, \pi/7, 2\pi/7, 4\pi/7$ and π. In the case $K = 0$, $B(K, \tau)$ is very stable and gives the ground-state energy. For other momenta the error bar increases exponentially as τ increases. But $B(K, \tau)$ approaches $E_1(K)$ even at $\tau J = 1 - 2$. We put $L = 4096, \Delta\tau = 0.5/J$ and $r = 8$. We take the time average over $6000\Delta\tau$.

B) AHC with S=1/2, N=32.

This system is too big for the exact diagonalization. But we can calculate $E_1(K)$ solving numerically the Bethe ansatz equation as was done by des Cloiseaux and Pearson[7]. The results by the PMC and the Bethe ansatz are shown in Fig.3.

C) AHC with S=1, N=32.

In Fig.2 it is shown that the limiting value of $B(K, \tau)$ gives the correct excitation energy at momentum K for the $S = 1$, $N = 14$ chain. In Fig.3 the elementary excitation obtained by the above method coincides with the Bethe ansatz solutions for the $S = 1/2$, $N = 32$ chain.

The Fourier transform of $\rho(l)$ is the structure factor $S(K)$:

$$S(K) \equiv \sum_{l=1}^{N} \exp(ilK)\rho(l). \tag{24}$$

Using the variational wave function $\sum_l S_l^z \exp(ikl)|0, 1>$, we can calculate the upper limit of the elementary excitation:

$$\epsilon(K) \leq g(K) \equiv 2J(1 - \cos K)[-\rho(1)]/S(K). \tag{25}$$

The elementary excitation for the $S = 1$, $N = 32$ chain shown in Fig.4 satisfies this variational relation. It has apparently the gap. The elementary excitation is not symmetric with respect to the inversion at $K = \pi/2$. At $K = \pi$ the gap is $\Delta = 0.4J$. On the other hand, the gap at $K = 0$ is about 2Δ.

Nightingale and Blöte[5] (NB) determined the lowest energy of $S^z = 0, 1$ and 2 using the PMC method. In Table 2 we showed $E_1(K)$ obtained by our method. In Fig.4 it is compared with the upper bound calculated in ref.13. the quantity $E_1(0)$ is the ground-state energy and $E_1(\pi)$ is the first excited energy. The value of $E_1(\pi)$, namely $-44.46J$, coincides with NB's lowest energy $-44.4364(40)J$ in the $S_z = 1$ subspace.

Table 1: Lowest-energy eigenvalues of the cyclic chain with $N = 32$ and $D/J = 0, 0.2$ in the subspace of total $S_Z = 0$ and total momentum Q in units of J by the projector Monte Carlo method.

$QN/(2\pi)$	$D/J = 0.2$	$D/J = 0.0$
0	-40.743	-44.875
1	-40.062	-43.93
2	-39.573	-43.56
3	-39.157	-43.18
4	-38.795	-42.84
5	-38.493	-42.55
6	-38.268	-42.35
7	-38.139	-42.23
8	-38.091	-42.22
9	-38.154	-42.26
10	-38.324	-42.42
11	-38.551	-42.67
12	-38.854	-42.99
13	-39.221	-43.38
14	-39.557	-43.81
15	-39.879	-44.22
16	-40.048	-44.46

The energy gap is about $0.4J$. It is noteworthy that the spectra for the $S = 1/2$ and $S = 1$ cases are completely different. The former has the shape $c|\sin k|$. From Fig.4 we find that the spectrum is asymmetric with respect to the axis $K = \pi/2$. It seems that the gap at $K = 0$ is twice as large as that at $K = \pi$:

$$\lim_{K \to 0}[E_1(K) - E_1(0)] = 2[E_1(\pi) - E_1(0)]. \qquad (26)$$

This may be explained by regarding the low-momentum excited state as a scattering state of two excitations with momenta close to π.

4   Anisotropic Spin-1 Chain

For the anisotropic system, the following two dynamical structure factors are important:

$$S_\parallel(Q,\omega) = \int_{-\infty}^{\infty} < S_Q^z(t)S_{-Q}^z(0) > \exp(-i\omega t)dt$$

$$= \sum_n | < n|S_Q^z|0 > |^2 \delta(\omega - (E_n - E_0)/\hbar),$$

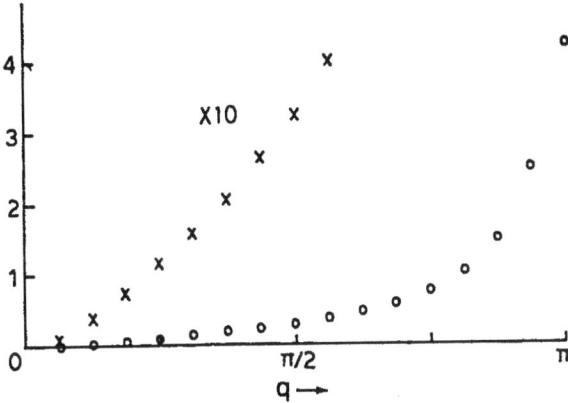

Figure 1: Structure factor $S(K)$ for the $S = 1$, $N = 32$ chain at the ground state. Circles represents its net values. Crosses denote the net values multiplied ten times. For small K, $S(K)$ behaves as K^2.

$$S_\perp(Q,\omega) = \int_{-\infty}^{\infty} < S_{-Q}^-(t)S_Q^+(0) > \exp(-i\omega t)dt$$

$$= \sum_n | < n|S_Q^-|0 > |^2 \delta(\omega - (E_n - E_0)/\hbar).$$

Here, $|0>$ and E_0 are the ground-state ket vector and its eigenvalue, respectively and $< n|$'s are the bra vectors of the eigenstates with the momentum $Q = 2\pi \times \text{integer}/N$. We put

$$S_Q^z \equiv N^{-1/2} \sum_l S_l^z \exp(-iQl),$$

$$S_Q^\pm \equiv N^{-1/2} \sum_l (S_l^x \pm iS_l^y) \exp(-iQl).$$

The $S_z = 0$ singlet branch should appear in $S_\|(Q,\omega)$ and the $S_z = \pm 1$ doublet branch should appear in $S_\perp(Q,\omega)$. Renard et al. observed both branches in the region near $Q = \pi$ for NENP[25-26]. It was difficult to observe the spectrum in other regions for the usual NENP sample because of the neutron scattering by protons. Recently a deuterium-enriched sample was prepared[29]. Ma et al. observed the dynamical structure factor in almost all regions of the momentum except $|Q|/\pi < 0.3$. For the ground state of the $S = 1$ chain, it is expected that a considerable part of the weight $S_\|(Q,\omega)$ has a peak at $\omega = (E_{Q,S_z=0} - E_0)/\hbar$. Here, $E_{Q,S_z=0}$ is the lowest energy in the subspace of the momentum Q and total $S_z = 0$. The symbol E_0 denotes $E_{Q=0,S_z=0}$. As shown in Table 1, we can calculate $E_{Q,S_z=0}$ using the PMC method. This situation is different from the case of the $S = 1/2$ Heisenberg antiferromagnet, where the lowest energy state is merely the lower edge of the continuous spectrum of $S_{\perp,\|}(Q,\omega)$. The structure factor $S_\perp(Q,\omega)$ has a peak at $\omega = (E_{Q,S_z=1} - E_{g.s.})/\hbar$, where $E_{Q,S_z=1}$ is the lowest energy in the subspace with $S_z = 1$ and the momentum

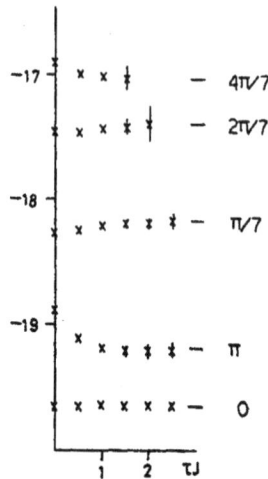

Figure 2: Function $B(K,\tau)/J$ for the $S=1, N=14$ chain at $K=0, \pi/7, 2\pi/7, 4\pi/7, \pi$. The values of $E(K)/J$'s are plotted as horizontal bars. the MC results and the exact diagonalization results are compared. The coincidence of $E(K)$ and $B(K,\tau)$ at $\tau J=1 \sim 2$ is very good. But statistical error grows exponentially as τ becomes large.

Q. At $D=0$, the values of $E_{Q,S_z=0}$ and $E_{Q,S_z=1}$ are the same and these levels form a triplet state. These excited states have the total spin 1. As shown in Fig.5, the results for $N=14, 16$ and $N=18$ coincide with those for $N=32$ obtained by the projector Monte Carlo method.

We find that the low-lying excitation almost exhausts the sum rule, exceeding 90 % except in the region of $Q \simeq 0$. This fact accords with the dominance of long-lived excitations, which was found experimentally in neutron scattering experiments[29].

It is shown from experiment that NENP has the anisotropy, and it is expected that $D/J=0.2$. Near $Q=\pi$ this discrepancy was observed in the neutron scattering experiment. If Q is far from π, it is difficult to recognize the discrepancy because the difference between these levels becomes narrower than the resolution width. In ref.29 a sharp peak is observed at $\pi \geq Q > 0.3\pi$. We calculate $E_{Q,S_z=0}$ and $E_{Q,S_z=1}$ for $N=14, 16, 18$ using the numerical diagonalization method. The largest vector has about 2.4×10^6 complex elements. For $N=32$ we calculate $E_{Q,S_z=0}$ using the projector Monte Carlo method. Numerical results are given in the Table 1, and are also plotted in Fig.5. From Fig.5 we can see that the elementary excitation spectrum curve at $N=32$ is almost the same as that at $N=18$. For $D/J=0.2$ we have two gaps at $Q=\pi$, $\Delta_{\parallel}^{\pi} \equiv E_{\pi,S_z=0} - E_0 = 0.69J$ and $\Delta_{\perp}^{\pi} \equiv E_{\pi,S_z=1} - E_0 = 0.29J$. This fact accords with the neutron scattering experiments and numerical calculations[30]. For a long time we have believed that the $S_z=0$ branch is always higher than the

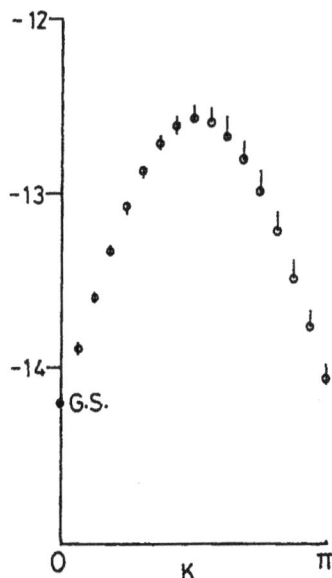

Figure 3: $E_1(K)/J$ for the $S = 1/2, N = 32$ AHC. Circles represents the results obtained by solving Bethe ansatz equation[7]. Bars shows the results by the MC calculation.

$S_z = \pm 1$ branch. But, as shown in Fig.2, this is a special phenomenon near $Q = \pi$. In a wide region $E_{Q,S_z=0}$ is lower than $E_{Q,S_z=\pm1}$. We find that there are also two gaps at $Q = 0$. If we assume that the lowest-energy states near $Q = 0$ are given by the scattering states of two elementary excitations near $Q = \pi$, we should have:

$$\Delta_\parallel^0 = 2\Delta_\perp^\pi, \quad \Delta_\perp^0 = \Delta_\perp^\pi + \Delta_\parallel^\pi. \tag{27}$$

Here, we used the conservation law of total S_z and total momentum. The obtained numerical results do not contradict this estimate. The splitting of the two branches should also be observed in the low-momentum region because the discrepancy between the two increases.

5   Summary and Discussion

We reviewed a new Monte Carlo method which calculates the excitation energy as a function of momentum. This method may be generalized if the Hamiltonian has some other symmetry. We are able to get the lowest-energy state in each subspace if an appropriate diagonal operator is found. By this method, the elementary excitation spectra of the $N = 32$ AHC with spin 1/2 and 1 are obtained. The exact diagonalization method is almost impossible to be applied to such long chains. The result of the $S = 1/2$ case coincides with des Cloiseaux and Pearson's theory[7]. In the $S = 1$ case, the spectrum of the elementary excitation has an energy gap.

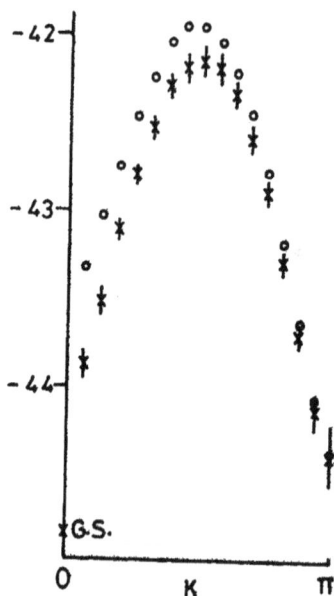

Figure 4: $E(K)/J$ for the $N = 32, S = 1$ chain. The spectrum has a gap at $K = \pi$. The value of the gap is about $0.4J$ and coincides with NB's calculation[5]. Small circles represents the upper bound of $E(K)/J$ given by the variational method. This upper bound was calculated using the structure factor and variational relation.

The energy gap is about $0.4J$. This coincides with Haldane's prediction[10] and NB's numerical calculation[5]. This excitation spectrum satisfies the variational relation with the structure factor $S(K)$, as shown in Fig.4. Calculated with the use of the exact diagonalization method, the low-lying excitation has a strong δ-function peak in $S(Q,\omega)$. It is almost certain that the elementary excitation near $Q = \pi$ has a finite weight in the limit $N \to \infty$. Its weight is more than 90 % of the total weight in almost all momenta. At finite temperature the δ-function peak should become a peak of a finite width because of the interaction between elementary excitations. At low temperature, however, the width is very narrow. It must be proportional to the number of excitations and of the order of $\exp(-\Delta_1^\tau/k_BT)$. On the other hand, there remains the possibility that the lowest-energy state is the lower edge of the continuum of scattering states near $Q = 0$. If this is true, there should be a transition from the δ-function peak to the lower edge of the scattering states at some momentum.

The Monte Carlo method reviewed in this paper is useful also for other quantum

Excitation spectrum of D/J=0.2 chain

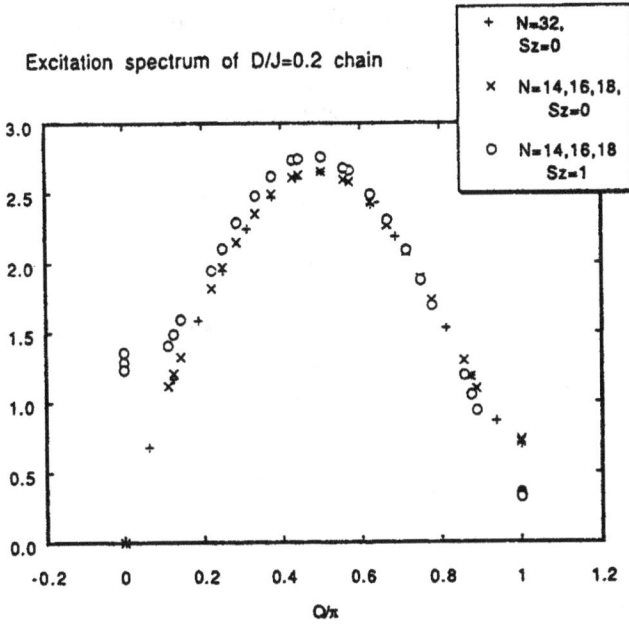

Figure 5: Excitation spectrum of the $D/J = 0.2$ chain. The elementary excitations from the ground state are shown with given S^z and total momentum. The system is a periodic chain with $D/J = 0.2$. The singlet branch is calculated for $N = 32$ by the quantum Monte Carlo method, while others by the Lanczos method. The singlet branch is higher than the doublet branch near the region $Q = \pi$, but becomes lower at $|Q/\pi| < 0.7$. The symbols + denote the results for $N = 32, S_z = 0$, while × those for $N = 14, 16, 18, S_z = 0$, and o those for $N = 14, 16, 18, S_z = 1$.

systems[31].

The author is grateful to Dr. Dorota Lipowska for critical reading of manuscripts.

References

1. M. Suzuki, Prog. Theor. Phys. **56**, 1457 (1976). M. Barma and B.S. Shastry, Phys. Lett. **61A** 15 (1977); Phys. Rev. **B18** 3351 (1978). J.E. Hirsch, R.L. Sugar, D.J. Scalapino and R. Blankenbecler, Phys. Rev. **B26** 5033 (1982). K.Sogo and M. Uchinami, J. Phys. **A19** 493 (1986). M.Uchinami, Phys. Lett. A **127** 151 (1988).
2. R. Blankenbecler and R.L. Sugar, Phys. Rev. **D27** 1304 (1983).
3. J.H. Hetherington, Phys. Rev. A **30** 2713 (1983).
4. K. Nomura and M. Takahashi, J. Phys. Soc. Jpn. **57** 1424 (1988).
5. M.P. Nightingale and H.W.J. Blöte, Phys. Rev. B **33** 659 (1986).
6. D.M. Ceperley and M.H. Kalos, Monte Carlo Methods in Statistical Physics, Edited by K. Binder (Springer, Berlin, 1979).
7. J. des Cloiseaux and J.J. Pearson, Phys. Rev. **128** 2131 (1962).

8. E.H. Lieb, T. Schultz and D.J. Mattis, Ann. Phys. NY **16** 407 (1961).
9. I. Affleck and E.H. Lieb, Lett. Math. Phys. **12**, 57 (1986).
10. F.D.M. Haldane, Phys. Rev. Lett. **50** 1153 (1983); Phys. Lett.**93A** 464 (1983).
11. R. Botet and R. Jullien, Phys. Rev. **B27**,613(1983); M. Kolb, R. Botet and R. Jullien, J. Phys. A **16**, L673 (1983); R. Botet, R. Jullien and M. Kolb, Phys. Rev. **B28**, 3914(1983).
12. U. Glaus and T. Schneider, Phys. Rev. B **30**,215(1984).
13. J.B. Parkinson and J.C. Bonner, Phys. Rev. **B32**,4703 (1985).
14. A. Moreo, Phys. Rev. **B35**, 8562(1987).
15. H. Betsuyaku, Phys. Rev. B **36**,799 (1987).
16. K. Saito, S. Takada and K. Kubo, J. Phys. Soc. Jpn **56**,3755(1987).
17. M. Takahashi, Phys. Rev. B **38** 5188 (1988).
18. K. Nomura, Phys. Rev. B **40**, 2421(1989); S. Liang, Phys. Rev. Lett. **64**, 1597(1990).
19. M. Takahashi, Phys. Rev. Lett. **62** 2313 (1989).
20. W.J.L. Buyers, R.M. Morra, R.L. Armstrong, P. Gerlach and K. Hirakawa, Phys. Rev. Lett. **56**, 371 (1986).
21. R.M. Morra, W.J.L. Buyers, R.L. Armstrong and K. Hirakawa, Phys. Rev. **B38**, 543(1988).
22. M. Steiner, K. Kakurai, J.K. Kjems, D. Petitgrand and R. Pynn, J. Appl. Phys. **61**, 3953 (1987).
23. Z. Tun, W.J.L. Buyers, R.L. Armstrong, K. Hirakawa and B. Briat, Phys. Rev. **B42**, 4677 (1990).
24. Z. Tun, W.J.L. Buyers, A. Harrison and J.A. Rayne, Phys. Rev. B **43**, 13331 (1991).
25. J.P. Renard, M. Verdaguer, L.P. Regnault, W.A.C. Erkelens, J. Rossa-Mignod and W.G. Stirling, Europhys. Lett. **3** 945 (1987).
26. J.P. Renard, M. Verdaguer, L.P. Regnault, W.A.C. Erkelens, J. Rossa-Mignod, J.Ribas, W.G. Stirling and C. Vettier, J. Appl. Phys. **63**, 3538 (1988).
27. L.P. Regnault, J. Rossa-Mignod, J.P. Renard, M. Verdaguer, and C. Vettier, Physica B **156** & **157**, 247 (1989).
28. T. Delica, K. Kopinga, H. Leshke and K.K. Mon, Europhys. Lett. **15** 55 (1991).
29. S. Ma, C. Broholm, D.H. Reich B.J. Sternlieb and R.W. Erwin ,Phys. Rev. Lett.**69**, 3571(1992).
30. O. Golinelli, Th. Jolicœur and R. Lacaze, Phys. Rev. **B45**,9798 (1992).
31. K. Hida, J. Phys. Soc. Jpn. **60** 1347 (1991), **61** 1013 (1992).

THE DECOUPLED CELL METHOD OF QUANTUM MONTE CARLO CALCULATION

Shigeo HOMMA

Physics Laboratory, Faculty of Engineering
Gunma University, Kiryu 376, Japan

Abstract

The decoupled cell method(DCM) and modified decoupled cel-
l method(mDCM) are introduced as new methods of quantum Monte
Carlo calculations. Both methods are applied to the one-dimensional
XY-model. The results are compared with the exact ones.

1. INTRODUCTION

In recent years Monte Carlo methods have been widely used in studying the equilibrium and nonequilibrium properties of many body systems.
They produced quite good estimates of the thermodynamic quantities of
classical statistical systems such as the hard sphere gas model [1], the Ising
model of ferro- and antiferromagnetism [2], and more realistic systems [3].
These methods are based on the idea, first put forward by Metropolis and
his coworkers [4], of generating the required probability distribution in the
configuration space as the limit of a Markov chain. This Markov chain can
be any one so long as 1) the transition probability $W(A \to B)$ from any
configuration A to any configuration B satisfies the condition of the detailed
balance at equilibrium:

$$P(A)W(A \to B) = P(B)W(B \to A), \tag{1}$$

and 2) the Markov chain is irreducible and recurrent. Here $P(A)$ (or $P(B)$)
is the probability of A or B in the required probability distribution. A realization of the Markov chain is performed once the quantity W is given. In

order to give W, we only need to know the ratio of the probabilities $P(A)$ and $P(B)$.

In a classical system we can usually find a suitable pair of configurations A and B for which this ratio can be calculated easily. However, this method is not directly applicable to quantum many- particle systems because of the noncommutativity between the local interaction operators. This results in the difficulty of defining the appropriate Markov chain in quantum mechanical cases. Accordingly a number of attempts have been proposed and tested to overcome the difficulties inherent in quantum mechanical systems. The first method is to use Monte Carlo simulation to evaluate thermodynamic quantities as a series expansion in terms of the Hamiltonian of a quantum system. This approach to quantum Monte Carlo calculation was initiated by Handscomb [5], and followed by many authors [6-10]. They used Monte Carlo calculation to estimate the expansion coefficients for one- and two-dimensional quantum spin systems. The second approach to quantum Monte Carlo calculation, first put forward by Suzuki [11], and used extensively by him and his coworkers [12], is to transform a d-dimensional quantum system into a mathematically equivalent $(d + 1)$-dimensional classical system by invoking the generalized Trotter formula (Suzuki-Trotter formula). Thus the non-commutativity problem in quantum systems is overcome and the classical Metropolis method of Monte Carlo calculation is applicable to the equivalent classical systems. The third approach to quantum Monte Carlo, called the Decoupled Cell Method (DCM), was introduced by Homma, Matsuda and Ogita as a direct extension of the classical Metroplis method [13,14].

The purpose of this report is to give a detailed account of the DCM and its modified version (mDCM) and to show the results obtained by applying it to low-dimensional quantum spin systems. In the next section we give the basic idea of the DCM and the procedures to perform Monte Carlo calculation by this method. In section 3 the results obtained by applying the DCM to the one-dimensional XY-model ($s = 1/2$) are shown and compared with the rigorous analytical results [15]. We also present the results obtained for the antiferromagnetic triangular lattice. In section 4 we describe the Modified Decoupled Cell Method (mDCM), which improves on the DCM by removing its inherent difficulty of breakdown of the detailed balance in the low temperature region. Here we also compare the results by the mDCM and the DCM. The last section is devoted to discussion.

2. THE DECOUPLED CELL METHOD

To explain the basic idea of the DCM let us consider a quantum spin system ($s = 1/2$) whose Hamiltonian is given by H. The state of the i-th site can be specified by a variable $s = \pm 1/2$, and the state of the total system by an N-dimensional state vector $|S> = |s_1, s_2, ..., s_N>$ whose i-site state is s_i. The probability of the state S in the canonical distribution is now given by

$$P(S) = < S| \exp(-\beta H)|S > /Z, \qquad (2)$$

where $\beta = (1/kT)$ and Z stands for the partition function of a system. As in the classical Metroplis method, the condition of irreducibility and re-currency of the Markov chain is satisfied by assigning positive transition probabilities between states that are different from each other only at one site $i(i = 1, 2, ..., N)$. The problem in the quantum mechanical case is how to obtain adequate transition probabilities consistent with the condition 1) described in the previous section.

Let $L_i(\nu)$ be a set of sites whose distance from the i-th site does not exceed a certain integer ν and $\overline{L}_i$ be a set of all sites not belonging to $L_i(\nu)$. We call such $L_i(\nu)$ the decoupled cell (DC) of radius ν with its center at the i-site. Let S_i denote the state of $L_i(\nu)$ excepting the i-site and $\overline{S}_i$ denote the state $\overline{L}_i$. The state of the total system can then be written as $S = (s_i, S_i, \overline{S}_i)$. The transition probability between $S = (s_i, S_i, \overline{S}_i)$ and $S' = (-s_i, S_i, \overline{S}_i)$ can be obtained from Eq.1 if one knows the value of

$$q(S) = \frac{P(S)}{P(S')} = \frac{< S| \exp(-\beta H)|S >}{< S'| \exp(-\beta H)|S' >}. \qquad (3)$$

Let $H(\nu, i)$ be the Hamiltonian of a DC which is obtained from H by deleting all the terms containing operators of $\overline{L}_i$. The basic ingredient of the DCM is to approximate Eq.3 by

$$q^{(\nu)}(S_i) = \frac{< s_i, S_i| \exp(-\beta H(\nu, i))|s_i, S_i >}{< -s_i, S_i| \exp(-\beta H(\nu, i))| - s_i, S_i >}. \qquad (4)$$

This approximation is based on the presumption that [The dependence of q on $\overline{S}_i$ gradually decreases by increasing ν]. The right hand side of Eq.4

can be obtained readily by solving the eigenvalue problem of $H(\nu, i)$ by computer. Let $(E_i, \phi_i; i = 1, 2, .., f = 2^{2\nu+1})$ be eigenvalues and eigenfunctions of $H(\nu, i)$. With their use Eq.4, $q^{(\nu)}(S_i)$ can be rewritten as

$$q^{(\nu)}(S_i) = \frac{\sum_n |<s_i, S_i|\phi_n>|^2 \exp(-\beta E_n)}{\sum_n |<-s_i, S_i|\phi_n>|^2 \exp(-\beta E_n)}. \tag{5}$$

The transition probability W_{DC} defined in the DCM is given by

$$W_{DC}(-s_i \rightarrow s_i) = max[1, q^{(\nu)}(S_i)]. \tag{6}$$

The physical meaning of replacing Eq.3 by the approximation Eq.4 was fully discussed by Matsuda et al. [16]. If we use Eq.6 as the transition probability from the state $S' = (-s_i, S_i, \overline{S}_i)$ to the state $S = (s_i, S_i, \overline{S}_i)$ in Monte Carlo calculation, we are able to obtain the Markov chain of the given quantum mechanical system using the Metroplis algorithm. Thus the DCM gives a natural extension of the classical Monte Carlo method to quantum systems. The detailed computational procedure of the DCM is given in references [14,17] . In the next section we apply the DCM to the one-dimensional (1D) XY- model for which exact solution was obtained, and to the antiferromagnetic XY model ($s = 1/2$) on the triangular lattice, in which the competition of frustration and quantum nature of the model system is of great interest.

3. ONE-DIMENSIONAL XY-MODEL AND ANTIFERROMAGNETIC TRIANGULAR LATTICE

The Hamiltonian of the 1D XY- model is given by

$$H = -2J\sum_i(s_i^x s_{i+1}^x + s_i^y s_{i+1}^y), s_i = s_{i+N}, \tag{7}$$

where N is the total number of lattice sites. Corresponding to the physical quantities which we calculate we choose a direction of a diagonal representation for the spin operator s. Here there are two possibilities for the diagonal representation.

(I) Taking the representation in which s_i^z is diagonal, we can calculate the following physical quantities:

Internal energy: $\epsilon = -N^{-1}4J\sum_i < s_i^z s_{i+1}^z >$,

Spin pair correlation function: $C^z(r) = N^{-1}\sum_i < s_i^z s_{i+r}^z >$,

In-plane susceptibility: $\chi_{in} = N^{-1}lim_{H\to 0}[< \sum_i s_i^z > /H]$,

where $< \Omega >$ denotes the canonical average of Ω .

(II) Taking the representation in which s_i^z is diagonal we can calculate The zero field susceptibility:
$$\chi_0 = (NkT)^{-1} < (\sum_i s_i^z)^2 >,$$

Spin pair correlation function: $C^z(r) = N^{-1}\sum_i < s_i^z s_{i+r}^z >$.

We show in Figs.1 and 2 the results, obtained by applying the DCM to the 1D XY-model, of ϵ and χ_0 [14,17] in conjunction with the exact results by Katsura [15]. With the increase in the size of a DC, the obtained results generally approach the exact one except for lower temperatures. In particular, the calculated values of ϵ give a negative specific heat below a certain demarcation temperature $T(d)$. This difficulty might be attributed to the breakdown of the detailed balance in the low temperature regions, arising from the finiteness of a DC. In the next section we modify the DCM to remove this difficulty.

The application of the DCM to the antiferromagnetic quantum XY- model ($s = 1/2$) on the triangular lattice (XYAFT) has also been performed, because of its intrinsic properties of frustration. The Hamiltonian of XYAFT is given by

$$H = -2J \sum_{<i,j>} (s_i^x s_j^x + s_i^y s_j^y), \tag{8}$$

where the sum $< i, j >$ is over all the nearest neighbour pairs on the lattice.

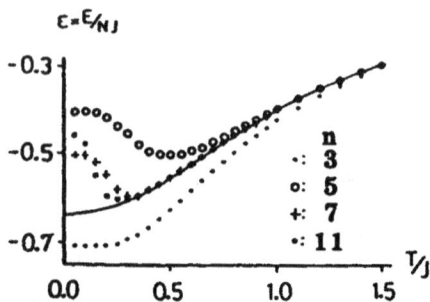

Fig.1 The internal energy ϵ per lattice site for 1D XY-model by DCM. The solid line is the exact result by Katsura [15]. The numbers in the figure represent the number of lattice sites in DC.

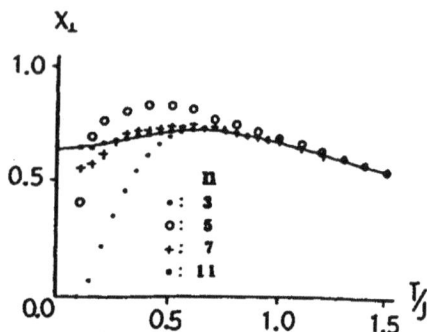

Fig.2 The zero field susceptibility along z-axis for 1D XY-model by DCM. The solid line is the exact result by Katsura [15]. The numbers in the figure represent the numbers of lattice sites in DC.

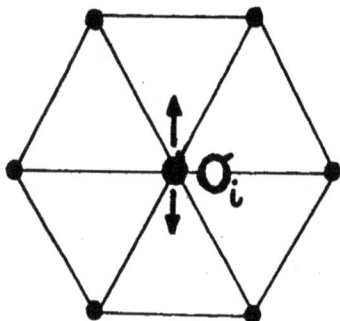

Fig.3 A size of a decoupled cell, where $\sigma_i = 2s_i$.

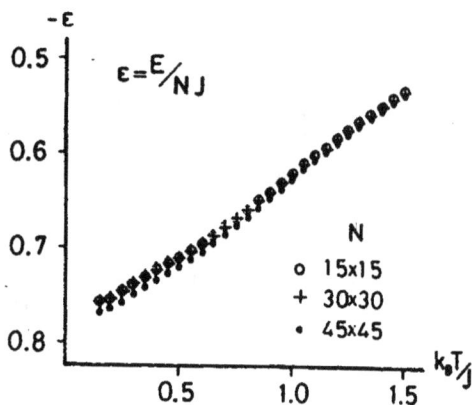

Fig.4 The internal energy ϵ per spin for XYAFT. N denotes the size of a system.

In this case the size of the DC used is depicted in Fig.3, where the central spin s_i makes a transition from $s_i = s$ to $-s$ with all the other spins in the cell fixed. In the system defined above we have two possibilities of the diagonal representations, that is, (1) to take s^x as a diagonal, and (2) to take s^z as a diagonal 2x2 matrix. In the first case we can calculate 1) internal energy,

$$N = 45 \times 45$$

$$\circ \quad C_2^x = \langle \hat{\sigma}_A^x \hat{\sigma}_{A'}^x \rangle$$

$$+ \quad C_1^x = \langle \hat{\sigma}_A^x \hat{\sigma}_B^x \rangle$$

$$\bullet \quad C_1^x = \langle \hat{\sigma}_A^x \hat{\sigma}_{B'}^x \rangle$$

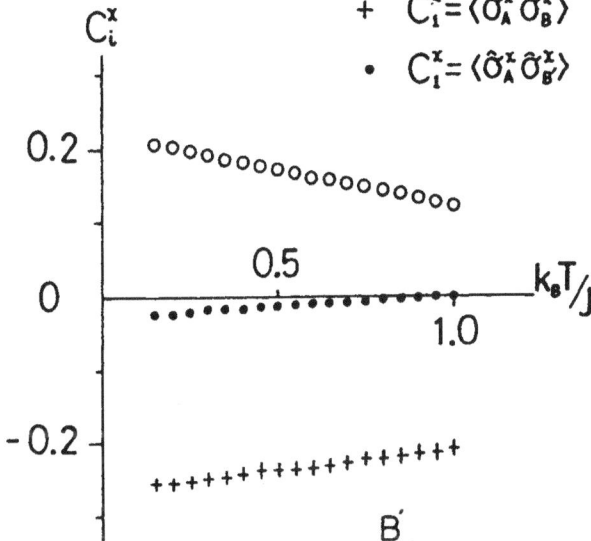

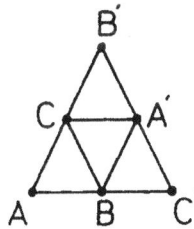

Fig.5 The temperature dependence of C_r^x for the first, second and third neighbours. Here $N = 45 \times 45$.

2) magnetization along x-axis, 3) magnetic susceptibility χ_x, and 4) pair correlation function of x-component of the quantum spin. And in the second case we can calculate 5) magnetization along z-axis, 6) magnetic susceptibility χ_z, and 7) pair correlation function of z-component of the spin. From these calculations we are able to know the possible spin orderings and a sublattice structure of XYAFT and the interrelation between the quantum effect and the frustration. We show in Figs.4, 5 and 6 the results for the internal energy, and spin pair correlation functions defined by

$$C_r^\alpha = 4 < s_i^\alpha s_{i+r}^\alpha >, \quad \alpha = x, z. \tag{9}$$

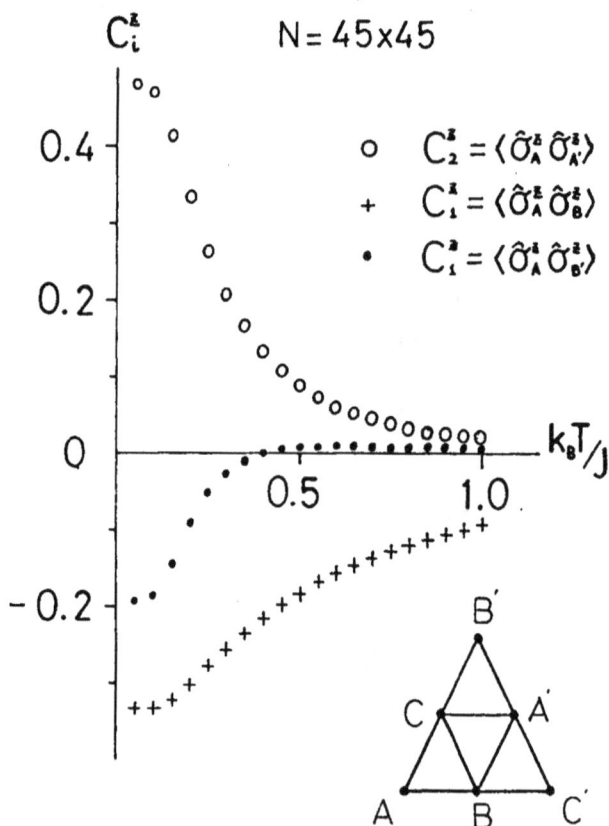

Fig.6 The temperature dependence of C_r^z for the first, second and third neighbours. Here $N = 45 \times 45$.

For C_r^z we see that by decreasing temperature its absolute value increases indicating a possibility of a certain spin ordering in the z-component of spin operator. In order to confirm whether it is true or only an artifact arising from the finiteness of the DC, we have to increase the size of the DC. Besides the 1D XY-model and the triangular antiferromagnet (XYAFT), the DCM has been used extensively to other low-dimensional quantum spin systems. The results obtained there will be found in refs.18-25.

4. MODIFIED DECOUPLED CELL METHOD (mDCM)

As was pointed out in the previous section when we apply the DCM to the 1D XY-model, a negative specific heat appears in the low temperature region. This might be attributed to the breakdown of the detailed balance originating from the finiteness of a DC used there. In order to remove this difficulty there are two possibilities. The first is to enlarge the size of the DC, but it is limited by the capability of computer power. The second is to improve the DCM to recover the detailed balance, extending the basic concept of the DCM for a certain size of the DC. As was explained in section 2, the DCM could be regarded as a natural extension of the classical Monte Carlo method to quantum systems. However, it should be noted that the transition probability defined by Eq.6 does not satisfy the detailed balance. The probability $W_{DC}(s_i \rightarrow -s_i)$ is a function of the neighbouring spins in the cell, which includes a finite number of spins. However, when we calculate the transition probability of some other spin in the same cell, we introduce W_{DC} in the form of Eq.6 independently. If the cell Hamiltonians $H(\nu, i)$ commute with each other, the detailed balance is satisfied automatically as far as it is satisfied locally. Thus Eq.6 gives correct transition probabilities for classical systems. But in quantum systems where $H(\nu, i)$ do not commute with each other, the transition probability defined at each lattice site independently does not satisfy the detailed balance. The flip of a spin s_k causes a change of all the transition probabilities for which $H(\nu, i)$ includes s_k. Thus the changes cannot be reduced to that of $W_{DC}(s_i \rightarrow -s_i)$. From this point of view, in determining the transition probability it is important to include not only the DC whose central site is i, but all the DC which include the i-site. Here we reformulate the DCM taking into account the above considerations.

First we decompose a system into identical cells (DC), whose shape and size are given. The way of a decomposition is not unique, as it depends on the size and shape of the DC. With a decomposition, labeled by j and a k-th cell, we associate the cell Hamiltonian $H_n(j,k)$, where n is the number of lattice sites included in each cell. Then the Hamiltonian of the system is written as a sum of $H_n(j,k)$ as

$$H = \frac{1}{r(n)} \sum_j (\sum_k H_n(j,k)), \qquad (10)$$

where the sum over k means the sum over all cells on a lattice and that over j means the sum over all different decompositions. The symbol $r(n)$ denotes the number of different decompositions. Using Eq.10 the probability of a certain spin configuration $|S, s_i >$ is given by

$$P(S, s_i) = \frac{1}{Z} < S, s_i| \exp(-(\frac{\beta}{r(n)} \sum_j (\sum_k H_n(j,k))))|S, s_i >, \qquad (11)$$

where Z is the partition function of the system. The ket (bra)$|S, s_i > (< S, s_i|)$ represents the spin configuration of the system, in which the i-th spin is s_i, with $s_i = \pm(1/2)$. We approximate Eq.11, invoking the basic concept discussed above, as

$$P(S, s_i) = \frac{1}{Z} \Pi_j \Pi_k < \sigma(j,k)| \exp(-(\frac{\beta}{r(n)} H_n(j,k)))|\sigma(j,k) >, \qquad (12)$$

where $|\sigma(j,k) >$ represents a spin state of the cell (DC) labeled by (j,k). The transition probability $W(s_i \rightarrow -s_i)$ is defined by

$$W(s_i \rightarrow -s_i) = max[1, \frac{P(S, -s_i)}{P(S, s_i)}], \qquad (13)$$

where $P(S, \pm s_i)$ is defined by Eq.11. We approximate Eq.11 , by substituting Eq.12 into $P(S, \pm s_i)$ in Eq.13 , to obtain W_{DC} in the modified Decoupled Cell Method (mDCM) as

$$W_{DC}(s_i \to -s_i)$$

$$= max[1, \frac{\Pi_j \Pi_k < \sigma(j,k), -s_i| \exp(-\frac{\beta}{r(n)} H_n(j,k))|\sigma(j,k), -s_i >}{\Pi_j \Pi_k < \sigma(j,k), s_i| \exp(-\frac{\beta}{r(n)} H_n(j,k))|\sigma(j,k), s_i >}].(14)$$

The product over k in Eq.14 must be over all cells which include the i-site; the total number of such cells is equal to the number of spins in a cell, that is n. Here it must be noted that in the DCM only the cell whose center is i-site is taken into account in Eq.14 . If the Hamiltonian H of a system consists of only a nearest neighbour coupling, Eqs.13 and 14 coincide with those of classical ones in the classical limit. However it should be noted that in one-dimensional lattice when the Hamiltonian H includes the second-neighbour interaction beside the nearest neighbour interaction, each second-neighbour interaction is included in Eq.12 by a factor $\frac{n-2}{n-1}$, whereas the nearest neigh-bour interaction is included in Eq.12 by $\frac{n-1}{n-1} = 1$. Thus Eq.14 does not give a correct expression in the classical limit in this case (in one-dimensional lat-tice with a second-neighbour interaction). In two-dimensional lattice proper size and shape of a cell (DC) depend on the type of lattice and the range of interactions. For the detailed account of the possible decompositions of a certain two-dimensional lattice with nearest- and second-neighbour interac-tions, into DC of the given shape and size , the reader is referred to ref.26. In calculating the internal energy of a system we use the following expression;

$$\epsilon = \frac{<H>}{N}$$
$$= \frac{1}{Nr(n)} \sum_j (\sum_k < H_n(j,k) >), \qquad (15)$$

where <> means the average with respect to the cell Hamiltonian $H_n(j,k)$. Thus we complete the modification of the DCM (mDCM) [26]. We applied the mDCM thus formulated to the 1D XY-model ($s = 1/2$) and calculated the internal energy ϵ and the zero-field susceptibility χ_0 defined in section 3,

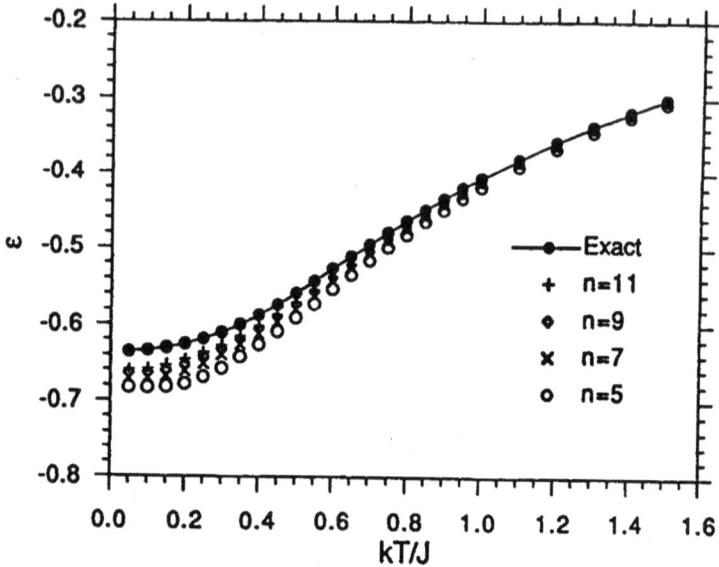

Fig.7 The internal energy ϵ per lattice site for the 1D XY-model by the mD-CM. The solid line is the exact result by Katsura [15]. The numbers in the figure represent the number of lattice sites in a DC.

increasing the size of a DC from 3 to 11 [26]. The number of lattice points used here was 128. In Fig.7 we show the result of the internal energy ϵ compared with the exact solution by Katsura [15]. As we can see there is no negative specific heat in the mDCM. Here we note that for the case $n = 11$ the agreement with the exact result is better for the DCM than the mDCM above $0.35kT/J$. In Fig.8 we show χ_0 obtained by the mDCM, where the solid line is the exact result by Katsura [15]. The results by the mDCM give better approximate values than those by the DCM.

The application of the mDCM to the antiferromagnetic $J_1 - J_2$ model on the square lattice ($s = 1/2$) has been done by Miyazawa and the present author, where the DC is square composed of 9 lattice sites. A preliminary calculation shows that, for the case $J_1 = J_2$, there appears a collinear state below $T_{col} = 0.3kT/J$. The extensive study of this model using the mDCM is in progress and will be published elsewhere [27].

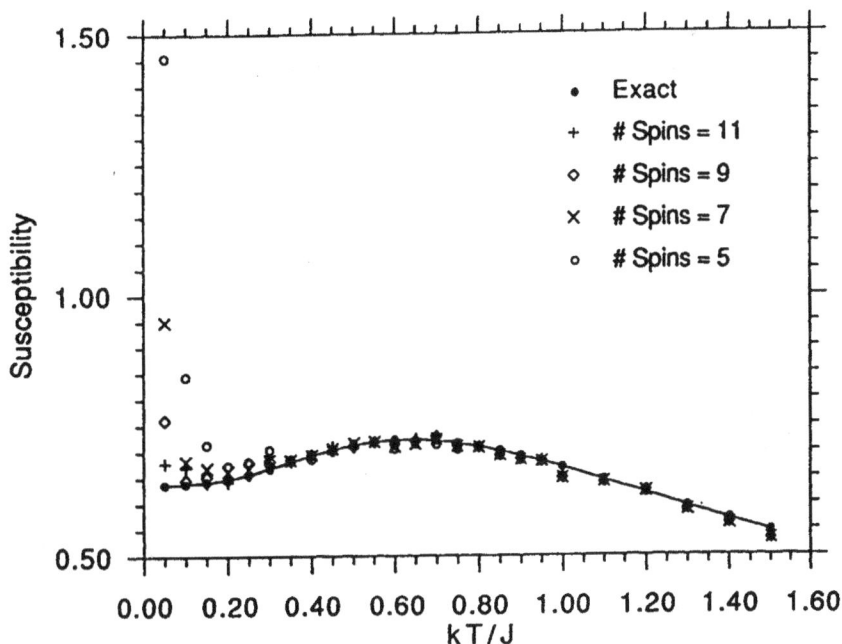

Fig.8 The zero-field susceptibility along z-axis for the 1D XY-model by the mDCM. The solid line is the exact result by Katsura [15]. The numbers in the figure represent the number of lattice sites in DC.

5. DISCUSSION

In this report we gave a brief description of the Decoupled Cell Method (DCM) and its modification (mDCM) as new methods of quantum Monte Carlo calculation. We, then, applied them to low-dimensional quantum spin systems and obtained the results that agree well with the exact ones for the 1D XY-model (s=1/2). As is easily seen from Eq.5, an advantage of the DCM and the mDCM is free from the negative-sign problem. Thus these methods are applicable to frustrated quantum spin systems such as the antiferromagnetic Heisenberg model on the triangular lattice, the $J_1 - J_2$ model on the square lattice or fermion systems such as the Hubbard model, where the negative sign problem is serious.

It is certain that by increasing the size of a DC we could diminish the error arising from its finiteness. The convergence of the DCM (mDCM) with increase in the size of a DC is the next problem to be solved analytically as well as numerically.

The present author expresses his thanks to Profs.H.Matsuda, N.Ogita and S.Miyazawa for discussions and to Dr.Dorota Lipowska for reading the manuscript.

REFERENCES

1. B.J.Alder and T.E.Wainwright, Phys. Rev. **127** (1962) 359
2. N.Ogita, A.Ueda, T.Matsubara, H.Matsuda and F.Yonezawa, J. Phys. Soc. Jpn. **26S** (1969) 145
3. O.G.Mouritsen, Computer Studies of Phase Transition and Critical Phenomena, (Springer -Verlag, Berlin, 1984)
4. N.Metropolis, A.W.Rosenbluth, M.N.Rosenbluth, A.H.Teller and E.Teller, J. Chem. Phys. **21** (1963) 1087
5. D.C.Handscomb, Proc. Cambridge Philos. Soc. **58** (1962) 594;**60** (1964) 115
6. S.Chakravarty and D.B.Stein, Phys. Rev. Lett. **49** (1982) 582
7. D.H.Lee, J.D.Joannoupoulos and J.W.Negele, Phys. Rev. **B30** (1984) 1599
8. J.W.Lyklema, Phys.Rev.Lett. **49** (1982) 66; Phys.Rev. **B27** (1983) 3108
9. D.H.Lee, J.D.Joannoupoulos and J.W.Negele and D.P.Landau, Phys. Rev. **B33** (1986) 450
10. S.Kadowaki and A.Ueda, Prog. Theor .Phys. **75** (1986) 451; **78** (1987) 224; **82** (1989) 493
11. M.Suzuki, Prog. Theor. Phys. **56** (1976) 1454
12. See, for example, Review by M.Suzuki in Quantum Monte Carlo Methods, ed. M. Suzuki (Springer-Verlag 1987)
13. S.Homma, H.Matsuda and N.Ogita, Prog. Theor. Phys. **72** (1984) 1245
14. S.Homma, H.Matsuda and N.Ogita, Prog. Theor. Phys. **75** (1986) 1058
15. S.Katsura, Phys. Rev. **57** (1962) 1508
16. H.Matsuda, K.Ishii, S.Homma and N.Ogita, Prog. Theor. Phys. **80** (1988) 583
17. S.Homma, H.Matsuda, N.Ogita and K.Sano, J. Phys. Soc. Jpn. **62** (1993) 880

18. S.Homma, K.Sano, H.Matsuda and N.Ogita, Prog. Theor. Phys. **S87** (1986) 127 and in Springer Series Solid State Science, 74,ed. M.Suzuki (Springer-Verlag, 1987) p.153

19. K.Sano, Prog. Theor. Phys. **77** (1987) 287

20. S.Homma, H.Matsuda, T.Horiki and N.Ogita, Prog. Theor. Phys. **80** (1988) 594

21. T.Horiki, S.Homma, H.Matsuda and N.Ogita, Prog. Theor. Phys. **82** (1989) 507

22. S.Homma, T.Horiki, H.Matsuda and N.Ogita, in Quantum Simulation of Condensed Matter Phenomena, ed. J. D. Doll and J.Gubernatis, (World Scientific, 1990) p.116

23. R.Creswick and C.Sisson, Mod. Phys. Lett. **B5** (1991) 907

24. C.Sisson and R.Creswick, in Computer Simulation in Condensed Matter Physics, ed. D.Landau, K.Man and H.Schuttler, (Springer-Verlag,1993)

25. C.Zeng and V.Elser, Phys. Rev. **B42** (1990) 8436

26. S.Miyazawa, S.Miyashita, M.S.Makivic and S.Homma, Prog.Theor. Phys. **89** (1993) 1167

27. S.Miyazawa and S.Homma, in preparation

Decoupled Cell Monte Carlo Study of the Critical Properties of the Spin-1/2 Ferromagnetic Heisenberg Model in Three Dimensions

Richard J. Creswick and Cynthia J. Sisson*
Department of Physics and Astronomy
University of South Carolina
Columbia, SC 29208

*current address: Department of Physics and Astronomy
Appalachian State University
Boone, NC 28608

Abstract

Decoupled Cell Monte Carlo calculations of the critical properties of the spin-1/2 Heisenberg model on cubic lattices are presented. Critical exponents for the ferromagnetic model are determined, establishing the utility of the Decoupled Cell Method for critical models in three dimensions. Our results are not in agreement with the ϵ-expansion, or Monte Carlo calculations for the classical Heisenberg model, but do obey the usual scaling laws.

1. Introduction

Monte Carlo simulations of quantum lattice models can be approached in several ways. Most methods make use of the Trotter-Suzuki theorem to represent the partition function of a d-dimensional quantum lattice model in terms of a d+1 dimensional classical model. While this method is quite general, and it has been applied successfully to many problems, it has drawbacks which have led to the search for other approaches. The most obvious difficulty is that in expanding the dimension of the

179

simulation from d to d+1 dimensions the size of the lattice one
can simulate is reduced. This is especially evident for systems
in three space dimensions and at low temperatures since the
number of discrete imaginary time steps grows as 1/T.

The second, and more difficult, problem with the imaginary time
approach is that the effective thermodynamic action, or free
energy functional, is not necessarily real. This can lead to
negative or even complex Boltzmann weights for some states, and
makes interpretation of these weights as (unnormalized)
probabilities difficult.

An alternate approach to calculating the transition probability
in a Monte Carlo process is the Decoupled Cell Method (DCM)
developed by Homma[1] and co-workers beginning in 1984. By using
the exact eigenstates of a local part, or cell, of the full
lattice Hamiltonian, the Boltzmann weight can be approximated as
a product of weights for each cell. These factors are all real
and positive, and the simulation is carried out in d dimensions,
so the two main difficulties of imaginary time methods are
avoided.

The method, however, is not exact except in the limit where the
cell comprises the whole lattice or the Hamiltonian reduces to a
sum of commuting local operators (e.g. the Ising limit of the
anisotropic Heisenberg model). On the other hand, if one analyzes
the DCM by looking at the cumulant expansion of the Boltzmann
weight, one can see that the DCM is exact in the high temperature
limit to order β^L where L is the characteristic size of the cell.
For low dimensional systems this implies that the DCM can give
very accurate results at finite temperatures. In the low
temperature limit the DCM results are determined by the ground
state and first few excited states of the cell, and are therefore

subject to finite cell-size effects[2].

Taken together, these observations led us to apply the DCM to the spin-1/2 Heisenberg model in three dimensions. In particular we are interested in the critical properties of the model, which occur at finite temperature where it is reasonable to expect the DCM to work well.

The Hamiltonian for the isotropic Heisenberg model is

$$H = -J\sum_{\langle ij \rangle} S_\alpha(i)\, S_\alpha(j) \tag{1}$$

where $\langle ij \rangle$ denotes a nearest neighbor pair of sites on a simple cubic lattice, and $\alpha=1,2,3$ labels the three spin components (sum over repeated indices implied). We implement the DCM on a cell consisting of the 8 sites (and 12 bonds) of an elementary cube.

Table 1 Theoretical Estimates for Critical Exponents					
method	α	β	γ	ν	η
ϵ-xpn[5]	-.130(21)	.368(4)	1.39(1)	.710(7)	.040(3)
MC[6]	-.118(18)	.36(1)	1.39(2)	.706(9)	.031(7)
series[3,4]	-.20(4)	.385(25)	1.43(1)	.735(15)	.05(6)
			1.388(2)		

We address two questions concerning the critical properties of the model. First, we wish to calculate the critical exponents and compare them to the values calculated by high-temperature series[3,4], the ϵ-expansion[5], and Monte Carlo simulation of the classical Heisenberg model[6] (table 1), and measured in experiments on isotropic quantum spin systems[7-10] (table 2).

In table 2 we show the estimates of the critical exponents

from high temperature series the ϵ-expansion, and Monte Carlo
calculations of the classical Heisenberg model. We see that these
results are consistent with each other and with the experimental
values. The ϵ-expansion is calculated for the n=3 classical
model, and the critical exponents calculated from the high
temperature expansion are extracted from the ferromagnetic
singularity in the series for the spin-1/2 Hamiltonian, (1).

Table 2
Experimental Values of Critical Exponents for
Isotropic Spin-1/2 Systems

Material	β	γ	ν
EuO	.368(5)[7]	1.30(2)[7]	.690(23)[8]
		1.396(30)[8]	
	.376(8)[8]		
EuS	.360(12)[8]	1.390(39)[8]	.720(27)[8]

2. Monte Carlo Results

We have performed Monte Carlo simulations on simple cubic
lattices of 8^3 to 24^3 spins. Each simulation is broken into as
many as 20 segments, depending on the size of the system. Each
segment is initialized by annealing, that is by equilibrating the
system in a descending sequence of temperatures starting above
the transition temperature and proceeding to the temperature of
the simulation. By starting above the transition temperature we
ensure that the order parameter assumes a statistically
independent value in each segment. The system is then
equilibrated for several thousand MCS/spin at the temperature of
the simulation. Finally, in the averaging phase, histograms are
constructed for the energy/spin and the z-component of the
magnetization, and the average z-z spin correlation is

calculated. After averaging for several thousand MCS/spin, the spin-configuration is stored and the process is repeated until we have accumulated several hundred thousand MCS/spin.

The susceptibility can be calculated from the magnetization histogram, but below the transition temperature the histogram is bimodal, so that simply calculating the second moment of the magnetization distribution would give incorrect results. To avoid this problem we assume that the magnetization distribution can be written as a sum of Gaussians centered at $\pm m_0$, which is exact in the infinite volume limit, and calculate both the second and fourth moments of the distribution. The spontaneous magnetization is then given by

$$m_0 = \left[3\langle m^2\rangle^2 - \langle m^4\rangle\right]^{\frac{1}{4}} \qquad (2)$$

and the susceptibility per spin is

$$\chi = N\beta\left(\langle m^2\rangle - m_0^2\right) \qquad (3)$$

where $\langle\cdots\rangle$ indicates an average over the measured magnetization distribution and $\beta=1/k_BT$.

The spontaneous magnetization can also be extracted from the correlation function by fitting the measured correlations to the form

$$G(z) = A\cosh\kappa\left(z - \frac{L}{2}\right) + m_0^2 \qquad (4)$$

While the values of m_0 determined by a least-squares fit to (4) depend on the particular functional form chosen for $G(z)$, they agree to less than a percent with the values determined from the magnetization histogram by (2).

3. Critical Exponents

We begin our study of the critical properties of the spin-1/2
Heisenberg model with the susceptibility of the ferromagnetic
model. The finite-size scaling form of the susceptibility is

$$\chi(t,L) = L^{\frac{\gamma}{\nu}} f_{hh}(tL^{\frac{1}{\nu}})$$ (5)

In (5), t is the reduced temperature,

$$t = \frac{T - T_c}{T_c}$$ (6)

and L is the linear size of the system. The universal function
f(x) has a single maximum at x=x₀, so that the peak in the
measured susceptibility scales with the size of the system as

$$L^{\frac{\gamma}{\nu}} .$$

Figure 1 is a log-log plot of
χ_{max} vs L. The slope of the
least-squares straight line
gives the exponent
γ/ν=2.02(2).

In figure 2 we show the three-
parameter scaling of the
susceptibility . The critical
temperature and exponents were
chosen to achieve the best
convergence of the data to a
single curve. The quality of

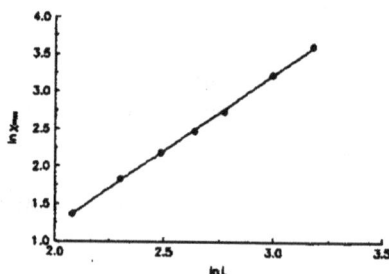

Fig. 1 Scaling of the maximum in
the susceptibility with size of
the lattice.

the fit was judged by eye, and therefore these values are
suggestive only.

The spontaneous magnetization evaluated at the critical temperature scales as

$$m(T_c, L) = L^{-\beta/\nu} f_h(0) \qquad (7)$$

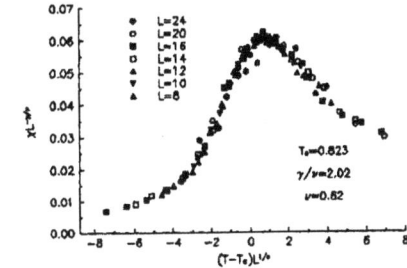

Fig. 2 Three-parameter scaling plot of the susceptibility.

In figure 3 we show the finite-size scaling of $m(T_c)$ and we find $\beta/\nu = 0.49(2)$. These results are consistent with our results for γ/ν through the scaling law

$$\frac{\gamma}{\nu} + 2\frac{\beta}{\nu} = d \qquad (8)$$

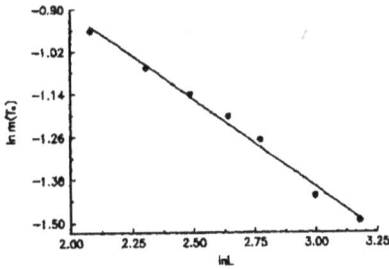

Fig. 3 Scaling of the spontaneous magnetization at T_c with size of the system.

Note that all these results are consistent with a small but negative value of η, which disagrees with the ϵ-expansion.

Three-parameter scaling of the magnetization is shown in figure 4 Again we find that the data collapses nicely onto a single curve.

We next examine the specific heat, which has the scaling form

$$C(t, L) = L^{\frac{\alpha}{\nu}} f_{tt}(t L^{\frac{1}{\nu}}) \qquad (9)$$

If we analyze the Monte Carlo data for the specific heat as we

did for the susceptibility we
find that the maximum in the
specific heat scales with the
size of the system as

$$C_{max} = L^{\frac{\alpha}{\nu}} f_{tt}(x_m) \qquad (10)$$

In figure 5 we plot the log of
C_{max} vs. log of L. We find a
good fit to a straight line
with $\alpha/\nu = 0.36(1)$.

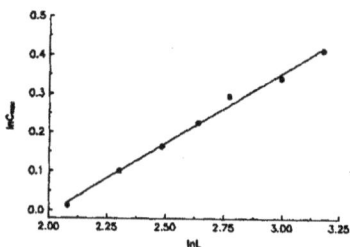

Fig. 4 Three-parameter scaling of
the spontaneous magnetization.

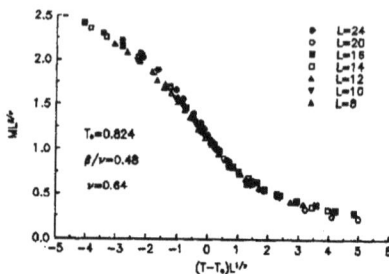

Fig. 5 Finite size scaling of
the maximum in the specific
heat.

A positive value of α is in
disagreement with the other
theoretical calculations shown in
table 1, and with the
experimentally determined values
of ν, which predict that α/ν
should be in the neighborhood of -
0.25 to -0.10. We should point
out, however, that in their work
on the classical Heisenberg model,
Peczak et al[4] also found that
their Monte Carlo data could be
interpreted in terms of a positive specific heat exponent with a
value of $\alpha = 0.21(4)$, which is consistent with our value. If in
fact the specific heat exponent is negative, then the non-
analytic component of the specific heat is a cusp superimposed on
a smooth background. This would make it very difficult to extract
scaling behavior from the Monte Carlo data.

However, if we take our Monte Carlo data at face value, it is

difficult to justify a negative
value for the specific heat
exponent. In figure 6 we show the
three-parameter scaling of the
specific heat data and find a
reasonably good collapse of the
data on to a single curve; no
such data collapse is possible
with a negative value for α.

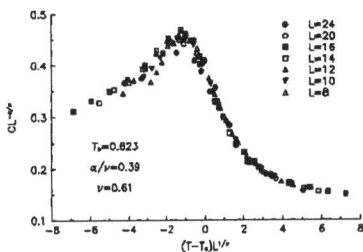

Fig. 6 Three parameter scaling
of the specific heat

In order to resolve the issue of
the sign of the specific heat
exponent we have used the "fourth-cumulant" method of Binder[8] to
determine the correlation-length exponent, ν, from the
magnetization data. According to the scaling hypothesis, the
ratio

$$C_4(tL^{1/\nu}) = \frac{\langle M^4 \rangle - 3 \langle M^2 \rangle^2}{\langle M^2 \rangle^2} \tag{11}$$

depends on the size of the system only through its argument. Near
the critical temperature we then have

$$C_4(tL^{1/\nu}) \approx C_4(0) + AL^{1/\nu}t \tag{12}$$

A plot of $C_4(T)$ vs T will then be a straight line, and the slopes
of these straight lines will scale with the size of the system as

$$\ln\left|\frac{dC_4(T_c)}{dT}\right| = B + \frac{1}{\nu}\ln L \tag{13}$$

In figure 7 we show the data for the fourth cumulants and in
figure 8 the plot of the log of the slope of $C_4(T)$ versus the log
L. A least-squares fit to a straight line gives $1/\nu=1.67(6)$. the

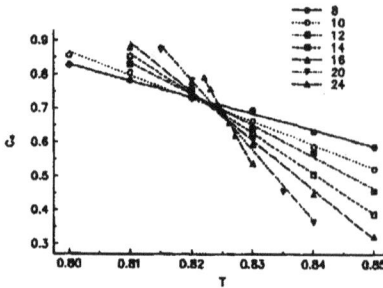

Fig. 7 Temperature dependence of the Binder fourth cumulant order parameter for systems of various size.

hyperscaling relation $\alpha = 2 - \nu d$ gives $\alpha = 0.20(6)$, in good agreement with the value determined directly from the specific heat.

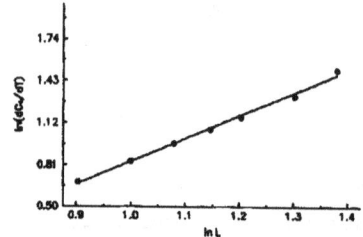

Fig. 8 Scaling of the slope of the Binder parameter with lattice size.

4. The Critical Temperature

The temperature at which the susceptibility or specific heat attains its maximum, T_m, (the rounding temperature) scales with the size of the system as

$$T_m = T_c \left(1 + x_m L^{-\frac{1}{\nu}} \right) \quad (7)$$

In figure 9 we show the finite size scaling of the rounding temperature taken from the susceptibility. Taking these two results

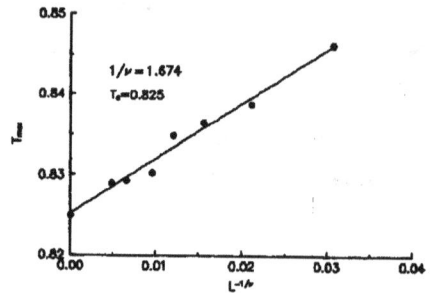

Fig. 9 Scaling of the rounding temperature with system size.

together, our best estimate is $T_{c,m}/J = 0.824(2)$. This result agrees reasonably well with the estimate from high-temperature series[2], T_c/J=0.840(4). We should point out that the value of non-universal quantities like the critical temperature depend to some extent on the choice of cell. The fair agreement of the DCM values for T_c with the high temperature series estimate indicates that the simple 8-site cubic cell we have employed gives a reasonably faithful representation of the Heisenberg model.

5. Conclusions

We have established the utility of the DCM method for quantum lattice models, especially in three dimensions. Our results for the critical exponents are collected in table 3.

Table 3			
Monte Carlo Results for the Critical Exponents of the Spin-1/2 Heisenberg Model			
	α/ν	β/ν	γ/ν	ν
DCM	0.36(1)	0.46(1)	2.02(2)	0.60(2)
MC(classical)[5]	-0.17(2)	0.516(3)	1.696(7)	0.706(9)
ϵ-expn[4]	-0.18(3)	0.517(6)	1.690(3)	0.735(15)

Our value of the specific heat exponent disagrees significantly with the results of the ϵ-expansion and high-temperature series, and we find a value of ν which is significantly smaller than the generally accepted value ν=0.7.

A direct comparison between the DCM results for the critical exponents and those of the ϵ-expansion and classical Monte Carlo can be seen in figures 10 and 11. The solid lines in each case are the relevant scaling laws. It is clear that the DCM results are in good agreement with the scaling laws, but in substantial disagreement with the results for the classical Heisenberg model.

In evaluating the results of our DCM calculations, the following points should be kept in mind. (1) The agreement of the exponents with the scaling laws indicates that by using the DCM we are studying the critical properties of some effective model (though possibly not the Heisenberg model). The critical temperatures are in reasonable agreement with those found by other methods. (2) Since the DCM uses exact eigenstates of the cell Hamiltonian, the probability of flipping is explicitly rotationally invariant. In this sense the DCM preserves the SU(2) symmetry of the spin-1/2 Heisenberg model.

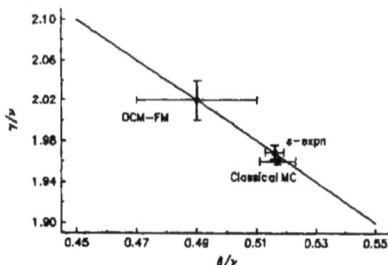

Fig. 10 Comparison of the critical exponents γ/ν and β/ν calculated by DCM, ϵ-expansion and classical Monte Carlo. The straight line is the scaling law $\gamma/\nu + 2\beta/\nu = d$.

There are two possible explanations for the discrepancy between the critical exponents calculated by DCM and those of the classical Heisenberg model. The first is that the approximations made in the DCM lead to incorrect values for the critical exponents. This leaves open the question of whether the classical and spin-1/2 Heisenberg models fall into the same universality class. However, if one takes the point of view that the universality class is determined solely by the dimension of space and the number of components of the order parameter, then we run into difficulty with point (3) above.

The second possibility is that the DCM gives a faithful picture

of the critical properties of
the spin-1/2 model, and
therefore the spin-1/2 model
is not in the same
universality class as the
classical Heisenberg model. We
do not believe that the
calculations presented here
constitute a proof of this
statement, but they certainly
call the matter into question.

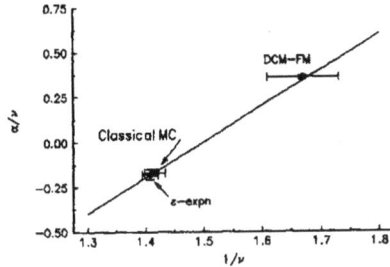

Fig. 11 Comparison of the critical exponents α/ν and $1/\nu$ calculated by DCM, ϵ-expansion and classical Monte Carlo. The straight line is the scaling law $\alpha/\nu = 2/\nu - d$.

The issue should be settled by
more precise calculations of
the exponents with larger cells, which we are currently carrying
out.

References

1. S. Homma, H. Matsuda, and N. Ogita, Prog. Theor. Phys. **72** 1245 (1984)
   S. Homma, H. Matsuda, and N. Ogita, Prog. Theor. Phys. **75** 1058 (1986)
   S. Homma, K. Sano, H. Matsuda, and N. Ogita, Prog. Theor. Phys. Supp. **87** 127 (1986)
   S. Homma, H. Matsuda, T. Horiki, and N. Ogita, Prog. Theor. Phys. **80** 594 (1988)
   T. Horiki, S. Homma, H. Matsuda, and N. Ogita, Prog. Theor. Phys. **82** 507 (1989)

2. C.J. Sisson, PhD Thesis, University of South Carolina, unpub. (1993)

3. G.S. Rushbrooke, G.A. Baker, and P.J. Wood, in <u>Phase Transitions and Critical Phenomena</u>, C. Domb and M.S. Green, eds., Vol 3 (Academic Press, 1974)

4. S. McKenzie, C. Domb, and D.L. Hunter, J. Phys. A **15** 3899 (1982)

5. J.C. Le Guillou and J. Zinn-Justin, J. Physique Lett. **46** L-137 (1985)

6. P. Peczak, A.M. Ferrenberg, and D.P. Landau, Phys. Rev. **B43**, 6087 (1991)

7. C.C. Huang, R.S. Pindale, and J.J. Ho, Sol. St. Comm. **14** 559 (1974)

8. J. Als-Neilsen and O.W. Dietrich, Phys Rev Lett. **27** 741 (1971)

9. K. Binder, Z. Phys. B **43**, 119 (1981)

VARIATIONAL MONTE CARLO STUDIES
OF CORRELATED ELECTRONS

Hiroyuki Shiba
Department of Physics, Tokyo Institute of Technology,
Oh-okayama, Meguo-ku, Tokyo 152, Japan

Abstract

A review is given on some of recent applications of the variational Monte Carlo method to correlated electrons. The systems discussed in this article include the Hubbard model, the $t - J$ model and the Kondo lattice, which represent various aspects of correlated electrons.

1 Introduction

Theoretical treatments of correlated electrons, itinerant electrons in particular, are difficult and challenging for computational physics. The quantum Monte Carlo method,[1] if it works, would be the best unbiased approach. Because of the negative-sign problem,[2] however, this method is not necessarily reliable except for such special cases as the high-temperature region or the half-filled case.

The variational Monte Carlo (VMC) method is complemetary to the quantum Monte Carlo method; it is free from such difficulty, but is approximate in general. The essence of the VMC method is to describe the ground state of the system with a suitable variational wave function. Even when a variational wave function is given, it is almost hopeless, in the case of quantum-mechanical many-boby systems, to carry out the calculation of the energy *etc.* analytically without any approximation. The spirit of the VMC is to perform those difficult calculations numerically, but in a controlled way.

The first successful application of the VMC method was made on the ground state of many Bose particles of ^{4}He by McMillan.[3] An extension of this approach to many fermion systems was carried out by Ceperley *et al.*,[4] who prepared the basic formulation for later development. Actually, prior to this development, Gutzwiller[5] had initiated the celebrated variational theory of the Hubbard model. However, it took some time for researchers[6-9] to notice that a combination of the VMC method with the Gutzwiller-type theory must be a fruitful route to study correlated electrons in solids. The present author reviewed in 1988 the initial achievement of the VMC method applied to the Hubbard model.[10] Therefore, to avoid an overlap we mainly discuss the development after 1988 in the present article.

Let us mention at this point that there are several ways in using the VMC method. The first is, needless to say, to find a reasonable approximate wave function for the ground state (and excited states). The second is to use it to check if a given wave function is an eigenstate of a Hamiltonian or not. In fact, if the trial wave function happens to be an eigenstate of the Hamiltonian, statistical fluctuations should vanish completely.[4,10] The third is to use a reasonable variational wave function as a starting point of the quantum Monte Carlo method or other methods. We shall see some examples for the first and second cases in this paper. The third case will show up in some articles in this volume.

2   Hubbard Model

The Hubbard Hamiltonian is the simplest model for correlated electrons given by

$$H = -t \sum_{(ij)\sigma} c_{i\sigma}^{\dagger} c_{j\sigma} + U \sum_{j} n_{j\uparrow} n_{j\downarrow}, \tag{1}$$

where the summation over (ij) is taken over nearest-neighbor pairs; the other notations are standard ones. The Gutzwiller theory[5] on the Hubbard Hamiltonian is to assume for the ground-state wave function

$$|\Psi_G\rangle = \prod_{j} [1 - (1-g)n_{j\uparrow} n_{j\downarrow}]|\Phi\rangle = g^D |\Phi\rangle \tag{2}$$

where g $(0 \leq g \leq 1$ for $U \geq 0)$ is a variational parameter corresponding to a local two-body correlation. $D = \sum_{j} n_{j\uparrow} n_{j\downarrow}$ represents the total number of doubly occupied sites and $|\Phi\rangle$ is the noninteracting Fermi sea. To discuss more general ordered states like the ferromagnetic and antiferromagnetic states, one generalizes $|\Phi\rangle$ by replacing the Fermi sea with a suitable Hartree-Fock-type wave function.[10]

A fascinating point of this theory lies in the simplicity of the wave function. For the half-filled case the single parameter g can cover both the insulating and metallic states: the wave function (2) with $g = 0$ corresponds to the insulating state, whereas the case with $g \neq 0$ is a correlated metal. Initially it was hoped from an approximate treatment[5,11] that this simple wave function is capable of describing the Mott transition for the half-filled case, since $g = 0$ corresponding to the Mott insulating state was concluded for U larger than a threshold value U_c. However, a more reliable analysis based on the VMC method[10] and exact analytic studies on the one-dimensional case[12] revealed new aspects, showing that some of the conclusions by Gutzwiller, Brinkman and Rice[5,11] are not correct. Incidentally, as far as the two approaches overlap, the results from the VMC and Vollhardt *et al.*'s exact theory are consistent with each other. This is a nice example showing the reliability and usefulness of the VMC method. It is now established by those studies that the Gutzwiller wave function is not sufficient to discuss the metal-insulator transition, as far as the dimension of the system is finite. Since then, this observation has led some researchers to go beyond (2) to improve the wave function. Here we wish to describe some efforts along this direction.

There are several ways to convince oneself that the wave function (2) is insufficient to describe the Mott insulating state for the half-filled case. The insulating state corresponding to $g = 0$ results in vanishing expectation value of energy, but this cannot be correct. Clearly, virtual processes which give a lowering of energy due to the second-order perturbation of hopping, are lacking in (2). To take into account this effect, $|\Psi_G\rangle$ in (2) is generalized to[13]

$$|\Psi\rangle = e^{-hT}|\Psi_G\rangle = e^{-hT}e^{-\zeta D}|\Phi\rangle, \qquad (3)$$

where T represents the kinetic energy in (1) and h is a new variational parameter. ζ is related to g as $e^{-\zeta} = g$. Actually, the wave function (3) with $g = 0$ was first proposed by Baeriswyl[14] to describe the large-U limit of the half-filled-band Hubbard model. The variational wave function (3) contains two parameters, g and ζ, for both of which the energy has to be minimized.

A VMC study based on (3) was carried out by Otsuka[13] for one-dimensional and two-dimensional Hubbard models. In contrast to the Jastrow-type wave function, on which most VMC studies are based, the wave function (3) requires a different algorithm for the VMC method. This can be seen easily, if we take an average of a physical quantity A in terms of (3):

$$\langle A\rangle = \frac{\langle\Phi|e^{-\zeta D}e^{-hT}Ae^{-hT}e^{-\zeta D}|\Phi\rangle}{\langle\Phi|e^{-\zeta D}e^{-2hT}e^{-\zeta D}|\Phi\rangle}. \qquad (4)$$

This form is similar to what one encounters in the quantum Monte Carlo simulation, in which the Suzuki-Trotter decomposition of $e^{-\tau H}$ ($H = T + UD$) leads to a product of exponential operators like (4). Therefore the technique developed for the quantum Monte Carlo method was used by Otsuka.[13] Since there are only a few exponential operators, the negative-sign problem does not arise in (4).

The ground-state energy obtained by this method is shown in Fig.1 for both one-dimensional (1D) and 2D (square lattice) cases. We note immediately that as far as the energy for the half-filled case is concerned, this theory provides an excellent and satisfactory result within the variational approach not only for the large-U region but also in intermediate or weak-U region. In particular, the resulting energy for 1D is extremely close to the exact solution; the energy for 2D is also expected to be reasonable. One interesting question is how the value of g for the lowest energy depends on U/t. This is an essential point to discuss the Mott transition within the wave function (3). Unfortunately, the g value for the lowest energy state has not been determined with a sufficient accuracy in ref.13 because of an extremely weak dependence of energy on g in the small-g region. We expect physically that the insulating state with $g = 0$ should be the lowest energy state for a large U and speculate that the minimum stays at $g = 0$ for 1D and 2D square lattices as far as U/t is finite. A confirmation is needed on this point as well as on the U dependence of g.

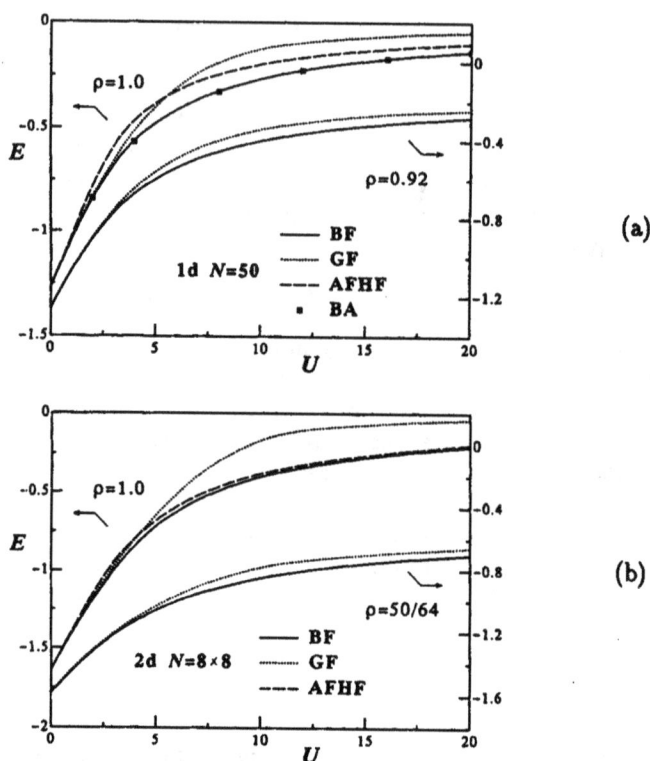

Figure 1: A comparison of the total energies obtained by various methods: (a) 1D half-filled Hubbard model and (b) 2D half-filled Hubbard model on the square lattice.[13]. The symbols GF, BF, AFHF, and BA mean the VMC with the Gutzwiller function (2), the VMC with the wave function (3), the antiferromagnetic Hartree-Fock approximation and the Bethe-Ansatz solution, respectively. ρ is the electron number per site and the unit of energy is t.

Two remarks on the half-filled case may be in order. First, it would be interesting to study carefully whether the wave function (3) can describe the Mott transition for a nonbipartite lattice, since in the case of bipartite lattices including the 2D square lattice the Mott insulating state is always accompanied with the antiferromagnetism. Second, the wave function of the form (3) appears to satisfy the criterion for describing the Mott insulating state that Millis and Coppersmith[15] have pointed out.

In contrast to the half-filled case, the nature of the ground state in non-half-filled cases is not clear except for 1D. It is very likely that in non-half-filled cases the wave function (3) is not sufficient;[13] intersite correlations are presumably important. One way to take into account such intesite correlations is to use the Jastrow-type correlation factor

$$\prod_{(jl)}\prod_{\sigma\sigma'}[1-(1-f_{jl})n_{j\sigma}n_{l\sigma'}] \qquad (5)$$

with a suitable f_{jl} instead of the local correlation in (2). Such an extension was done by Fazekas and Penc[16] and by Yokoyama and Shiba.[17] Although a partial success has been achieved in fact, more work is needed on non-half-filled cases. We shall see in the next section that the Jastrow-type correlation factor (5) plays an important role for the 1D $t - J$ model.

3 $t - J$ Model

The $t - J$ model is the following Hamiltonian within the subspace having no double occupancy

$$H = -t \sum_{(ij)\sigma} (c_{i\sigma}^{\dagger} c_{j\sigma} + \text{H.c.}) + J \sum_{(ij)} (\vec{S}_i \cdot \vec{S}_j - \frac{1}{4} n_i n_j), \qquad (6)$$

where $\vec{S}_j$ and n_j are spin and density operators at site j, respectively.

The phase diagram of the $t - J$ model in the space of J/t and the electron density is very interesting, because it is expected to contain a superconducting phase somewhere. The local constraint that no double occupancy is allowed at any site is easy for the VMC in contrast to other approaches.

The phase diagram of the 1D $t - J$ model has been explored extensively by various methods:[18] the exact diagonalization method,[19] the exact solution for $U/t \to 0$ and for the supersymmetric case $(J/t = 2)$[20] and the VMC method.[21,22]

Here we wish to review the VMC studies on the 1D $t - J$ model. Hellberg and Mele[21] proposed an interesting theory in which a long-range Jastrow-type correlation factor of the form (5) is multiplied to the Gutzwiller-projected Fermi sea:

$$|\Psi_{\text{HM}}\rangle = \prod_{(jl)} \prod_{\sigma\sigma'} [1 - (1 - f_{jl}) n_{j\sigma} n_{l\sigma'}] |\Psi_{\text{G}}(g = 0)\rangle$$

$$f_{jl} = |e^{i2\pi j/N} - e^{i2\pi l/N}|^{\nu} \qquad (7)$$

where ν is the only variational parameter controlling the correlation. Positive values of ν correspond to a repulsive correlation, whereas negative values mean an attractive correlation. Minimizing the energy with respect to ν, Hellberg and Mele obtained the phase diagram, which is shown in Fig.2. Note that the phase separation occurs for $\nu < -1/2$. The major part of the phase diagram is surprisingly similar to what Ogata et al.[19] obtained by a combination of the exact diagonalization and the conformal field theory. In the region with $0 > \nu > -1/2$ the superconducting correlation is the most dominant, while for $\nu > 0$ the magnetic correlation has the slowest decay. Notice also that for the supersymmetric case $J/t = 2$ the optimal value of ν is close to 0, in which case $f_{jl} = 1$ holds so that the wave function (7) is nothing but the Gutzwiller-projected Fermi sea $|\Psi_{\text{G}}(g = 0)\rangle$. Actually this was pointed out also by Yokoyama and Ogata.[22] Figure 3 shows a comparison of the momentum distribution fuction and spin and charge correlation functions for $J/t = 2$. A nice feature of the long-range correlation (8) is that it can describe the Luttinger liquid, to which the system (6)

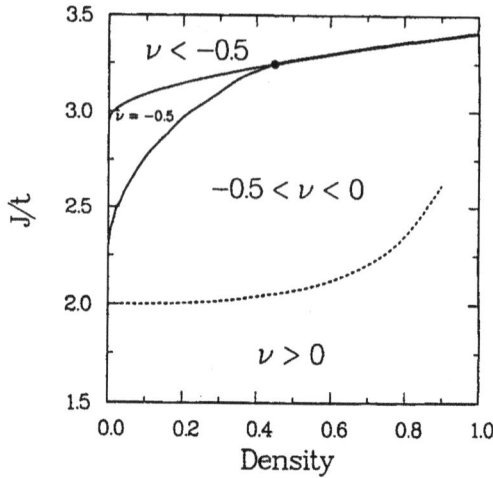

Figure 2: The phase digram of the 1D $t - J$ model determined by the variational wave function (7).[21] The optimal value of ν for each region is shown.

is believed to belong. In fact, Kawakami and Horsch[23] and Hellberg and Mele[24] found analytically (and numerically[21,24]) how the exponent of the power-law decay of correlation functions, which is characteristic in the Luttinger liquid, is related to ν. It seems therefore that the variational wave function (7) captures essential physics of the 1D $t - J$ model.

As is the case for most strongly correlated systems in 2D, the 2D $t - J$ model on the square lattice is very controversial and its phase diagram has not been determined completely. The high-temperature expansion[25] was applied to determine the phase diagram, the phase separation line in particular. It was suggested also in this work that a ferromagnetic phase may exist even for $J > 0$. Stimulated by Anderson's suggestion[26] that the Luttinger liquid may be realized also in 2D, Varenti and Gros[27] applied the VMC method based on the wave function similar to (7) and obtained a result supporting the Luttinger liquid. However, we believe more work is needed on this delicate issue.

One amusing development related to the $t - J$ model is that the Gutzwiller-projected Fermi sea $|\Psi_G(g = 0)\rangle = \prod_j(1 - n_{j\uparrow}n_{j\downarrow})|\Phi\rangle$ was shown[28] to be the ground state of a slightly generalized version of (6), namely, the 1D supersymmetric $t - J$ model with long-range exchange and transfer:

$$H = - \sum_{i \neq j, \sigma} t_{ij} c_{i\sigma}^{\dagger} c_{j\sigma} + \sum_{i \neq j} J_{ij}(\vec{S}_i \cdot \vec{S}_j - \frac{1}{4}n_i n_j) \tag{8}$$

where $t_{ij} = J_{ij} = t(\pi/N)\sin^{-2}[\pi(i - j)/N]$ $(t > 0)$ and the periodic boundary condition for the N-site chain is imposed. Again H is restricted within the subspace with no double occupancy and the total number of electrons N_c $(\leq N)$ is arbitrary. Kuramoto

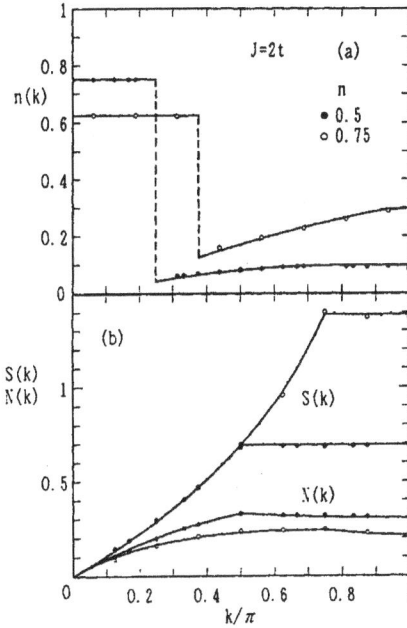

Figure 3: A comparison of (a) the momentum distribution function $n(k)$ and (b) the spin $(S(k))$ and charge $(N(k))$ correlation functions in the 1D supersymmetric $t - J$ model.[22] The solid lines represent the results for the Gutzwiller-projected Fermi sea, while open and filled circles show the exact-diagonalization results up to 16 sites. n represents the electron number per site.

and Yokoyama[29] proved that a set of wave functions generated from $|\Psi_G(g = 0)\rangle$ by introducing spin and charge currents and spin polarization

$$|\Psi_G(g = 0; N_\uparrow, J_\uparrow; N_\downarrow, J_\downarrow)\rangle = \prod_j (1 - n_{j\uparrow} n_{j\downarrow}) |\Phi(N_\uparrow, J_\uparrow; N_\downarrow, J_\downarrow)\rangle \qquad (9)$$

are also exact eigenstates of the Hamiltonian (8). Here $|\Phi(g = 0; N_\uparrow, J_\uparrow; N_\downarrow, J_\downarrow)\rangle$ represents a product of Slater determinants, in which the states within $\pi(-N_\sigma + 1 + J_\sigma)/N \leq k \leq \pi(N_\sigma - 1 + J_\sigma)/N$ are occupied for spin σ. They showed that the expectation value of H

$$\frac{\langle \Psi_G(g = 0; N_\uparrow, J_\uparrow; N_\downarrow, J_\downarrow) | H | \Psi_G(g = 0; N_\uparrow, J_\uparrow; N_\downarrow, J_\downarrow) \rangle}{\langle \Psi_G(g = 0; N_\uparrow, J_\uparrow; N_\downarrow, J_\downarrow) | \Psi_G(g = 0; N_\uparrow, J_\uparrow; N_\downarrow, J_\downarrow) \rangle} \qquad (10)$$

does not have any statistical fluctuations, when Monte Carlo samplings are carried out.[4,10] This is an "experimental" way to show that (9) is an eigenfunction of the Hamiltonian (8). The result is consistent with that due to the "asymptotic Bethe-Ansatz",[30] which gives an additional support to the VMC.

4   Kondo Lattice

The third system we take up is the Kondo lattice

$$H = \sum_{k\sigma} \varepsilon_k c_{k\sigma}^\dagger c_{k\sigma} + J \sum_{j\sigma\sigma'} \vec{S}_j \cdot \vec{s}_{\sigma\sigma'} c_{j\sigma}^\dagger c_{j\sigma'}, \tag{11}$$

where each site has a localized spin with $S = 1/2$, which is exchange-coupled with conduction electrons having the dispersion ε_k. Although conduction electrons are noninteracting with each other, the coupling with localized spins indirectly induces correlations among them. We assume $J > 0$ (*i.e.* antiferromagnetic), which is interesting from the viewpoint of quantum effects.

One important problem concerning the Kondo lattice is the Luttinger theorem: [31] whether one should count *localized* electrons producing $S = 1/2$ to determine the Fermi surface or not. Let us construct a spin-singlet ground state including localized spins. Such a state can be made by forming an overlapping singlet cloud at each site and making compatible with the Fermi statistics as[32]

$$|\Psi\rangle = \prod_k^{(N-N_c)/2} c_{k\uparrow} c_{k\downarrow} \prod_j (f_{j\uparrow}^\dagger \tilde{c}_{j\downarrow}^\dagger - f_{j\downarrow}^\dagger \tilde{c}_{j\uparrow}^\dagger)|0\rangle, \tag{12}$$

where $|0\rangle$ represents the vacuum; N_c and N are the total number of conduction electrons and lattice points, respectively; $N_c \le N$ is assumed for definiteness. $f_{j\sigma}^\dagger$ is the creation operator of localized electron with spin σ at site j. $\tilde{c}_{j\sigma} = N^{-1/2} \sum_k e^{ikr_j} a(k) c_{k\sigma}$ is the Kondo cloud around the site j, where $a(k)$ is the form factor of the cloud, which should be determined variationally. In Eq.(12) $N - N_c$ conduction electrons are taken away from the top of the conduction band ε_k.

The wave function (12) is written with a mixed representation of real and k-spaces. To avoid this inconvenience, let us rewrite (12) by introducing explicitly the projection operator $\hat{P}$ onto the subspace in which each site is always occupied one f-electron representing the localized spin:

$$\hat{P} = \prod_j [n_{j\uparrow}^f (1 - n_{j\downarrow}^f) + n_{j\downarrow}^f (1 - n_{j\uparrow}^f)]. \tag{13}$$

where $n_{j\sigma}^f$ is the number of f-electron at site j. First we note that

$$\prod_j^N [f_{j\uparrow}^\dagger \tilde{c}_{j\downarrow}^\dagger - f_{j\downarrow}^\dagger \tilde{c}_{j\uparrow}^\dagger]|0\rangle = \prod_j^N [\tilde{c}_{j\uparrow}^\dagger f_{j\uparrow} + \tilde{c}_{j\downarrow}^\dagger f_{j\downarrow}] \prod_m^N f_{m\uparrow}^\dagger f_{m\downarrow}^\dagger |0\rangle$$

$$= \lambda^{-N} \hat{P} \prod_\sigma \prod_j^N (\lambda + \tilde{c}_{j\sigma}^\dagger f_{j\sigma}) \prod_m^N f_{m\uparrow}^\dagger f_{m\downarrow}^\dagger |0\rangle \tag{14}$$

holds. λ is arbitrary because of the presence of $\hat{P}$. Now we use operators in k-space to rewrite (14) further. Then Eq.(14) is reduced to

$$\lambda^{-N} \hat{P} \prod_{k\sigma} (\lambda f_{k\sigma}^\dagger + a(k)^* c_{k\sigma}^\dagger)|0\rangle. \tag{15}$$

Here $f_{k\sigma}^\dagger$ is the Fourier transform of $f_{j\sigma}^\dagger$. Substituting this into (12), we find finally

$$|\Psi\rangle = \hat{P} \prod_\sigma \prod_k^{(N_c+N)/2} [\lambda f_{k\sigma}^\dagger + a(k)^* c_{k\sigma}^\dagger]|0\rangle, \qquad (16)$$

where the product in k is taken over the lowest $(N_c + N)/2$ states. Because of $\hat{P}$ the f-electrons are completely localized as they should be. A constant factor irrelavant to calculating expectation values has been dropped in Eq.(16). Thus we have shown that (12) is equivalent to (16); the latter is in essence a *Gutzwiller-projected mixed-band* wave function.

The VMC method, for which the local constraint $\hat{P}$ does not cause any technical problem, was applied to the wave function (16).[32] Since (16) has a form of mixed band, the form factor $a(k)$ was assumed to be given by

$$a(k) = \frac{2\tilde{V}}{\tilde{\varepsilon}_f - \varepsilon_k + \sqrt{(\tilde{\varepsilon}_f - \varepsilon_k)^2 + 4\tilde{V}^2}} \qquad (17)$$

where $\tilde{\varepsilon}_f$ and $\tilde{V}$ are variational parameters determined by the VMC. This formulation was applied to the simplest one-dimensional model: $\varepsilon_k = -2t\cos(k)$.

Figure 4 shows the momentum distribution function $n(k)$ of conduction electrons. The jump of $n(k)$ occurs at the Fermi wavenumber corresponding to the "large Fermi surface;" in other words, the number of localized spins should be counted to determine the Fermi surface. The smallness of the jump suggests a heavy mass of those conduction electrons due to the Kondo coupling with localized spins. On the other hand, the gradual but major change of $n(k)$ occurs at the Fermi wavenumber of the "small Fermi surface." As far as the ground state is a singlet including localized spins, the "large Fermi surface" should remain true in general. Quite recently Ueda *et. al.*[33] have presented a possibility of the "large Fermi surface," stabilizing the singlet state with a frustration. This may be regarded as a support to the above VMC study.

5 Summary

In this paper we reviewed some applications of the variational Monte Carlo method to correlated electrons: the Hubbard model, the $t - J$ model and the Kondo lattice. Due to difficulty of strongly correlated electrons, what has been achieved so far by the VMC is limited, but the VMC method has its own merits compared with other approaches. In particular it is extremely useful that the wave function is given explicitly in the VMC method, even though it is approximate in most cases.

The author is indebted to H. Otsuka for useful discusssions. This work is partly supported by Grant-in Aid for Scientific Research on Priority Area, Computational Physics as a New Frontier in Condensed Matter Research from the Ministry of Education, Science and Culture.

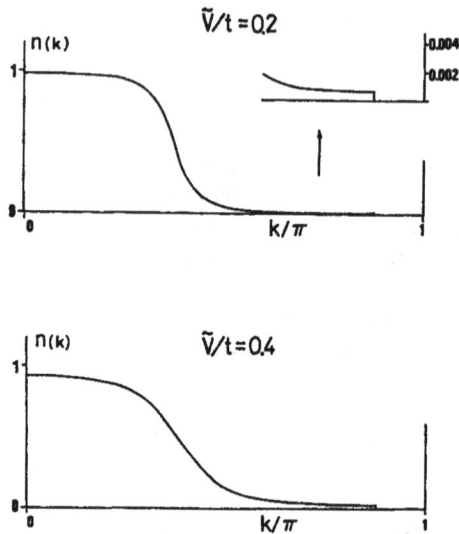

Figure 4: The momentum distribution function of the Kondo lattice determined by the VMC.[32] The conduction electron density N_c/N is chosen as 3/4. $\tilde{V}/t$ is 0.2 or 0.4 for the two cases; $\tilde{\varepsilon}_f/t$ is fixed at -0.7630.

References

1. See for instance, *Quantum Monte Carlo Methods in Equilibrium and Nonequilibrium Systems* (ed. M. Suzuki, Springer-Verlag, 1987).
2. See for instance N. Hatano and M. Suzuki: in this volume and M. Imada: in this volume.
3. W. L. McMillan: Phys. Rev. **138**, A442 (1965).
4. D. Ceperley, G. V. Chester and K. H. Kalos: Phys. Rev. **B16**, 3081 (1977).
5. M. C. Gutzwiller: Phys. Rev. Lett. **10**, 159 (1963); Phys. Rev. **134**, A1726 (1965).
6. T. A. Kaplan, P. Horsch and P. Fulde: Phys. Rev. Lett. **49**, 889 (1982).
7. H. Shiba: J. Phys. Soc. Jpn. **55**, 2765 (1986).
8. H. Yokoyama and H. Shiba: J. Phys. Soc. Jpn. **56**, 3582 (1987).
9. C. Gros, R. Joynt and T. M. Rice: Phys. Rev. **B36**, 381 (1987).
10. H. Shiba: *Two-Dimensional Strongly Correlated Electronic Systems* (ed. by Z. Z. Gan and Z. B. Su, 1989), p.161.
11. W. F. Brinkman and T. M. Rice: Phys. Rev. **B2**, 4302 (1970).
12. W. Metzner and D. Vollhardt: Phys. Rev. Lett. **59**, 121 (1987); F. Gebhard and D. Vollhardt: Phys. Rev. Lett. **59**, 147 (1987).
13. H. Otsuka: J. Phys. Soc. Jpn. **61**, 1645 (1992) and private communication.
14. D. Baeriswyl: *Nonlinearity in Condensed Matter* ed. by A. R. Bishop, D. K. Campbell, P. Kumar and S. E. Trullinger (Springer, 1987), p.187.

15. A. J. Millis and S. N. Coppersmith: Phys. Rev. **B43**, 13770 (1991).
16. P. Fazekas and K. Penc: Int. J. Phys. **B1**, 1021 (1988).
17. H. Yokoyama and H. Shiba: J. Phys. Soc. Jpn. **59**, 3669 (1990).
18. See for instance H. Shiba and M. Ogata: Prog. Theor. Phys. Suppl. **108**, 265 (1992)
19. M. Ogata, M. U. Luchini, S. Sorella and F. F. Assaad: Phys. Rev. Lett. **66**, 2388 (1991).
20. N. Kawakami and S. K. Yang: Phys. Rev. Lett. **65**, 2309 (1990).
21. C. S. Hellberg and E. J. Mele: Phys. Rev. Lett. **67**, 2080 (1991).
22. H. Yokoyama and M. Ogata: Phys. Rev. Lett. **67**, 3610 (1991).
23. N. Kawakami and P. Horsch: Phys. Rev. Lett. **68**, 3110 (1992).
24. C. S. Hellberg and E. J. Mele: Phys. Rev. Lett. **68**, 3111 (1992).
25. W. O. Putikka, M. U. Luchini and T. M. Rice: Phys. Rev. Lett. **68**, 538 (1992); W. O. Putikka, M. U. Luchini and M. Ogata: Phys. Rev. Lett. **69**, 2288 (1992).
26. P. W. Anderson: Phys. Rev. Lett. **64**, 1839 (1990); *ibid.* **67**, 2092 (1991).
27. R. Valenti and C. Gros: Phys. Rev. Lett. **68**, 2402 (1992).
28. Y. Kuramoto and H. Yokoyama: Phys. Rev. Lett. **67**, 1338 (1991).
29. H. Yokoyama and Y. Kuramoto: J. Phys. Soc. Jpn. **61**, 3046 (1992).
30. N. Kawakami: Phys. Rev. **B45**, 7525 (1992).
31. J. M. Luttinger: Phys. Rev. **119**, 1153 (1960).
32. H. Shiba and P. Fazekas: Prog. Theor. Phys. Suppl. **101**, 403 (1991)
33. K. Ueda, T. Nishino and H. Tsunetsugu: preprint.

QUANTUM MONTE CARLO SIMULATION OF MULTIBAND FERMION SYSTEMS AND ITS APPLICATION TO SUPERCONDUCTIVITY

Kazuhiko Kuroki and Hideo Aoki

Department of Physics, University of Tokyo, Hongo, Tokyo 113, Japan

Abstract

Application of the quantum Monte Carlo method to multiband fermion systems is described. The result for our recent model for superconductivity in a class of multiband systems comprising a metallic band interacting repulsively with an insulating band has unraveled the superconducting ground state as hallmarked by the pairing correlation function similar to that for the attractive Hubbard model. The similarity persists, surprisingly, well into the non-perturbative regime. This confirms the generic mechanism for superconductivity proposed for the system having a metallic band on top of another, which may be either weakly or strongly correlated charge-transfer insulator, Mott-Hubbard system, etc, and provides an example of the remarkable strength of the quantum Monte Carlo method in a non-perturbative regime that is otherwise difficult to probe.

1 Introduction

A wealth of new and unexpected phenomena that originate from strong electron correlations has been revealed from recent developments in both experimental and theoretical physics. Strongly correlated systems have turned out to be ubiquitous, and range from high-T_C superconductors,[1] heavy fermion systems or magnetic systems to fractional quantum Hall systems. To tackle these problems has raised a theoretical challenge, since conventional approaches such as perturbational or mean-field theories are of little use. This makes numerically implemented methods such as the quantum Monte Carlo method[2] invaluable.

In particular, the quantum Monte Carlo (QMC) method offers a way to probe systems with relatively large sizes, which is crucial in investigating the appearance of long-range orders such as magnetism or superconductivity. Specifically, if one wishes to propose a mechanism of superconductivity, the existence of effective attractions between electron by no means automatically implies superconductivity, but we have to demonstrate the existence of the off-diagonal long-range order as measured by the pairing correlation function. The QMC method provides a powerful tool to examine this.

The occurrence of superconductivity has been examined with QMC method for various models. For instance, existence of the off-diagonal long-range order in electron-phonon systems has been studied.[3,4] There is also a study on the Kosterlitz-Thouless transition to a superconducting state in the two-dimensional (2D) single-band attractive Hubbard model.[5,6] On the other hand, *electronic* mechanisms of superconductivity in repulsively interacting fermion systems have been of great interest, which has gained a special impetus after the discovery of high temperature superconductivity. Following a seminal suggestion by Anderson that strong electron correlations should play an essential role, QMC studies for the 2D repulsive Hubbard model have been performed extensively, but by now it has become clear that superconductivity does not arise in the usual (single-band repulsive) Hubbard model.[7,8,9,10]

Superconductivity in repulsively interacting *multiband* systems has also been studied by means of QMC method. Motivated by the structure of high T_C cuprates, many authors have investigated the models that take account of Cu3d and O2p orbitals.[7,11,12,13,14,15] For example, Cooper paring correlation function and susceptibility have been calculated for the so-called d-p model (or Emery's model) that considers one hybridised band of Cu3$d_{x^2-y^2}$ and O2p_σ orbitals.[7,11,12,13,14] Other authors have introduced models comprising O2p_π as well as O2p_σ orbitals to consider a pair-transfer term between them.[15] The superconductivity becomes enhanced due to the pair transfer, which idea was originally introduced by Suhl *et al*[16] and by Kondo.[17]

The present authors, on the other hand, have considered another class of multi-band systems, in which multiple bands having different symmetries (with no hybridisation) interact via direct Coulomb repulsion between electrons. As a typical case in this class of systems, we can show that the model Hamiltonian consisting of interacting metallic and insulating bands can be canonically transformed into the attractive Hubbard model in the perturbational regime.[18] Thus the superconductivity is unambiguously identified, since the attractive Hubbard model is shown to be superconductive.[5,6,19,20] The mechanism of superconductivity here is generic in the sense that the insulating band accompanying the metallic one may be either weakly or strongly correlated charge-transfer insulator, Mott-Hubbard system, etc.

In the non-perturbative regime for possibly higher T_C, which is of real interest, the QMC study has revealed that the similarity with the attractive Hubbard model persists, surprisingly, well into the non-perturbative regime. This provides an example of the strength of the QMC method in probing a non-perturbative regime. The purpose of the present article is to describe the QMC method and its application to the study of superconductivity in multiband systems.

One can go one step further to complement the QMC result with analytical approaches such as g-ology, and it has then turned out that the insulating band in our multiband model may be metallic ('nearly insulating'), which can be even desirable for higher T_C as we shall touch upon towards the end of this article.

We note that this class of model has a universality class different from those of hybridised models such as the d-p model or the periodic Anderson model. The essential ingredient in the latter class of models is the spin-spin coupling (the *d-p* model, for instance, reduces to a coupled spin-fermion system by a canonical transformation), as contrasted with the charge-charge coupling between unhybridised bands in our models. Our class of systems is also distinct from the Suhl-Kondo systems, in which the interband transfer of pairs rather than the interband repulsion between electrons is considered.

2 Quantum Monte Carlo method for multiband fermion systems

The basic idea of the QMC method we have adopted has been conceived by Blankenbecler *et al*,[21] and was first applied to the single-band Hubbard model by Hirsch.[22] At first it was difficult to explore low-temperature properties due to a numerical instability in the method. Later developments in the algorithm with the use of a stabilization procedure[8,9,23,24,25] enabled us to study ground states and low-temperature properties. The ground-state formalism with the canonical ensemble has been applied to single-band models,[8,9,10,24,25] to multiband models,[18,26,27] and to electron-phonon systems.[3,4] A finite-temperature method with the grand canonical ensemble has also been used to investigate low-temperature properties of single-band [6,25] and multiband systems.[11,12,13,14,15]

Here, we recapitulate the ground-state formalism for multiband systems, which is a straightforward extension of the single-band version. Let us write the Hamiltonian of the system as

$$H = H_0 + H_1,\qquad(1)$$

where H_0 is the kinetic energy (one-body) term, while H_1 is the interaction (two-body) term, which may include on-site, off-site, intraband, and interband interactions. We denote the typical value of the hopping and interaction parameters as t and V, respectively. The ground-state energy of the system may be written as

$$E_{\rm GS} = -\lim_{\beta\to\infty}\frac{1}{\Delta\beta}\log\frac{\rho(\beta+\Delta\beta)}{\rho(\beta)},\qquad(2)$$

$$\rho(\beta) = \langle\phi_T|e^{-\beta H}|\phi_T\rangle,\qquad(3)$$

where β is $1/kT$ and $|\phi_T\rangle$ is a trial state which is assumed to be non-orthogonal to the ground state. To evaluate eqn.3, we first decompose $\exp(-\beta H)$ using the Trotter-Suzuki formula,[28]

$$e^{-\beta H} = \left(e^{-\Delta\tau H}\right)^L,\qquad \Delta\tau = \beta/L\qquad(4)$$

$$e^{-\Delta\tau H} = e^{-\Delta\tau H_0/2}e^{-\Delta\tau H_1}e^{-\Delta\tau H_0/2} + O((\Delta\tau t)^3(V/t)).\qquad(5)$$

Here, the number of decomposition, L, should be large so that $\Delta\tau = \beta/L$ is sufficiently small. The error in the decomposition (the last term in eqn.5, which we shall ignore) increases with the strength of the interaction, (V/t), so that larger L has to be taken for larger interactions. Each $e^{-\Delta\tau H_0/2}e^{-\Delta\tau H_1}e^{-\Delta\tau H_0/2}$ is called a Trotter slice.

When we assume for simplicity that each interaction term is bilinear in number operators, the interaction terms can be linearized with the discrete Hubbard-Stratonovich transformation,[29] which takes the form

$$\exp(-\Delta\tau V n_{i\sigma}^\nu n_{j\sigma'}^\mu) = \frac{1}{2}\sum_{s=\pm 1}\exp[\lambda s(n_{i\sigma}^\nu - n_{j\sigma'}^\mu) - \frac{\Delta\tau V}{2}(n_{i\sigma}^\nu + n_{j\sigma'}^\mu)], \qquad (6)$$

with $\cosh(\lambda) = \exp(\Delta\tau V/2)$ for $V > 0$, and

$$\exp(-\Delta\tau V n_{i\sigma}^\nu n_{j\sigma'}^\mu) = \frac{1}{2}\sum_{s=\pm 1}\exp[\lambda s(n_{i\sigma}^\nu + n_{j\sigma'}^\mu) - \lambda'(n_{i\sigma}^\nu + n_{j\sigma'}^\mu)], \qquad (7)$$

with $\exp(-\Delta\tau V) = \cosh(2\lambda)/\cosh^2(\lambda)$ and $\lambda' = \ln(\cosh\lambda)$ for $V < 0$, where $n_{i\sigma}^\nu$ is the number operator of an electron at site i in band ν with spin σ. This transformation is applied to all the interaction terms in $e^{-\Delta\tau H_1}$ in each Trotter slice. Let K be the number of two-body interaction terms contained in the Hamiltonian. For instance, if we have

$$H_1 = V\sum_i \sum_{\sigma,\sigma'} n_{i\sigma}^\alpha n_{i\sigma'}^{\beta_1} + U\sum_j n_{j\uparrow}^{\beta_2} n_{j\downarrow}^{\beta_2}, \qquad (8)$$

where α, β_1 and β_2 denote different bands, we have four kinds of Stratonovich variables per site (corresponding to $n_{i\uparrow}^\alpha n_{i\uparrow}^{\beta_1}, n_{i\uparrow}^\alpha n_{i\downarrow}^{\beta_1}, n_{i\downarrow}^\alpha n_{i\uparrow}^{\beta_1}, n_{i\downarrow}^\alpha n_{i\downarrow}^{\beta_1}$) for the V-term, and one kind for the U-term. If there are N unit cells (γN orbitals in total with $\gamma = 3$ here) in the system, $K = 4N + N = 5N$ in this example. Then we can express $\exp(-\Delta\tau H_1)$ in the form,

$$\exp(-\Delta\tau H_1) = \sum_{s_1=\pm 1}\sum_{s_2=\pm 1}\cdots\sum_{s_K=\pm 1} u(s_1, s_2, \cdots, s_K), \qquad (9)$$

where $u(s_1, s_2, \cdots, s_K)$ is written as a product of exponential terms of linear combination of number operators. Eqn. 3 is then transformed into

$$\langle\phi_T|\exp(-\beta H)|\phi_T\rangle = \sum_{\{s_i(l)\}}\langle\phi_T|U(\{s_i(l)\};\beta)|\phi_T\rangle, \qquad (10)$$

where l specifies the Trotter slice. $U(\{s_i(l)\};\beta)$, which is a product of terms of the form $\exp[\text{one-body operator}]$, can be represented by matrices in a single-particle basis. The trial state, $|\phi_T\rangle$, may also be expressed in terms of coefficient matrices of $\gamma N \times M$ for an M-electron system, so that the matrix corresponding to $\langle\phi_T|U(\{s_i(l)\};\beta)|\phi_T\rangle$ can be obtained by a multiplication of matrices. It is shown that the value of $\langle\phi_T|U(\{s_i(l)\};\beta)|\phi_T\rangle$ can be evaluated as a determinant of the corresponding matrix.

Now the argument in the logarithmic term in eqn.2 is evaluated with the importance sampling technique as follows. First we rewrite the expression as

$$\frac{\rho(\beta+\Delta\beta)}{\rho(\beta)} = \frac{\sum_{\{s_i(l)\}}\langle\phi_T|U(\{s_i(l)\};\beta+\Delta\beta)|\phi_T\rangle}{\sum_{\{s_i(l)\}}\langle\phi_T|U(\{s_i(l)\};\beta|\phi_T\rangle}$$

$$\equiv \frac{\sum_{\{s_i(l)\}} |\Omega(\{s_i(l)\}; \beta)| \frac{\Omega(\{s_i(l)\}; \beta + \Delta\beta)}{|\Omega(\{s_i(l)\}; \beta)|}}{\sum_{\{s_i(l)\}} |\Omega(\{s_i(l)\}; \beta)|} \bigg/ \frac{\sum_{\{s_i(l)\}} |\Omega(\{s_i(l)\}; \beta)| \mathrm{sign}\Omega(\{s_i(l)\}; \beta)}{\sum_{\{s_i(l)\}} |\Omega(\{s_i(l)\}; \beta)|}$$

(11)

where $\Omega(\{s_i(l)\}; \beta) = \langle \phi_T | U(\{s_i(l)\}; \beta) | \phi_T \rangle$. We can then evaluate the average on the right-hand side of eqn.11 from a series of configurations of Stratonovich variables generated according to the probability,

$$P(\{s_i(l)\}; \beta) = \frac{|\Omega(\{s_i(l)\}; \beta)|}{\sum_{\{s_i(l)\}} |\Omega(\{s_i(l)\}; \beta)|},$$

(12)

to take the average of $\Omega(\{s_i(l)\}; \beta + \Delta\beta)/|\Omega(\{s_i(l)\}; \beta)|$ and $\mathrm{sign}\Omega(\{s_i(l)\}; \beta)$. We can generate the configurations by flipping the Stratonovich variables one by one according to the probability,

$$
\begin{aligned}
&p(s_i(l) = +1 \rightarrow s_i(l) = -1) \\
&= \frac{P(\{\cdots, s_i(l) = -1, \cdots\}; \beta)}{P(\{\cdots, s_i(l) = +1, \cdots\}; \beta) + P(\{\cdots, s_i(l) = -1, \cdots\}; \beta)} \\
&= \frac{\Omega(\{\cdots, s_i(l) = -1, \cdots\}; \beta)}{\Omega(\{\cdots, s_i(l) = +1, \cdots\}; \beta) + \Omega(\{\cdots, s_i(l) = -1, \cdots\}; \beta)},
\end{aligned}
$$

(13)

which may be done without calculating $\sum_{\{s_i(l)\}} |\Omega(\{s_i(l)\}; \beta)|$. The energy is thus obtained from eqn.2. Other quantities such as one-body Green's function or correlation functions may also be obtained within the formalism.

For large β, which becomes necessary for ground-state calculations, the number of Trotter slices required to obtain $\langle \phi_T | U(\{s_i(l)\}; \beta) | \phi_T \rangle$ becomes large, and this causes a numerical instability since the product of a large number of matrices becomes ill conditioned. This difficulty can be circumvented by orthonormalising the resultant matrix once per several multiplications, and it is then possible to take sufficiently large β.[8,9,23,24,25] In addition, however, there is other problems when β is large.

For a fixed value of β, the number ($K \times L$) of the Stratonovich variables, $s_i(l)$, increases with the number of sites (γN), kinds of interactions, and the strength of interactions, as can be seen from above. Computation time required for one Monte Carlo step, which contains the update of all the Stratonovich variables, linearly increases with the number of Stratonovich variables. Moreover, the larger the number of Stratonovich variables, more Monte Carlo steps are needed for convergence. Thus, for large β, the tractable system size, and the kinds and strength of interactions are restricted especially for multiband models.

Another problem known as the negative sign problem is specific to fermion systems: the positive and negative components of $\mathrm{sign}\Omega(\{s_i(l)\}; \beta)$ tend to be equally weighted for large β and/or large values of interactions. Since $\mathrm{sign}\Omega(\{s_i(l)\}; \beta)$ appears in the denominator in eqn.11, we have to evaluate $\sim 0/0$, which is numerically unstable. In one-dimensional (1D) systems this problem does not arise as far as the number of electrons satisfies the closed-shell condition. This remains the case for multiband 1D systems when the hybridisation between the bands is absent, while in the hybridised case such as the d-p model the sign problem arises even in 1D, since an introduction of the interband transfer effectively makes the system two dimensional (with two strands).

Despite these limitations, the QMC method is powerful, since tractable system sizes are much larger than those in the exact diagonalization method. For instance, the tractable size up to now for the 2D single-band Hubbard model is a 16×16 system in QMC[8] as contrasted with a 4×4 system in the exact diagonalization. As for multiband models, we could still manage QMC calculations for a system as large as a 75-orbital 1D chain (with 25 unit cells, see Fig.2 in section4), which requires 20 hours of CPU time on HITAC S-820. As we have mentioned in Introduction, large system sizes are crucial since we have to identify the behaviour of correlation functions at large distances. Indeed, the present study provides an example of exploiting the QMC method in identifying long-range order in finite systems in that we actually establish a mapping of the proposed repulsive model to the attractive Hubbard model, which is known to superconduct.

3   Mechanism of superconductivity in repulsively interacting multiband systems

In this section, we introduce our model, and characterise its properties analytically in certain limits to facilitate interpretation of the QMC results.

We start with a model (Fig.1), in which a carrier band(α) interacts repulsively with a charge-transfer (CT) insulator (β) that consists of a band from alternating orbitals (β_1 and β_2) with a level offset. The Hamiltonian is given, in standard notations, as

$$
\begin{aligned}
H =\ & -t_\alpha \sum_{\langle i,i'\rangle} \sum_\sigma (c_{i\sigma}^{\alpha\dagger}c_{i'\sigma}^\alpha + \text{h.c.}) - t_\beta \sum_{\langle i,j\rangle}\sum_\sigma (c_{i\sigma}^{\beta_1\dagger}c_{j\sigma}^{\beta_2} + \text{h.c.}) \\
& +V\sum_i \sum_{\sigma,\sigma'} n_{i\sigma}^\alpha n_{i\sigma'}^{\beta_1} + U\sum_j n_{j\uparrow}^{\beta_2} n_{j\downarrow}^{\beta_2} \\
& +\sum_\sigma (\varepsilon^\alpha \sum_i n_{i\sigma}^\alpha + \varepsilon^{\beta_1}\sum_i n_{i\sigma}^{\beta_1} + \varepsilon^{\beta_2}\sum_j n_{j\sigma}^{\beta_2}).
\end{aligned}
\tag{14}
$$

Here t_α is the transfer energy between adjacent α orbitals, t_β the transfer energy between adjacent β_1 and β_2 orbitals, and $\varepsilon^\alpha, \varepsilon^{\beta_1}, \varepsilon^{\beta_2}$ are respective energy levels. An α electron interacts with a β_1 electron on the same site with a repulsion V. We also assume an on-site repulsion, U, within β_2 orbitals, while the intraband repulsion within the α band is neglected. We consider no hybridization between α and β orbitals, which are assumed to possess different symmetries.

Let us first consider the case of weakly correlated CT insulator, in which U is small compared with $\Delta\varepsilon = \varepsilon^{\beta_1} - \varepsilon^{\beta_2}$, and an average of two β electrons per unit cell make the β band an insulator. In the highly insulating case of large $\Delta\varepsilon$, we can make

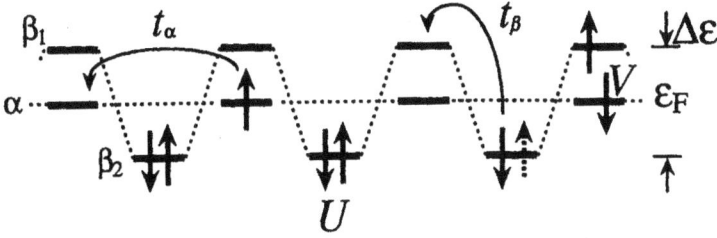

Figure 1: A multiband model in which the β band is a weakly correlated CT insulator. Dotted lines represent electron-transfer terms. The case of $\Delta\varepsilon > 0$ is shown here.

a canonical transformation that eliminates the β electron hopping in the lowest order in $(t_\beta/\Delta\varepsilon)$ with an appropriate generator, S. We are then left with a Hamiltonian with effective interactions between α electrons in the leading order, which reads, up to a constant,

$$
\begin{aligned}
H_{\text{eff}} &= e^{iS} H e^{-iS} \\
&\simeq -t_\alpha \sum_{\langle i,i'\rangle} \sum_\sigma (c^{\alpha\dagger}_{i\sigma} c^\alpha_{i'\sigma} + \text{h.c.}) \\
&\quad -2z_1 t_\beta^2 \left(\frac{1}{\Delta\varepsilon_{\text{eff}} + 2V} + \frac{1}{\Delta\varepsilon_{\text{eff}}} - \frac{2}{\Delta\varepsilon_{\text{eff}} + V} \right) \sum_i n^\alpha_{i\uparrow} n^\alpha_{i\downarrow},
\end{aligned}
\tag{15}
$$

where z_1 denotes the number of β_2 orbitals neighbouring a β_1. The effective interaction arising from the interband repulsion is thus negative definite, and we end up with the single-band attractive Hubbard model (AHM). We note that the situation does not depend on the sign of the interband interaction, V, and the same formula applies for $\Delta\varepsilon < 0$ if we replace $\Delta\varepsilon$ and V with $-\Delta\varepsilon$ and $-V$, respectively.

From this, the α-band correlation functions in our model should coincide with those of the AHM at least in the limit of large $\Delta\varepsilon$. We can in fact extend our analysis to show that, as $\Delta\varepsilon$ becomes smaller, the pairing correlation in the multiband model remains essentially the same as that in the AHM as far as $\Delta\varepsilon$ is sufficiently large, except that the former is multiplied at large distances by an overall prefactor, $\eta(\simeq 1)$.[18]

The AHM has been extensively studied from the Bethe ansatz exact solution in 1D[19] and by the QMC method in 2D.[5,6] The results show that AHM is superconductive except for the half-filled band for which superconductivity is degenerate with the charge density wave (CDW) state. Thus, our model may be considered to be superconductive at least in the large $|\Delta\varepsilon|$ limit.

4 Quantum Monte Carlo results

Now the real problem is the occurrence of superconductivity in the non-perturbative regime, since the perturbational regime has only small effective attractions. This has in fact motivated us to perform QMC calculations for the present model with realistic parameters. We have restricted our calculation to 1D systems since (i) we can explore larger systems to study the behaviour of the correlation function at larger distances, and (ii) the negative sign problem in QMC calculations does not arise in 1D as long as we keep the filling in each band closed-shell, since the hybridization between the bands is absent in our model.

In Fig.2(a) we show the pairing correlation function in real space subtracted by its value for the non-interacting case with $t_\beta/\Delta\varepsilon$ as large as 1/3. We have fitted the results to the QMC result for AHM of the same sample size by adjusting the negative U. In the fitting, we have multiplied the pairing correlation in the AHM for $|i-j| \neq 0$ with $\eta(\simeq 1)$ as mentioned above. The pairing correlation function in our model is indeed in dramatic agreement with that of AHM. Similar agreement is obtained for the CDW correlation function (Fig.2(b)). Thus the property originally confirmed by a perturbational argument turns out to extend, surprisingly, well into the non-perturbative regime, which is reminiscent to the single-band case, where the repulsive half-filled (non-half-filled) Hubbard model retains similarity to the Heisenberg model (*t-J* model) for finite values of U.

The effective attraction increases as we make the β band less insulating, and an obvious question is whether the superconductivity is retained then. We show in Fig.3 the QMC result for the pairing correlation function for various values of $|\Delta\varepsilon|$, which controls the insulating property (excitation gap) of the β band. The case of $\Delta\varepsilon < 0$ is chosen here, since the effective attraction becomes stronger than the case of $\Delta\varepsilon > 0$ as seen from a comment below eqn.15. We consider two (one up and one down spin) electrons doped in the α band, since the AHM has a special property that the pairing correlation for this number of carriers becomes a constant in real space, which facilitates an accurate comparison of the correlation function in our model to that of the AHM. The result shows that the correlation function for the present model remains flat for $|i-j| \neq 0$ for $|\Delta\varepsilon|$ decreased down to 2 with a fixed interband interaction, $V = 1$. The effective attraction, which is measured by the value of the correlation function at $|i-j| = 0$, increases for smaller $|\Delta\varepsilon|$.

So far we have presented the results for the model with weakly correlated CT insulator. Remarkably, we can show that the present mechanism of superconductivity is applied to systems with various kinds of insulating bands: the insulating band can be a strongly correlated one such as the strongly correlated CT insulator (with $U \gg |\Delta\varepsilon|$) depicted in Fig.4, or the Mott-Hubbard insulator. Superconductivity has been again shown from both the canonical transformation into an effective AHM as in eqn.15, and the QMC result for the pairing correlation in the non-perturbative regime. The QMC result (Fig.5) for the case in which the β band is a strongly correlated CT insulator shows that the pairing correlation again resembles that of

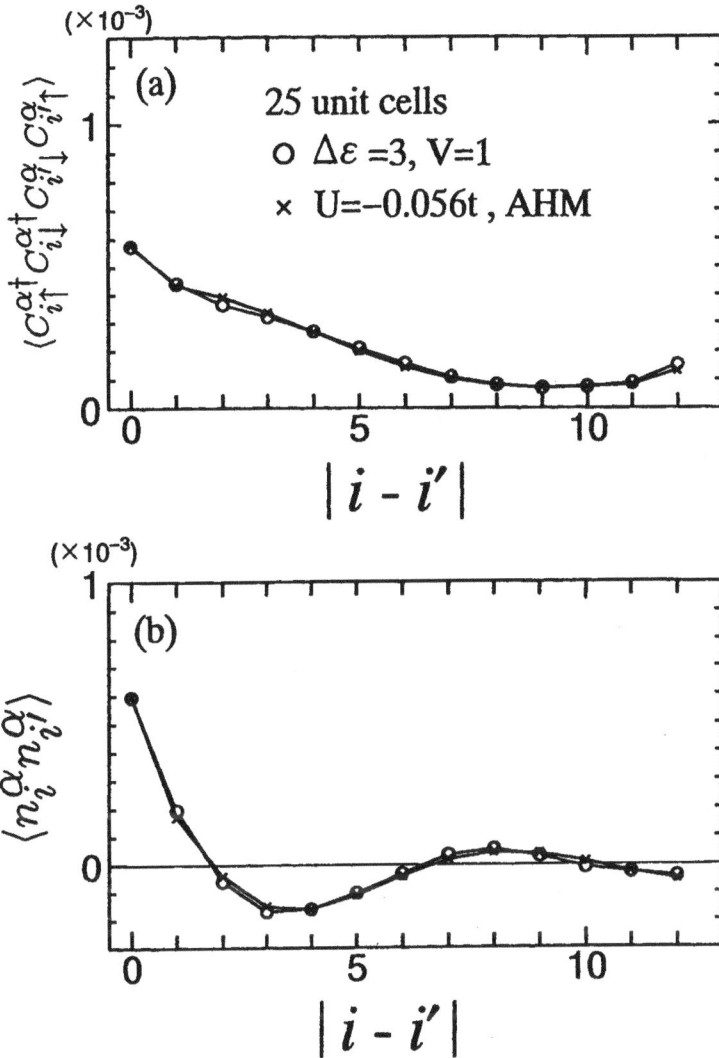

Figure 2: QMC results for the correlation function (subtracted by the value for the non-interacting case) against distance for the α-band pairing(a) and the α-band CDW(b). The model is depicted in Fig.1 in a system with 25 unit cells (75 orbitals) with 6 electrons doped in the α band on top of 50 β electrons so that the β_2 orbitals are fully filled. The result is shown for $\Delta\varepsilon = 3$, $V = 1$, $t_\alpha = t_\beta = 1$, which is compared with the result ($\times$) for the single-band attractive Hubbard model with $U = -0.056t$.

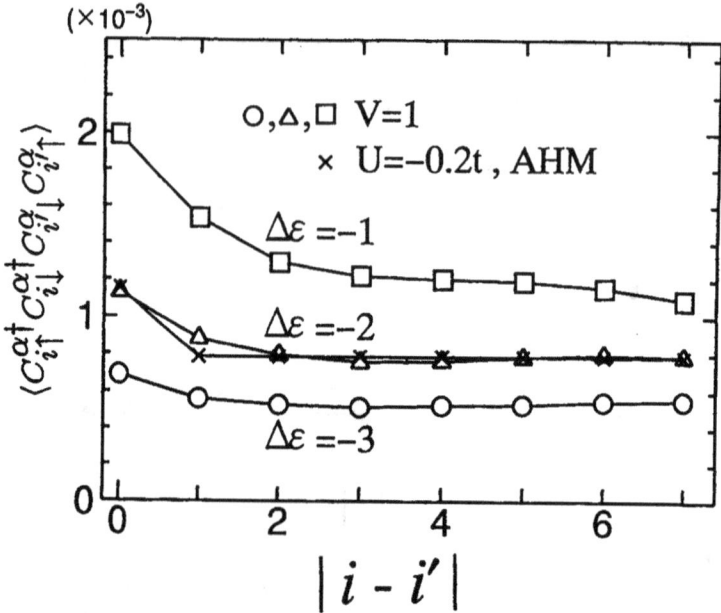

Figure 3: The dependence of the pairing correlation on $\Delta\varepsilon$ is shown with 2 α electrons on top of 30 β electrons in 15 unit cells (45 orbitals). We vary $\Delta\varepsilon$ from $-3(\bigcirc)$, $-2(\triangle)$ to $-1(\square)$ with the fixed $V = t_\alpha = t_\beta = 1$. The result for the single-band attractive Hubbard model with $U = -0.2t$ best-fit for $\Delta\varepsilon = -2$ is represented by $\times$.

AHM for sufficiently large U. For $U = 0$ the similarity is completely degraded, while the attraction continues to increase.

The similarity to AHM in this case is determined by a balance of the intraband repulsion U, the gap $|\Delta\varepsilon|$, and the interband repulsion V. To see this, we show in Fig.6(a) and (b) the pairing correlation for $|\Delta\varepsilon| = 2$ and 3, fixing U at $U = 4$ (a) or 3 (b), and $V = 1$. For $U = 3$ and $|\Delta\varepsilon| = 2$, i.e., when both U and $|\Delta\varepsilon|$ are small, the similarity is degraded, while in other cases it is retained as seen from the flatness of the pairing correlation.

So far we have given results for the filling for which the β band is insulating. When the β band is doped with carriers, the pairing correlation in the α band deviates from an AHM-like behaviour, but the effective attraction increases, as seen in Fig.7. This is reminiscent of the degradation of similarity to AHM in Fig.6(b), where both U and $|\Delta\varepsilon|$ are small.

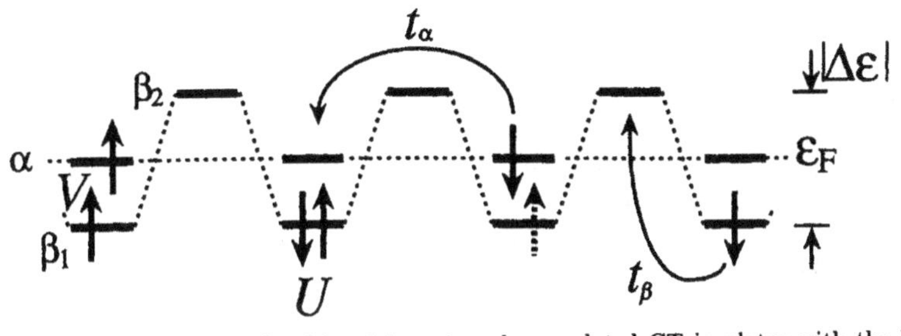

Figure 4: A model in which the β band is a strongly correlated CT insulator with the half-filled lower β orbitals. The case of $\Delta\varepsilon < 0$ is shown here.

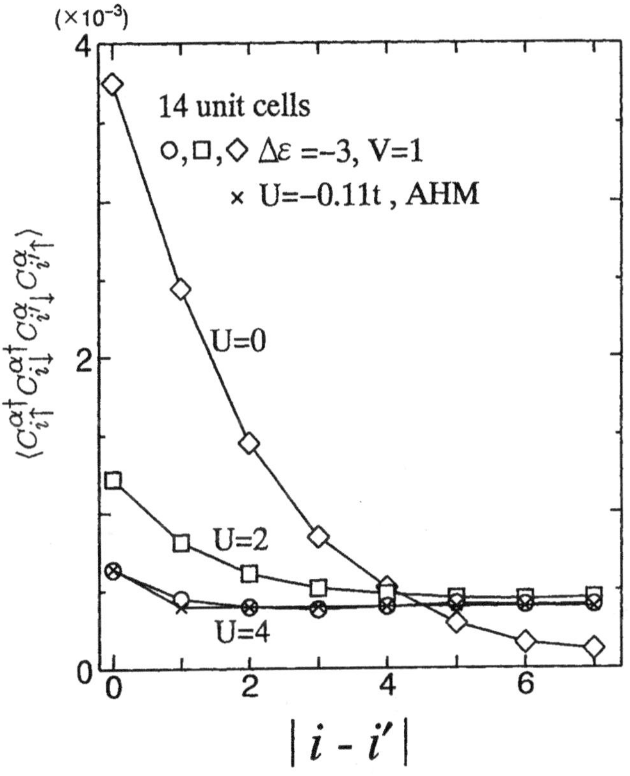

Figure 5: U-dependence of the pairing correlation in the model with strongly correlated β band (Fig.4) with 2 α electrons on top of 14 β electrons in 14 unit cells (42 orbitals). We vary U from 4($\bigcirc$), 2($\square$) to 0($\Diamond$) with the fixed level offset, $\Delta\varepsilon = -3$, and $V = t_\alpha = t_\beta = 1$. For comparison the result ($\times$) for the single-band attractive Hubbard model with $U = -0.11t$ is shown.

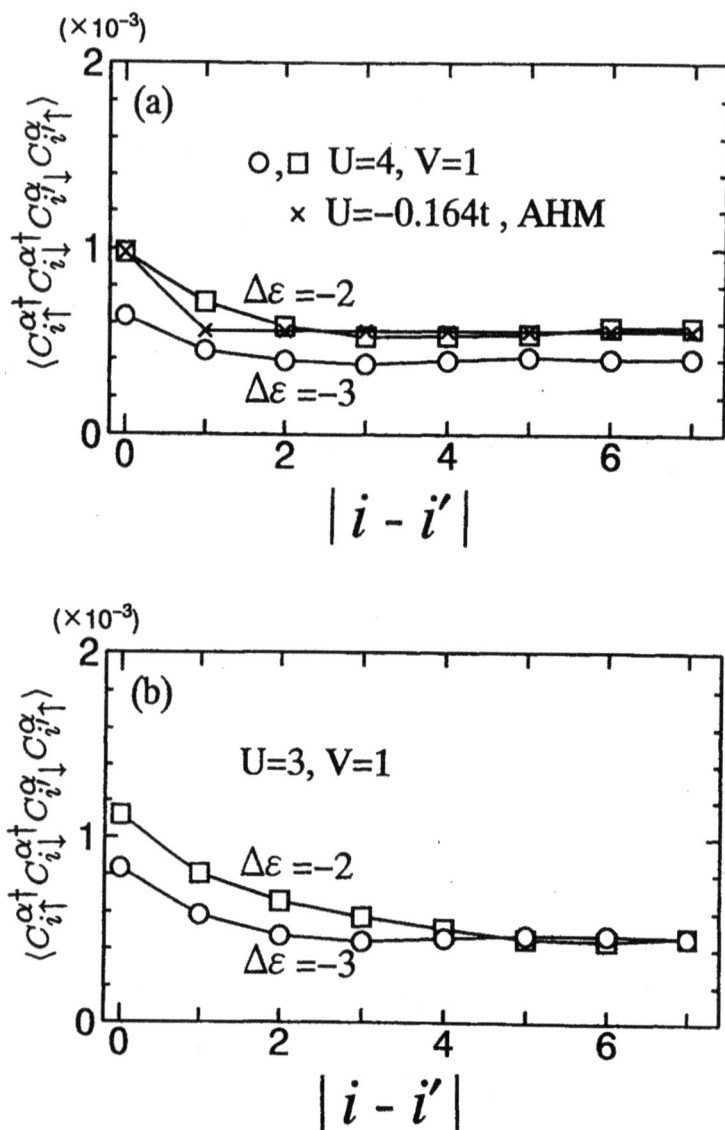

Figure 6: The change in the pairing correlation when $\Delta\varepsilon$ is varied from $-3(\bigcirc)$ to $-2(\square)$ for 2 α electrons on top of 14 β electrons in 14 unit cells in the model with strongly correlated β band Fig.4). U is fixed at $U = 4$(a) or $U = 3$(b) with $V = 1$, while $\times$ in (a) represents the result for the single-band attractive Hubbard model with $U = -0.164t$.

$(\times 10^{-3})$

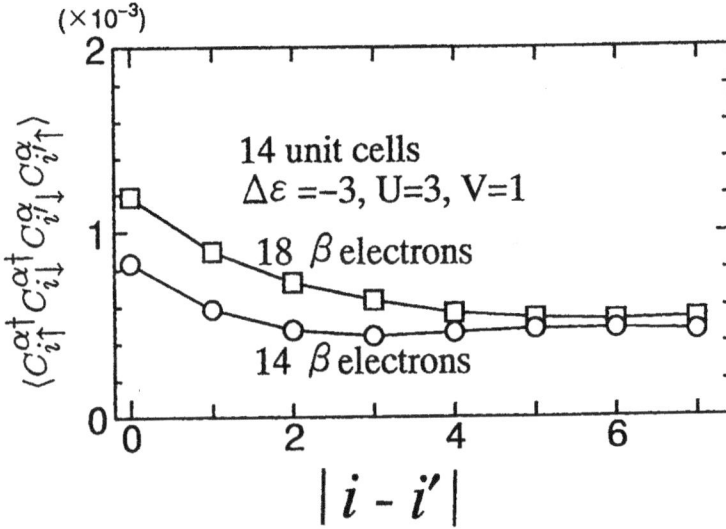

Figure 7: QMC result ($\square$) for the pairing correlation in the model with strongly correlated β band with 14 unit cells when both α and β bands are doped. Here we have two electrons in α band while the β orbitals is also doped with 4 electrons (to 14+4=18 β electrons) for $\Delta\varepsilon = -3$, $U = 3$, and $V = t_\alpha = t_\beta = 1$. The case with 14 electrons is also shown ($\bigcirc$) for comparison.

5 Concluding remarks

In the present study, we have employed the QMC method to establish a generic mechanism for superconductivity in a class of multiband models consisting of a metallic band interacting repulsively with an insulating band. The mapping of the models to the single-band attractive Hubbard model is done by comparing the pairing correlation functions in the non-perturbative regime, where the attraction becomes relatively large. The similarity to AHM is eventually degraded when the gap and/or the repulsion within the β band becomes too small, or when the β band is doped, but the effective attraction continues to increase.

Since the similarity to AHM is only a sufficient condition for superconductivity, the intriguing question we can next ask is whether another superconductivity takes over when the similarity is lost. If a multiband system cannot be mapped to a single-band system, which is generally the case, superconductivity has to compete with not only intraband but also interband orders and phase separations. Thus the problem becomes highly non-trivial.

We can explore this problem with the two-band g-ology (a bosonization and renormalization approach for interacting fermion gas in 1D), which gives results consistent with the QMC results, and also gives an evidence for the existence of a superconductivity that cannot be renormalised into a single-band model.[30] In this sense the QMC method and other methods can be just complementary,

We can apply our models to the cuprate superconductors. For them we can employ the *strongly correlated CT insulator* among the present class of multiband models, for which we assign the $Cu3d_{x^2-y^2}$-$O2p_\sigma$ band as the insulating β_2-β_1 band, while orbitals having a different symmetry from that such as $O2p_\pi$, $O2p_z$ (holes), or Cu4s (electrons) may be assigned as the α orbitals. Possibility of these orbitals being relevant has been suggested by some of the experimental results[31,32] and first-principles calculations.[33,34,35] In real systems, the direct on-site repulsion will supersede the effective attraction. This problem can be readily circumvented by considering off-site bare repulsion between β and α orbitals, which induces effective *off-site* attractions between carriers in the α orbitals, leading to an extended pairing superconductivity.[18]

In conclusion, we have shown that the QMC method is extremely powerful for exploring fascinations in the multiband fermion systems. The present example also indicates that a deeper insight into properties of strongly correlated fermion systems may be attained by combining the QMC method with other analytical or numerical methods.

Acknowledgments

The numerical calculations were done on HITAC S-820/80 in the University of Tokyo Computer Centre. This work was in part supported by a Grant-in-aid for Specially Promoted Research from the Ministry of Education, Science, and Culture, Japan.

References

1. J.G. Bednorz and K.A. Müller, Z. Phys. B **64**, 189(1986).
2. See for a general review, *Quantum Monte Carlo* ed. by M. Suzuki (Springer, Berlin, 1987).
3. A. Muramatsu and W. Hanke, Phys. Rev. B **38** 878 (1988).
4. M. Frick, I. Morgenstern, and W. von der Linden, Z. Phys. B **82**, 339 (1991).
5. R.T. Scalettar et al, Phys. Rev. Lett. **62**, 1407 (1989).
6. A. Moreo and D.J. Scalapino, Phys. Rev. Lett. **66**, 946 (1991).
7. M. Imada, J. Phys. Soc. Jpn. **57**, 3128 (1988).
8. M. Imada and Y. Hatsugai, J. Phys. Soc. Jpn. **58**, 3752 (1989).
9. W.von der Linden, I. Morenstern, and H. de Raedt, Phys. Rev. B **41**, 4669 (1990).
10. N. Furukawa and M. Imada, J. Phys. Soc. Jpn. **61**, 3331 (1992).
11. G. Dopf, A. Muramatsu, and W. Hanke, Phys. Rev. B **41**, 853 (1990); Phys. Rev. Lett. **68**, 353 (1992).

12. M. Frick, P.C. Pattnaik, I. Morgenstern, D.M. Newns, and W. von der Linden, Phys. Rev. B **42**, 2665 (1990).
13. Y. Asai, Physica C **185-189**, 1497 (1991).
14. R.T. Scalettar, D.J. Scalapino, R.L. Sugar, and S.R. White, Phys. Rev. B **44**, 770 (1991).
15. Y. Asai, in *Proc. Int. Workshop on Electronic Properties and Mechanisms in High-T_c Superconductors*, ed. by K. Kadowaki and T. Oguchi (Elsevier, 1991), p.1633.
16. H. Suhl, B.T. Mattis and L.R. Walker, Phys. Rev. Lett. **3**, 552 (1959).
17. J. Kondo, Prog. Theor. Phys. **29**, 1 (1963).
18. K. Kuroki and H. Aoki, Phys. Rev. Lett. **69**, 3820 (1992); Phys. Rev. B **48**, to be published.
19. N. N. Bogoliubov and V. E. Korepin, Int. J. Mod. Phys. B **3**, 427 (1989).
20. R. Micnas, J. Ranninger and S. Robaszkiewicz, Rev. Mod. Phys. **62** 113 (1990), and references therein.
21. R. Blankenbecler, D.J. Scalapino, and R.L. Sugar, Phys. Rev. D **24**, 2278 (1981).
22. J.E. Hirsch, Phys. Rev. B **31** 4403 (1985).
23. G. Sugiyama and S.E. Koonin, Ann. Phys. **168**, 1 (1986).
24. S. Sorella, E. Tosatti, S. Baroni, R. Car, and M. Parinello, Int. J. Mod. Phys. B **1**, 993 (1988).
25. S.R. White *et al*, Phys. Rev. B **40**, 506 (1991).
26. K. Kuroki and H. Aoki, Solid State Commun. **73**, 563 (1990); H. Aoki and K. Kuroki, Phys. Rev. B **42**, 2125 (1990).
27. K. Kuroki, H. Aoki, and Y. Takada, J. Phys. Soc. Japan **61**, 1161 (1992).
28. M. Suzuki, Prog. Theor. Phys. **56**, 1454 (1976).
29. J.E. Hirsch, Phys. Rev. **B 28**, 4059 (1983).
30. K. Kuroki and H. Aoki, to be published.
31. A. Fujimori *et al*, Phys. Rev. B **42**, 325 (1990).
32. Y. Yoshinari, PhD thesis, Univ. of Tokyo, 1991.
33. Y. Guo, J.M. Langlois, W.A. Goddard III, Science **239**, 896 (1988).
34. H. Kamimura and M. Eto, J. Phys. Soc. Jpn. **59**, 3053 (1990).
35. A. Svane, Phys. Rev. Lett. **68**, 1900 (1992).

QUANTUM MONTE CARLO IN THE INFINITE DIMENSIONAL LIMIT

M. Jarrell, H. Akhlaghpour and Th. Pruschke[†]
Department of Physics
University of Cincinnati
Cincinnati, Ohio 45221

[†] Institut für Theoretische Physik
Universität Regensburg
93040 Regensburg, Germany

e-mail: jarrell@wanderer.phy.uc.edu

Abstract

A new Quantum Monte Carlo algorithm is used to solve models of strongly correlated systems in the limits of infinite spatial dimensions and lattice size. A detailed description of this algorithm is given, concentrating on the methods used to reduce systematic errors and to measure the two-particle properties. The algorithm is then applied to the Hubbard model, which is shown to retain the qualitative features expected in the limit of finite dimensions. These include a Mott gap and antiferromagnetism for the half-filled model. Away from half filling the Mott gap disappears, replaced by a Kondo-like peak in the density of states associated with screening of the local spin. At low temperatures and low doping, a transition to an incommensurate magnetic state is also observed. In the paramagnetic state of the lightly doped model, we find anomalies in the resistivity and inverse NMR rate which are strikingly similar to those characteristic for the cuprate high temperature superconductors ($\rho \sim T$ and $1/T_1 \sim T$ at high temperatures).

1 Introduction

Many different materials are thought to be described by models of strongly correlated electrons. For example, The Hubbard model[1] is thought to at least qualitatively describe some of the properties of transition metal oxides, and possibly high temperature superconductors [2]; the Periodic Anderson model (PAM) seems to be the generic model for Heavy Fermion systems and the Anderson insulators; the Holstein model incorporates the essential physics of strongly interacting electrons and phonons. All of these model Hamiltonians contain at least two major ingredients: a local interaction term and a non-local hopping term. For example, for the Hubbard model reads

$$H = -\frac{t^*}{2\sqrt{d}} \sum_{\langle j,k\rangle\sigma} (c_{j\sigma}^\dagger c_{k\sigma} + c_{k\sigma}^\dagger c_{j\sigma}) + \sum_j \epsilon(n_{j\uparrow} + n_{j\downarrow}) + U(n_{j\uparrow} - 1/2)(n_{j\downarrow} - 1/2), \quad (1)$$

where $c_{j\sigma}^\dagger$ ($c_{j\sigma}$) creates (destroys) an electron at site j with spin σ, $n_{i\sigma} = c_{i\sigma}^\dagger c_{i\sigma}$, and $t^* = 1$ sets our unit of energy.

However, except for special limits, even such simplified models like (1) cannot be solved exactly. For example, for the Hubbard model, no exact solution exist except in one dimension, where the knowledge is in fact rather complete [3]. The Periodic Anderson model is only solvable in the limit where the orbital degeneracy diverges[4], and the Holstein model is only solvable in the Eliashberg-Migdal limit where vertex corrections may be neglected. Clearly a new approach to these models is needed if nontrivial exact solutions are desired.

Recently, Metzner and Vollhardt suggested such a new approach [5, 6, 7] based on an expansion in $1/d$ about the point $d = \infty$ to study these strongly correlated lattice models. The main simplification of this limit is that the single-particle Green's functions $G(R, i\omega_n)$ fall off very quickly in this limit, $G(R, i\omega_n) \propto d^{R/2}$, so that nonlocal diagrams in the irreducible self energy (and two-particle irreducible vertices) may be neglected. Thus, the self energy for the single-particle Green's function and the irreducible vertex functions have no momentum dependence and are functionals of the local Green's function [5, 8]. The many-body problem may then be solved by mapping it onto an auxiliary impurity problem [9] in a time-dependent field (that mimics the hopping of an electron onto a site at time τ and off the site at a time τ'). The effective action for the impurity problem is [10]

$$S_{eff} = \sum_\sigma \int_0^\beta d\tau \int_0^\beta d\tau' c_\sigma^\dagger(\tau) G_0^{-1}(\tau - \tau') c_\sigma(\tau') + S_{int} \quad (2)$$

where G_0^{-1} is the "bare" Green's function that contains *all of the dynamical information of the other sites of the lattice* and S_{int} describes the local interactions. The interacting Green's function is determined by[10, 11]

$$G^{-1}(i\omega_n) = G_0^{-1}(i\omega_n) - \Sigma(i\omega_n), \quad (3)$$

at each Matsubara frequency $\omega_n = (2n + 1)\pi T$. The impurity problem is mapped onto the infinite-dimensional lattice by self-consistently equating the impurity Green's function with the local Green's function of the lattice

$$G(i\omega_n) = \sum_{\mathbf{k}} G(\mathbf{k}, i\omega_n) = \frac{1}{\sqrt{\pi}} \int_{-\infty}^{\infty} dy \frac{\rho(y)}{i\omega_n - \epsilon - \Sigma_n - y}, \quad (4)$$

where $\rho(y) = \exp(-y^2)/\sqrt{\pi}$ is the non-interacting density of states (DOS). Thus, given some way of solving the problem described by Eq. 2, the set of Eq. 2-4 form a closed nontrivial solution of correlated lattice problems in the limit $d = \infty$.

Both the local and momentum-dependent two-particle Green's functions may also be calculated with this approach. Quite generally, the two-particle propagators with the same symmetry as the lattice, and which involve no net frequency transfer, $\chi_q(i\omega_n, i\omega_m)$ are related to the corresponding local propagator $\chi_{ii}(i\omega_n, i\omega_m)$ by the following matrix form

$$\chi_q^{-1} = \chi_{q0}^{-1} + \chi_{ii}^{-1} - \chi_{ii0}^{-1}. \tag{5}$$

Here , χ_{q0} is the momentum dependent two-particle propagator without vertex corrections (i.e. just the simplest bubble diagram involving the fully dressed single-particle propagator). It may be calculated with the methods developed by Müller-Hartmann[6, 12] at any value of $x(\mathbf{q}) = 1/d \sum_l cos(q_l)$ which defines a set of the equivalent points in the infinite-dimensional Brillouin zone. χ_{ii0} is the local undressed two-particle propagator, which may be constructed from the local single-particle propagator, and χ_{ii} is measured in the Monte Carlo process[11], as described below.

2 Monte Carlo Algorithm

The dynamics of the impurity problem are identical to that of an impurity embedded in a host metal described by G_0 [[8]–[10],[11]].Thus, given G_0, the impurity problem may be solved by using the quantum Monte Carlo (QMC) algorithm of Hirsch and Fye[13] (an alternative derivation of this algorithm is presented in the appendix). In the QMC the problem is cast into a discrete path formalism in imaginary time, τ_l, where $\tau_l = l\Delta\tau$, $\Delta\tau = \beta/L$, and L is the number of times slices. The values of L used ranged from 40 to 160, with the largest values of L reserved for the largest values of β since the time required by the algorithm increases like L^3. Experience shows that no severe sign problem occurs in the QMC process. Since the bare Green's function G_0^{-1} in Eq. 3 is not *a priori* known, the QMC algorithm must be iterated to determine a self-consistent solution for the Green's function of the infinite-dimensional lattice. The procedure [11] is to begin with a bare Green's function G_0^{-1}, use the QMC algorithm to determine the self energy Σ, calculate the lattice Green's function from Eq. 4, and determine a new bare Green's function from Eq. 3. This process is iterated until convergence is reached.

The details of the (Hirsch-Fye[13]) impurity algorithm are reproduced in the appendix. For the remainder of this section, we will discuss the modifications necessary to apply this algorithm to the infinite-dimensional limit. The main difficulty is that the Hirsch-Fye algorithm requires an imaginary-time path integral technique which only produces data for $G(\tau)$ at a *discrete* set of points in Euclidean time $0 < \tau < \beta$; whereas, Eqs. 3-4 require either the Matsubara frequency Green's function or the corresponding self energy. This involves a numerical approximation of the integral

$$G(i\omega_n) = \int_0^\beta d\tau e^{-i\omega_n\tau} G(\tau). \tag{6}$$

Fourier transforming discretely sampled data presents some well known difficulties[14]. The principle difficulty is that Nyquist's theorem tells us that above some frequency $\omega_n = 1/2\Delta\tau$, unpredictable results are produced by conventional quadrature techniques. Typically this problem is overcome by fitting the discrete data $G(\tau)$ with a smooth cubic spline, and then performing the integral on the splined data[14]. Since the integral on the splined data may be sampled on a much finer grid than the original data, this process is referred to as over sampling.

However, a problem still remains at high frequencies, since the resulting $G(i\omega_n)$ goes quickly to zero for frequencies above the Nyquist cutoff $1/2\Delta\tau$. This presents a difficulty since causality requires that

$$\lim_{\omega_n \to \infty} G(i\omega_n) \sim \frac{1}{i\omega_n}, \qquad (7)$$

In order to maintain causality[15] of the Matsubara frequency Green's functions, we condition the Fourier transform with a perturbation theory result. That is, we write

$$G(i\omega_n) = G_{pt}(i\omega_n) + \int_0^\beta d\tau e^{-i\omega_n \tau} \left(G(\tau) - G_{pt}(\tau) \right). \qquad (8)$$

where G_{pt} is a Green's function obtained from perturbation theory, and the integral here is performed by the oversampling method described above.

There are two obvious advantages to this approach. First, the integral goes to zero for frequencies greater than the Nyquist frequency $1/2\Delta\tau$, so that the resulting Green's function has the same asymptotic behavior as the perturbation theory result, and is thus causal. Second, often, the perturbation theory result is asymptotically exact (i.e. results from a high temperature expansion etc.), and this then presents a way of appending exact QMC results at low frequency with asymptotically exact perturbation theory results at high frequency. The flow chart for the resulting algorithm is shown in Fig. 1.

Once convergence of the algorithm in Fig. 1 is reached, the physical properties of the system are calculated with the Monte Carlo. A variety of two-particle properties may also be calculated in the QMC approach since the irreducible vertex function is also local. For most quantities, this is straight-forward; however, the two-particle Green's functions $\chi_{ii}(i\omega_n, i\omega_m)$ are difficult to measure efficiently. For example, consider the opposite-spin particle-hole propagator

$$\chi_{ii}(i\omega_n, i\omega_m) = \int_0^\beta d\tau_1 \cdots d\tau_4 e^{[i\omega_n(\tau_1-\tau_2)-i\omega_m(\tau_3-\tau_4)]} \langle T_\tau C_\uparrow(\tau_1) C_\downarrow(\tau_2) C_\downarrow^\dagger(\tau_4) C_\uparrow^\dagger(\tau_3) \rangle. \qquad (9)$$

For a particular configuration of the Hubbard-Stratonovich fields, the Fermions are noninteracting, thus the expectation value indicated by the angle brackets above may be evaluated in two steps. First, using Wick's theorem, its value is tabulated for each field configuration $\{s_l\}$. Second we Monte Carlo average over these configurations. After the first step, the equation becomes

$$\chi_{ii}(i\omega_n, i\omega_m) = \left\langle \int_0^\beta d\tau_1 \cdots d\tau_4 e^{[i\omega_n(\tau_1-\tau_2)-i\omega_m(\tau_3-\tau_4)]} g_\uparrow(\tau_4, \tau_1) g_\downarrow(\tau_3, \tau_2) \right\rangle_{m.c.} \qquad (10)$$

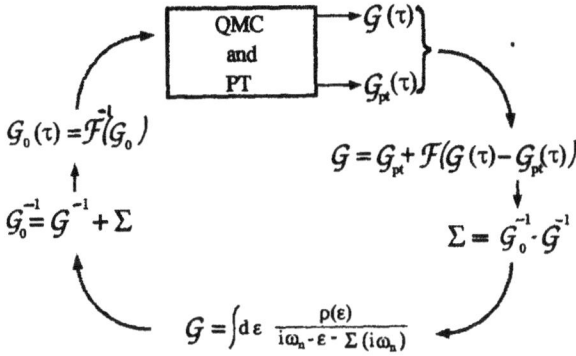

Figure 1: *Flowchart for the $d = \infty$ algorithm. The symbol $\mathcal{F}$ denote that a Fourier transform is to be performed by oversampling, and $\mathcal{F}^{-1}$ denotes its inverse.*

where the *m.c.* subscript means that the Monte Carlo average is still to be performed.

To measure this on the computer, the integrals must be approximated by sums. Since the Green's functions change discontinuously when the two time arguments intersect, the best integral approximation that can be used here is the trapezoidal approximation. Using this, we will run into Green's functions with both time arguments the same $g(j,j)$. This is stored as $g(j^+,j)$ (i.e. it is assumed that the first time argument is slightly greater than the second), but in the sums we clearly want the equal time Green's function to be the average $\{g(j^+,j) + g(j,j^+)\}/2 = g(j^+,j) - 1/2$. If we call **g**, with $1/2$ subtracted from its diagonal elements, $\overline{\mathbf{g}}$, then

$$\chi_{ii}(j,k) = \left\langle \left(\sum_{n,n'} \Delta\tau \, e^{+i\pi n'(2j+1)} \overline{g}_\uparrow(n',n) \, \Delta\tau \, e^{-i\pi n'(2k+1)} \right) \right.$$
$$\left. \left(\sum_{m,m'} \Delta\tau \, e^{-i\pi m'(2j+1)} \overline{g}_\downarrow(m',m) \, \Delta\tau \, e^{+i\pi m'(2k+1)} \right) \right\rangle_{m.c.} \tag{11}$$

This measurement may be performed efficiently if each term in parenthesis is tabulated first and stored as a matrix, and then the direct product of the two matrices taken as the estimate of χ_{ii}. When done this way, the time required for this measurement scales like $\sim L^3$ rather than $\sim L^4$ as would result from a straight-forward evaluation of the sums implicit in Eq. 10.

3   Application: The Hubbard Model

To illustrate the usefulness of this algorithm, we will apply it to the Hubbard model and show that the model in this limit retains most of the physics expected of the model in the finite-dimensional limit.

3.1   Magnetic Properties

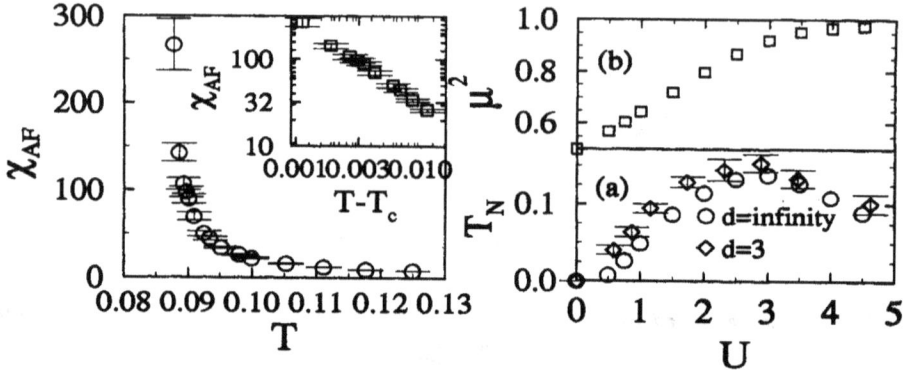

Figure 2: *LEFT: Antiferromagnetic susceptibility $\chi_{AF}(T)$, versus temperature T when $U = 1.5$ and $\epsilon = 0.0$. The logarithmic scaling behavior is shown in the inset. The data close to the transition fit the form $\chi_{AF} \propto |T - T_N|^{\nu}$ with $T_N = 0.0866 \pm 0.0003$ and $\nu = -0.99 \pm 0.05$ consistent with the mean-field behavior expected for $d = \infty$.*

Figure 3: *RIGHT: (a) Antiferromagnetic T_N and (b) $\mu^2 = \langle (n_\uparrow - n_\downarrow)^2 \rangle$ (calculated at $T = T_N$) as functions of U for the half-filled model ($\epsilon = 0$). In (a) the circles represent the values obtained from the present approach while the diamonds are data for a three-dimensional Hubbard model extracted from ref.[17].*

At half filling and large U, the system can gain hybridization energy if spins on adjacent sites are antiferromagnetically aligned. Thus, as the temperature falls, antiferromagnetism is expected in the half-filled model. This is shown in Fig. 2. Furthermore, since each lattice site has an infinite number of near neighbors, we expect this transition to be mean-field-like, $\chi_{AF} \sim 1/|T - T_c|$. This behavior is illustrated in the inset. The antiferromagnetic transition temperature T_N obtained within the current approach is plotted as a function of U in Fig. 3a. For small values of U, where the local spin moment is also small, we find that T_N is exponentially small, consistent with perturbation theory [16]. For very large values of U, where the spin moment and antiferromagnetic exchange have saturated to their maximum values, one expects that the transition temperature will fall monotonically with increasing U, $T_N \sim 1/U$ [16]. This is because the net antiferromagnetic exchange $dJ \sim t^{*2}/U$ also decreases with increasing U. Thus, one expects a peak in $T_N(U)$ for some intermediate value of U as seen in Fig. 3a. For comparison we included in Fig. 3a $T_N(U)$ for the $d = 3$ Hubbard model as calculated by Scalettar et al. [17]. The shape and order of magnitude compare very well, although the $d = 3$ data always have slightly larger values. At least part of this discrepancy is a consequence of the different analytic structures of the free DOS in $d = 3$ and $d = \infty$. In Fig. 3b the unscreened squared magnetic moment $\mu^2 = \langle (n_\uparrow - n_\downarrow)^2 \rangle$, calculated at the transition $T = T_N$, is plotted versus U when $\epsilon = 0$. For the half-filled model μ^2 ranges from $\mu^2 = 0.5$ in the uncorrelated limit ($U = 0$), to $\mu^2 = 1$ in the strongly correlated limit ($U \to \infty$). Note that the peak in $T_N(U)$ occurs near the point where μ^2 begins to saturate to one.

Away from half filling the transition temperature falls quickly with doping, $\delta = 1- < n >$. Once the doping exceeds a certain value, the transition is to an incommensurate state, as indicated by the divergence of the susceptibility $\chi(x(\mathbf{q}), T)$ at a value of $x > -1$. This behavior is illustrated in Fig. 4 when $U = 4$. The transition temperature may then be determined by extrapolating the inverse of the peak in the susceptibility, $1/\chi_{max}$, as shown in the inset. In Fig. 5, the corresponding transition temperatures are plotted as a function of doping. The open symbols indicate that the first divergence of the susceptibility was antiferromagnetic ($x = -1$); whereas, the filled symbols indicate that the susceptibility diverges into an incommensurate state (in the case $\delta = 0.2455$, the extrapolation lead to a negative transition temperature, indicating the lack of a transition at this doping).

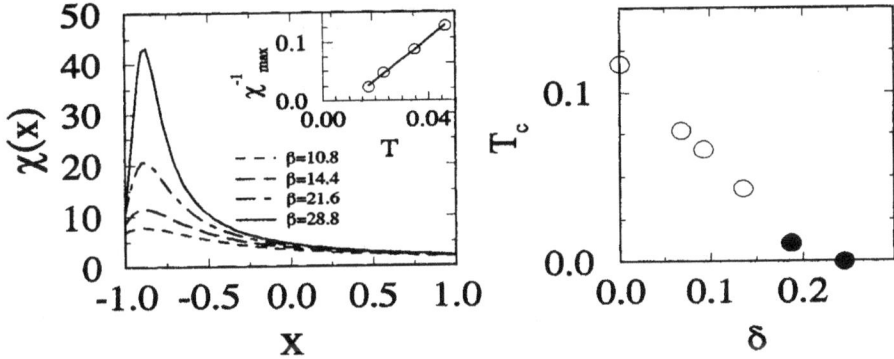

Figure 4: *LEFT: Susceptibility, $\chi(x(\mathbf{q}), T)$ versus $x(\mathbf{q}) = 1/d \sum_l \cos(q_l)$ when $U = 4$ and $\delta = 0.1788$. The scaling behavior of the peak susceptibility χ_{max} is shown in the inset. The data close to the transition fit the form $1/\chi_{max} = 1/\chi(x = -0.875, T) \propto T - T_c$ with $T_c = 0.011$, consistent with the mean-field behavior expected for $d = \infty$.*

Figure 5: *RIGHT: Transition temperature T_c versus doping when $U = 4$. The open symbols indicate that the first divergence of the susceptibility was antiferromagnetic ($x = -1$); whereas, the filled symbols indicate that the susceptibility diverges into an incommensurate state. (In the case $\delta = 0.2455$, the inverse peak susceptibility extrapolated to a negative transition temperature, indicating the lack of a transition at this doping.)*

3.2 Single-Particle Density of States

In this section we will discuss the single-particle properties of the $d = \infty$ Hubbard model. The most important quantity in this connection is the single-particle density of states (DOS) defined by $N(\omega) = -1/2\pi \sum_\sigma \mathrm{Im} G_{\sigma ii}(\omega + i0^+)$. Since the QMC method supplies us only with the imaginary time Green's function $G_{ii}(\tau)$, we used the maximum entropy procedure [18, 19, 20] for providing its analytic continuation. In most cases, the finite-U non-crossing approximation[21] was used to provide a default model for this procedure. However, for Figs. 6–7, a simple ($U = 0$) Gaussian density of states was used as a default.

Above the AF transition of the half-filled model, one expects to see a correlation induced Mott-Hubbard gap for sufficiently large values of U. This is shown in Fig. 7, where $N(\omega)$ is plotted for several values of U when $\beta = 7.2$, $\epsilon = 0$, and $\delta = 0$. As the

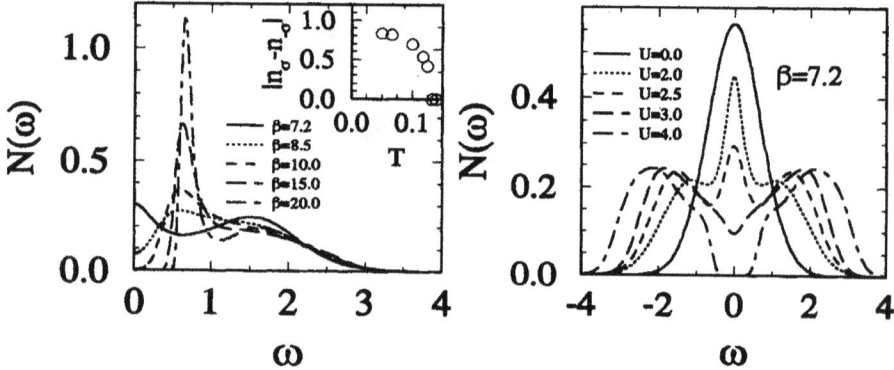

Figure 6: LEFT: *Evolution of the Neél gap in the DOS in the antiferromagnetic phase of the half-filled model when $U = 2.5$. In the inset, the ordered Neél moment is plotted versus temperature.*

Figure 7: RIGHT: *Evolution of $N(\omega)$ in the paramagnetic phase of the half-filled model for fixed inverse temperature $\beta = 7.2$.*

Coulomb repulsion U is increased from zero the broad central peak in the spectrum becomes narrower and is gradually suppressed while at the same time two side bands build up which are carrying the majority of the spectral weight. The central peak can be associated with a quasi-particle resonance, and the side peaks as usual with incoherent charge transfer on and off the site. As U continues to rise, the spectrum begins to develop a pseudogap at zero frequency when $U > 3.4 \approx U_C$. The "critical" value U_C increases with increasing temperature, and for $U > U_C$ the pseudogap grows linearly in U.

This algorithm[11] may be easily modified[22] to work below T_N in the antiferromagnetic phase of the half-filled model. As shown in Fig. 6, when $T < T_N$ a gap develops in the density of states due to the doubling of the unit cell. It is interesting to note that this gap develops continuously, so that when $T \lesssim T_N$, thermally induced states make the DOS finite at the Fermi surface. The width of this gap, measured from $\omega = 0$ to the first peak, increases monotonically with the ordered moment $|n_\downarrow - n_\uparrow|$ shown in the inset to Fig. 6.

Away from half filling, as shown in Fig. 8, the paramagnetic system becomes a heavy metal characterized by a narrow peak of width $\approx T_0$ in the single-particle density of states near the Fermi surface [23]. The development of this peak is associated with a screening of the local moments[23], as shown in the inset to Fig. 8. Hence, we associate the peak with the Kondo effect and its width T_0 with the Kondo scale. T_0 may then be determined by requiring $\chi_{ii}(T = 0) = 1/T_0$[24].

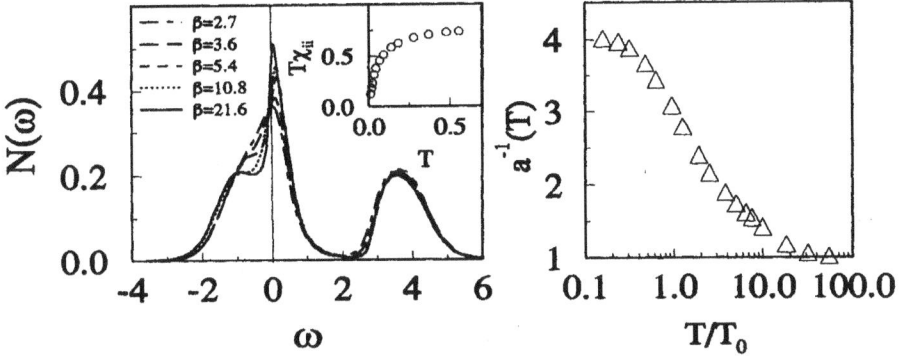

Figure 8: *LEFT: The evolution of the density of states when $U = 4$ and $\delta = 0.1878$. The development of a sharp peak at the Fermi surface is correlated with the reduction of the screened local moment $T\chi_{ii}(T)$, as shown in the inset.*

Figure 9: *RIGHT: The evolution of the finite temperature quasiparticle renormalization factor when $U = 4$ and $\delta = 0.1878$. At low temperatures, the mass enhancement $m^*/m = a^{-1}(T = 0) \approx 4$*

The development of the Kondo peak is also associated with an enhancement of the effective electron mass, as shown in Fig. 9. To examine whether the low-temperature phase of the $D = \infty$ Hubbard model is a Fermi liquid or not one may calculate the finite temperature quasi-particle renormalization factor as defined by Serene and Hess[25], $a^{-1}(T) = 1 - \text{Im}\Sigma(i\omega_o)/\omega_o$. In Fig. 9, the renormalization factor is plotted versus T/T_0 when $U = 4$ and $\delta = 0.1878$. Three different regions may be identified in this data: *The Fermi liquid regime, $T \ll T_0$.* As the temperature is reduced below T_0, $a^{-1}(T)$ saturates to a finite value, indicating the formation of a Fermi liquid. The zero temperature limit of $a^{-1}(T)$ may be obtained by extrapolation, and it gives the effective mass for the quasi-particles of this Fermi liquid $m^*/m = a^{-1}(0) \approx 4$. For fixed U, the effective mass appears to diverge continuously as half filling is approached[24]. *The high temperature regime, $T \gg T_0$.* The Kondo peak disappears [23] and $a^{-1}(T) \to 1$ as in the free metal. The physical properties no longer show scaling. *The crossover regime, $T \approx T_0$.* Here, $a^{-1}(T)$ increases in a roughly log-linear fashion. This behavior $a^{-1}(T) \sim \ln(T)$ has been identified [25] as a signal of marginal Fermi liquid behavior [26]. Let us stress, though, that this special behavior appears only as a crossover feature from a high- to a low-temperature regime and does not persist as an anomalous ground state!

3.3   Transport Anomalies

The Hubbard Hamiltonian [1] is usually considered the simplest model that contains at least some of the anomalous features of the high-T_c superconductors [27, 2]. These anomalies include a resistivity and the inverse NMR rate $1/T_1$ which increase linearly with temperature. We find that the infinite-dimensional Hubbard model exhibits strikingly similar anomalies.

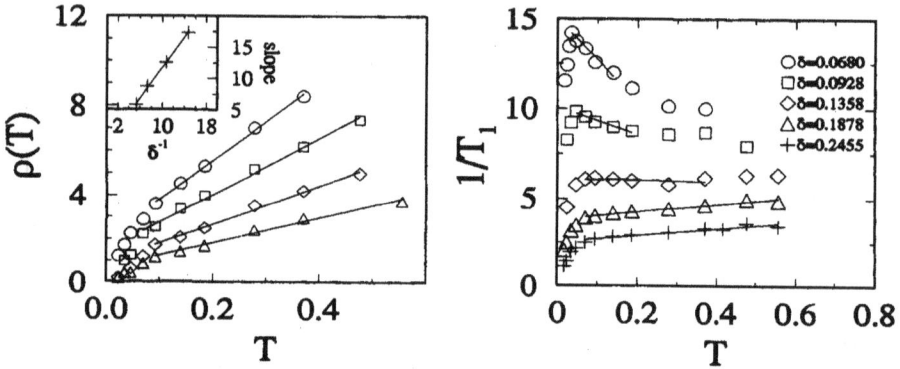

Figure 10: *LEFT: Resistivity versus temperature for several different dopings when U = 4. The lines are are linear fits to the anomalous resistivity. The slope determined by this fitting procedure increases linearly with $1/\delta$ as shown in the inset.*

Figure 11: *RIGHT: NMR relaxation $1/T_1$ vs. temperature for different dopings at $U = 4t^*$. The lines are linear fits to the anomalous data.*

The NMR relaxation rate $1/T_1$ is shown in Fig. 11 for several dopings vs. temperature. Remarkable is a linear region, which is most pronounced for larger dopings. The slope of this linear region shows in addition a sign change for dopings less than $\approx 15\%$. This quite anomalous behavior is also observed in the NMR data of the cuprate superconductors [28]

A similarly interesting behavior can also be found in the resistivity data collected in Fig. 10. For doping $\delta > 10\%$ one encounters a pronounced linear region, $\rho(T) \sim aT$. As shown in the inset to Fig. 10, the slope a of this linear resistivity depends on the doping δ like $a \sim \delta^{-1}$. Both results are again in agreement with experimental results for the cuprate superconductors [29]

It is interesting to speculate about the excitations responsible for these anomalies. Given that non-local dynamical correlations are suppressed in the infinite-dimensional limit, one may eliminate all excitations except local spin and charge excitations as candidates. Since the latter occur at relatively large energy scales ($\sim U$), charge fluctuations may also be eliminated, leaving only the local spin fluctuations, i. e. Kondo screening, as the scattering mechanism responsible for the transport anomalies. Transport anomalies due to Kondo screening are not unusual. They are responsible for the anomalous transport seen in dilute magnetic alloys, and heavy Fermion metals. However, since in the Hubbard model each lattice site is embedded in a band consisting only of similar correlated sites, the self consistency causes a strong normalization of the site's environment. This strong renormalization causes the transport in the Hubbard model to be different than that seen in the Periodic Anderson model.

4 Conclusion

Conventional methods of computational physics, such as QMC, are often restricted to relatively small systems, and display technical problems like the sign-problem. These difficulties make calculations and interpretations very difficult, especially in the interesting parameter regimes of low temperatures and strong electronic correlations. Our method of combining the simplifications arising in the limit of $d \rightarrow \infty$ with standard QMC techniques completely overcomes these difficulties. We are thus able to present a new approach that allows us to address consistently and (numerically) exactly physical properties of strongly correlated electron systems in a nontrivial limit over a wide range of parameters in the thermodynamic limit.

As an example, we applied this method to the single-band Hubbard model, Eq. 1. In contrast to various approximate methods, our solution shows all the expected features – an antiferromagnetically ordered ground-state close to half filling, and a metal-insulator transition at half filling – while retaining the essential dynamics induced by the local correlations. Especially this last feature led to new and unexpected results away from half filling, namely some sort of self-consistent "Kondo-effect" that yields quite anomalous transport properties.

We have also applied this method to the periodic Anderson[30] and Holstein[31] models, and demonstrated that they also retain the properties expected of their three-dimensional counterparts. We believe that this new approach represents a powerful method to solve strongly correlated lattice models and that most of our results will persist even for finite dimensionality.

We would like to acknowledge useful conversations with D.L. Cox, J. Freericks, B. Goodman, H. R. Krishnamurthy, J. Keller, M. Ma, R. Scalettar, and Matthew Steiner. This work was supported by the National Science Foundation grant number DMR-9107563, and by the Ohio Supercomputing Center. Two of the authors (MJ and TP) would also like to thank the hospitality of the Institute für Theoretische Physik of the University of Regensburg where part of this work was done.

1 Derivation of Quantum MC algorithm of Hirsch and Fye

The purpose of this section is to derive the Hirsch-Fye algorithm[13] using Grassmann algebra. We begin by splitting the Single Impurity Anderson Model (SIAM)

Hamiltonian, into bare and interacting parts, $H_{SIAM} = H_0 + H_I$, where

$$H_0 = \sum_{k,\sigma}(\epsilon_c + \epsilon_k)c_{k,\sigma}^\dagger c_{k,\sigma} + V\sum_{k,\sigma}(f_\sigma^\dagger c_{k,\sigma} + c_{k,\sigma}^\dagger f_\sigma) + \epsilon(n_\uparrow + n_\downarrow) \qquad (A1)$$

and

$$H_I = U(n_{f\uparrow} - 1/2)(n_{f\downarrow} - 1/2). \qquad (A2)$$

To obtain the Trotter-Suzuki decomposition for the partition function [32] we divide the interval $[0, \beta]$ into L sufficiently small subintervals such that $\Delta\tau^2[H_0, H_I]$ may be neglected. This leads to

$$Z = Tr e^{-\beta H} = Tr\prod_{l=1}^{L}e^{-\Delta\tau H} \approx Tr\prod_{l=1}^{L}e^{-\Delta\tau H_0}e^{-\Delta\tau H_I}. \qquad (A3)$$

The interacting part of the Hamiltonian may be further decoupled by mapping it to an auxiliary Ising field via a discrete Hirsch-Hubbard-Stratonovich[33] transformation,

$$e^{-\Delta\tau H_I} = e^{-\Delta\tau U(n_{f\uparrow}-1/2)(n_{f\downarrow}-1/2)} = 1/2e^{-\Delta\tau U/4}\sum_{s=\pm 1}e^{\alpha s(n_{f\uparrow}-n_{f\downarrow})} \qquad (A4)$$

where $cosh(\alpha) = e^{\Delta\tau U/2}$. Finally, one may cast Eq. (A3) into functional-integral form by using coherent states (Grassmann variables for Fermions). If we integrate out the host Fermionic degrees of freedom $\{c_{k,\sigma}\}$, then we end up with an action analogous to Eq. 2

$$S_{eff} = \Delta\tau^2\sum_{l,l',\sigma}f_{\sigma,l}^*G_0(l,l')f_{\sigma,l'} + S_{int}, \qquad (A5)$$

where

$$S_{int} = \sum_{l,\sigma}-f_{\sigma,l}^*(f_{\sigma,l} - f_{\sigma,l-1}) + \Delta\tau f_{\sigma,l}^*(\epsilon_f + U/2 + \frac{\alpha}{\Delta\tau}s_l\sigma)f_{\sigma,l} \qquad (A6)$$

and

$$G_0^{-1}(l,l') = \delta_{l,l'} - \delta_{l-1,l'}(1 - \Delta\tau\epsilon_k). \qquad (A7)$$

At this point the correspondence of the SIAM and the infinite-dimensional Hubbard model is clear. In both, G_0 contains the information about the host into which the impurity is embedded. The difference is that G_0 must be determined self-consistently for the lattice model.

We will now proceed to derive the Monte Carlo algorithm[13] sufficient for either the SIAM or the infinite-dimensional lattice problem. By integrating over $f_{\sigma,l}$ we can write down the partition function (neglecting a numerical prefactor), as

$$Z = \sum_{\{s_l\}}det(G_{\uparrow\{s_l\}}^{-1})det(G_{\downarrow\{s_l\}}^{-1}) \qquad (A8)$$

where

$$G_{\sigma,\{s_l\}}^{-1}(l,l') = \delta_{l,l'} - \delta_{l-1,l'}(1 - \Delta\tau\epsilon_f + \alpha s_l\sigma) - \Delta\tau^2 V^2 G_0(l,l') \qquad (A9)$$

and we sum over all configurations of Hubbard-Stratonovich field $\{s_l\}$. If we reexponentiate the above formula by defining $\mathcal{V}_{\sigma,\{s_l\}}(l) \equiv \Delta\tau(\epsilon_f + \alpha s_l\sigma/\Delta\tau)$, we can write it in a simple notation as

$$G_\sigma^{-1} = 1 + Te^{\mathcal{V}_\sigma} + \Delta\tau^2 V^2 G_0,$$

where T is $\delta_{l-1,l'}$ and $\mathcal{V}_\sigma \equiv \mathcal{V}_{\sigma,\{s_l\}}(l)$ for one special configuration. For another field configuration the only difference comes from $\mathcal{V}_\sigma$ such that $G_\sigma'^{-1} - G_\sigma^{-1} = T(e^{\mathcal{V}_\sigma'} - e^{\mathcal{V}_\sigma}) + O(\Delta\tau^{3/2})$ (note that α is of the order of $\Delta\tau^{1/2}$). On the other hand $T = (G_\sigma^{-1} - 1 - \Delta\tau^2 V^2 G_0)e^{-\mathcal{V}_\sigma}$ which results

$$G_\sigma'^{-1} - G_\sigma^{-1} = (G_\sigma^{-1} - 1)e^{-\mathcal{V}_\sigma}(e^{\mathcal{V}_\sigma'} - e^{\mathcal{V}_\sigma}) + O(\Delta\tau^{3/2}). \qquad (A10)$$

Multiplying from the left by G and from the right by G' and, ignoring terms $O(\Delta\tau^{3/2})$, we find

$$G'_\sigma = G_\sigma + (G_\sigma - 1)(e^{\mathcal{V}_\sigma'-\mathcal{V}_\sigma} - 1)G'_\sigma \text{ or } G_\sigma G_\sigma'^{-1} = 1 + (1 - G_\sigma)(e^{\mathcal{V}_\sigma'-\mathcal{V}_\sigma} - 1) \quad (A11)$$

The probability of having a configuration $\{s_l\}$ is $P_s \propto det(G_{\uparrow\{s_l\}}^{-1})det(G_{\downarrow\{s_l\}}^{-1})$; on the other hand the detailed balance requires $P_{s'}P_{s'->s} = P_s P_{s->s'}$ for all s'. We may satisfy this requirement by defining the probability of going from $\{s_l\}$ to $\{s_l'\}$ as $R/(1+R)$, where

$$R \equiv \frac{det(G'_\uparrow)det(G'_\downarrow)}{det(G_\uparrow)det(G_\downarrow)} \qquad (A12)$$

is the relative weight of two configurations. If the difference between two configuration is due to a flip of a single Hubbard Stratonovich field at the mth imaginary time slice then[13]

$$R = [1 + (1 - G_{m,m})(e^{-2\alpha\sigma s_m} - 1)]^{-1}. \qquad (A13)$$

Finally we can write down the evolution of the Green's function in the QMC time as[13]

$$G'_{i,j} = G_{i,j} + (G_{i,m} - \delta_{i,m}) \times (e^{-2\alpha\sigma s_m} - 1)\{1 + (1 - G_{m,m})(e^{-2\alpha\sigma s_m} - 1)\}^{-1}G_{m,j}. \quad (A14)$$

The QMC process precedes by sequentially proposing changes in each spin, accepting these changes with probability $P_{s->s'}$, and updating the Greens function with Eq. A14 when the change is accepted.

References

1. J. Hubbard, Proc. R. Soc. A**276**, 238(1963);
   M.C. Gutzwiller, Phys. Rev. Lett. **10**, 159(1963);
   J. Kanamori, Prog. Theor. Phys. **30**, 257(1963).

2. Proceedings of the international conference on "Itinerant-Electron Magnetism". Physica B+C**91** (1977); for a review about the theory of the Hubbard model see also D. Vollhardt, Rev. Mod. Phys. **56**, 99(1984).

3. E.H. Lieb and F.Y. Wu, Phys. Rev. Lett. **20**, 1445 (1968);
   H. Frahm nd V.E. Korepin, Phys. Rev. B**42**, 10533(1990);
   N. Kawakami and S.-K. Yang, Phys. Rev. Lett. **65**, 2309(1990).

4. G. Czycholl, J. Mag. Mag. Mat. **47,48**, 350 (1985).

5. W. Metzner and D. Vollhardt, Phys. Rev. Lett. **62**, 324 (1989).

6. E. Müller-Hartmann, Z. Phys. B **74**, 507 (1989).

7. P.G.J. van Dongen and D. Vollhardt, Phys. Rev. Lett. **65**, 1663 (1990).

8. H. Schweitzer and G. Czycholl, Z. Phys. B **77**, 327 (1990).

9. U. Brandt and Ch. Mielsch, Z. Phys. B **75**, 365(1989); **79**, 295(1990).

10. A. Georges and G. Kotliar, Phys. Rev. B**45**, 6479(1992).

11. M. Jarrell Phys. Rev. Lett. **69**, 168(1992).

12. E. Müller-Hartmann, Z. Phys. B **76**, 216(1989).

13. J.E. Hirsch and R.M. Fye, Phys. Rev. Lett. **56**, 2521 (1986).

14. W. H. Press, B. P. Flannery, S. A. Teukolsky, and W. T. Vetterling, *Numerical Recipes* (Cambridge University Press, Cambridge, 1986).

15. Obeying causality is crucial when Bayesian methods are used to analytically continue QMC data, since the form of the Bayesian prior is dictated by causality.

16. P.G.J. van Dongen, Phys. Rev. Lett. **67**, 757 (1991).

17. R.T. Scalettar, D.J. Scalapino, R.L. Sugar and D. Toussaint, Phys. Rev. B**39**, 4711(1989).

18. J.E. Gubernatis et al., Phys. Rev. B**44**, 6011 (1991).

19. R.K. Bryan, Eur. Biophys. J. **18**,165 (1990).

20. R.N. Silver, D.S. Sivia and J.E. Gubernatis, Phys. Rev. B**41**, 2380 (1989).

21. Th. Pruschke, D.L. Cox and M. Jarrell, Phys. Rev. B**47**, 3553 (1993).

22. A. George, G. Kotliar and Q. Si, Int. Jou. of Mod. Phys. **6**, 705(1992).

23. M. Jarrell and Th. Pruschke, Z. Phys. B **90**, 187(1993).

24. M. Jarrell and Th. Pruschke, submitted for publication.

25. J.W. Serene and D.W. Hess, Phys. Rev. B**44**, 3391(1991).

26. C.M. Varma, P.B. Littlewood, S. Schmitt-Rink, E. Abrahams and A.E. Ruckenstein, Phys. Rev. Lett. **63**, 1996(1989).

27. J.G. Bednorz and K.A. Müller, Z. Phys. B **64**, 189(1986).

28. T. Imai et al., Physica C **162-164**, 169 (1989);
    T. Imai, C. P. Slichter, A. P. Paulikas, and B. Veal, preprint.

29. H. Takagi et al. Phys. Rev. Lett. **69**, 2975(1992);
    Cristoph Quitmann, Thesis RWTH Aachen.

30. M. Jarrell, H. Akhlaghpour, and Th. Pruschke, Phys. Rev. Lett. **70**, 1670 (1993).

31. J. K. Freericks, M. Jarrell, and D. J. Scalapino. Phys. Rev. B, in print.

32. H. F. Trotter, Proc. Am. Math. Soc. **10** 545 (1959);
    M. Suzuki, Prog. Theor. Phys. **56**, 1454 (1976)

33. J.E. Hirsch,Phys. Rev. B**28**, 4059(1983), and **31**, 4403(1985).

ASPECTS OF THE SIGN PROBLEM

J H Samson

Department of Physics, Loughborough University of Technology,
Loughborough, Leics LE11 3TU, England
e-mail: j.h.samson@lut.ac.uk

Abstract

The interpretation of the sign problem is discussed with reference to auxiliary field quantum Monte Carlo calculations on the Hubbard and Heisenberg models. In many cases one must use a vector auxiliary field, in which case the sign problem can be interpreted geometrically in terms of a Berry phase. The classical effective Hamiltonian for the auxiliary field, obtained by integrating out finite-frequencies, is complex as a result of the negative-weight paths. The static approximation is the leading term in a $1/N$ expansion of this Hamiltonian.

1 Introduction

Monte Carlo (MC) simulation allows integration over a high-dimensional configuration space Γ. In classical statistical mechanics this is the classical phase space: a point $\mathbf{x} \in \Gamma$ might represent a configuration $\{\mathbf{S}_i\}$ of spins on a d-dimensional lattice. The task is to find the expectation value of a function A:

$$\langle A \rangle = \frac{\int_\Gamma A(\mathbf{x}) \, w(\mathbf{x}) d\mathbf{x}}{\int_\Gamma w(\mathbf{x}) d\mathbf{x}}, \tag{1}$$

where the weight $w(\mathbf{x})$ is typically the Boltzmann distribution. A uniform random sampling of points is extremely inefficient for large systems, as most of configuration space has exponentially small weight. The Metropolis algorithm therefore samples points $\mathbf{x}_k$ from the (positive-definite!) distribution $w(\mathbf{x})$, given a means of calculating the ratio $w(\mathbf{x})/w(\mathbf{y})$ for neighbouring configurations $\mathbf{x}$ and $\mathbf{y}$. Then the expectation value becomes

$$\langle A \rangle = \lim_{n \to \infty} \frac{1}{n} \sum_{k=1}^{n} A(\mathbf{x}_k). \tag{2}$$

In a quantum many-body system the eigenstates cannot in general be sampled directly, and less direct techniques must be used. A natural method of dealing with the statistical mechanics of quantum systems is *path integration* [1]. The configuration space Γ becomes the space of paths (with imaginary time suitably discretised) and the weight $w(\mathbf{x})$ the exponential of an action. Such integrals can be evaluated by MC integration *if the action is real* (and the weight therefore positive), as it is in the case of spinless bosons.

235

A serious difficulty plaguing such *quantum Monte Carlo* (qMC) calculations is that, in fermion and spin systems, the action is not real. The calculation may still proceed if points are sampled according to the modulus $|w(\mathbf{x})|$ of the weight, and the sign (or phase) $s(\mathbf{x})$ is absorbed into the integrand:

$$\langle A \rangle = \lim_{n \to \infty} \frac{\sum_{k=1}^{n} A(\mathbf{x}_k) s(\mathbf{x}_k)}{\sum_{k=1}^{n} s(\mathbf{x}_k)} \text{ , with } s(\mathbf{x}) = \frac{w(\mathbf{x})}{|w(\mathbf{x})|} . \tag{3}$$

This is feasible if the average sign is close to 1. However, in many cases the average sign tends to zero exponentially with decreasing temperature until sampling errors overwhelm the calculation. This chapter is devoted to interpreting this so-called *sign problem* or, more generally, *phase problem*, in the auxiliary-field qMC method. This approach uses the Hubbard-Stratonovich transformation (HST) [2] to eliminate electronic degrees of freedom, leaving an effective Hamiltonian for an auxiliary field related to an order parameter of the system. The system is then described by a functional integral over this imaginary-time-dependent field.

Section 2 discusses the application of the HST to the Hubbard and Heisenberg models. Although the HST has been widely used with a variety of uncontrolled approximations, notably in the field of itinerant magnetism, only in the early 1980's has it become possible to compute such integrals directly by qMC methods [3].

The sign problem is interpreted geometrically in section 3. In the functional integral, an auxiliary field transports the electron system along a path in state space. A wave function, transported around a path, develops a *Berry phase* in addition to the naïvely expected dynamical phase. This phase factor is responsible for the sign problem. At high temperatures there is insufficient imaginary time for electrons to follow the field adiabatically; they respond only to the time-averaged field, and the static approximation is recovered.

One aim of such work is to study the statistical mechanics and possible phase transitions of many-body systems. Section 4 considers the possibility of representing the collective degrees of freedom by integrating out the finite-frequency paths to leave a classical effective Hamiltonian for which the usual classical methods can be used. The paths of negative weight leave their mark here: this Hamiltonian is complex at low temperatures. In a $1/N$ expansion of the potential, the leading term is the static approximation, which can be unphysical at low temperatures.

The chapter concludes with a brief speculation on the significance of the sign problem.

2   The Hubbard-Stratonovich Transformation

2.1   General formulation

We wish to derive an expression for the grand partition function $Z = \mathrm{Tr} e^{-\beta(H-\mu n)}$ for a Hamiltonian for electrons with two-body interactions. The Hamiltonian, in an arbitrary single-particle basis $\{|\alpha\rangle\}$, labelled for example by site, orbital and spin indices, is

$$H = H_{band} + H_I \tag{4}$$

where
$$H_{band} = \sum_{\alpha\beta} t_{\alpha\beta} \, c_{\alpha}^{\dagger} c_{\beta} \qquad (5)$$

and
$$H_I = \frac{1}{2} \sum_{\alpha\beta\gamma\delta} v_{\alpha\beta,\gamma\delta} \, c_{\alpha}^{\dagger} c_{\beta} c_{\gamma}^{\dagger} c_{\delta}. \qquad (6)$$

This includes the Hubbard and Anderson models and, in a restricted sector of Hilbert space, the Heisenberg and Ising models. If the terms in the Hamiltonian commuted, it would be possible to diagonalise the Hamiltonian in terms of the eigenstates of the fermion bilinears $c_{\alpha}^{\dagger} c_{\beta}$. (This is indeed possible in the Ising model, where there is a site single-occupancy constraint, the bilinears in question are $n_{i\uparrow}$ and $n_{i\downarrow}$ and a MC simulation in state space is therefore possible.) In a practical calculation the interaction is local, so that the number of bilinears is proportional to the size of the system. One writes the Hamiltonian as a sum of a small number p (possibly 1) of terms H_r and uses the Trotter-Suzuki (TS) formula [4] to divide the imaginary-time interval $0 \leq \tau < \beta$ into a large number L of time-slices of width $\Delta\tau = \beta/L$:

$$\exp\left(-\beta H\right) = \exp\left(-\beta \sum_{r=1}^{p} H_r\right) = \lim_{L \to \infty} \left(\prod_{r=1}^{p} e^{-H_r \Delta\tau}\right)^{L}. \qquad (7)$$

The error for finite L, expressed as the norm of the difference of the operators, is $O(1/L)$. This transforms a d-dimensional lattice into a $d+1$-dimensional lattice with L layers in the time direction. If a Hamiltonian H contains two-particle interactions, matrix elements of the form $\langle m|e^{-\beta H}|n\rangle$ will be difficult to calculate while those of the form $\langle m|H|n\rangle$ can be calculated, where $\{|n\rangle\}$ is a many-body basis. If H only contains single-particle terms, both types of matrix element can be calculated. This suggests two ways to proceed. We can use the TS formula in such a way that the exponential can be linearised or diagonalised in each time-slice. One can then introduce a complete set of many-body states between each time slice (*world line qMC*) [5]; alternatively, one can introduce an overcomplete set parametrised by a continuous variable such as the direction of a spin (*coherent state integration*). The other method is to use the Hubbard-Stratonovich transformation (HST) to cast the Hamiltonian into single-particle form (*auxiliary field qMC*). In all cases the price is an extra (time) direction in the lattice.

This chapter is principally concerned with the auxiliary field method. In its original form, the method takes a TS decomposition with $H_1 = H_{band}$ and $H_2 = H_I$ and uses a Gaussian identity (the HST) within each time slice to replace the two-body interaction $v_{\alpha\beta,\gamma\delta} c_{\alpha}^{\dagger} c_{\beta} c_{\gamma}^{\dagger} c_{\delta}$ by a single-particle Hamiltonian in an imaginary-time-dependent auxiliary field with matrix elements $u_{\alpha\beta}(\tau)$. Following the standard derivation [6] we obtain the partition function as a functional integral

$$Z = \frac{1}{\mathcal{N}} \int \prod_{\alpha\beta} \mathcal{D}u_{\alpha\beta}(\tau) e^{-\beta V[u]} = \frac{1}{\mathcal{N}} \int \prod_{\alpha\beta} \mathcal{D}u_{\alpha\beta}(\tau) e^{-\beta V_0[u]} z[u], \qquad (8)$$

where
$$V_0[u] = -kT \int_0^{\beta} \sum_{\alpha\beta\gamma\delta} \frac{1}{2} u_{\alpha\beta}(\tau) \bar{v}_{\alpha\beta,\gamma\delta}^{-1} u_{\gamma\delta}(\tau) \, d\tau, \qquad (9)$$

with
$$z[u] = \mathrm{Tr}\, U[u;\beta] \qquad (10)$$

and
$$U[u;\tau_1] = \mathcal{T} \exp \int_0^{\tau_1} \sum_{\alpha\beta} \left(-t_{\alpha\beta} - u_{\alpha\beta}(\tau) + \mu\delta_{\alpha\beta} \right) c_{\alpha}^\dagger c_{\beta\tau} \, d\tau. \tag{11}$$

The "partition function" of electrons in the field, $z[u]$, is a fermion determinant [7,8]. To define the matrix v^{-1} we take $\alpha\beta$ and $\gamma\delta$ as the two indices of v and assume for simplicity of presentation that it is negative-definite. The imaginary-time-ordering symbol $\mathcal{T}$ orders factors with τ decreasing from left to right, and $\mathcal{N}$ is a (somewhat ill-defined) normalisation. The integrand $e^{-\beta V[u]}$, if it is positive-definite, can be interpreted as a probability density for the sample space $\{u_{\alpha\beta,l}\}$, where l labels the time slice, and the integral (8) can be evaluated by importance sampling. Note that, since $u(\tau)$ is an independent variable of integration at each time slice with no intrinsic dynamics, it is not a continuous function. This method has mostly been applied to lattice systems but has been applied to the long-ranged Coulomb interaction [9] (where u might be thought of as an electrostatic potential). The auxiliary field does not always have such a direct physical interpretation; however, its correlation functions are closely related to the correlation functions of the collective electronic degrees of freedom.

The sign problem now emerges to frustrate practical application of the above formalism. Unlike in a classical MC calculation, the integrand in Eq. 8 is not in general positive-definite (or even real). Indeed, the average value of the sign usually falls exponentially to zero at low temperatures [10], rendering any expectation values numerically unstable, although in some cases the sign may remain finite [11]. This is the well-known *sign problem* and its generalisation the *phase problem*. The interest is in the sign or phase of $z[u]$, since the factor $\exp(-\beta V_0)$ is positive. For certain Hamiltonians and fields (such as a one-band Hubbard model with scalar auxiliary fields discussed below) z is real, but it is complex if the field breaks time-reversal invariance. For each configuration u there is a conjugate configuration $\tilde{u}_{\alpha\beta}(\tau) = (u_{\beta\alpha}(\beta-\tau))^*$. The weight of $\tilde{u}$ is the complex conjugate of the weight of u as $z[\tilde{u}] = z[u]^*$. Since both a path and its conjugate appear in the integration with the same V_0, only the real part of $z[u]$ is needed; if u is a real symmetric matrix, the weight is real.

2.2   The Heisenberg Model

The auxiliary-field formulation has been applied to the Heisenberg model (but not for qMC purposes where much more efficient methods exist). We compare the world-line, coherent state and auxiliary field methods as applied to this model, as a simple example of the problems of the formalism in more general cases.

The spin-s Heisenberg Hamiltonian is
$$H = - \sum_{ij} J_{ij} \, \mathbf{S}_i \cdot \mathbf{S}_j. \tag{12}$$

We will allow for self-interaction of spins ($J_{ii} \neq 0$); although of no physical significance, it will be needed for convergence of the functional integral. As usual, qMC calculations start from the TS formula.

The *world-line* method uses a "checkerboard decomposition" to distribute the bonds in the Heisenberg Hamiltonian (12) between the terms H_r in the TS formula (7). By inserting a complete set of states between each factor, the method converts a d-dimensional Heisenberg model into a $d+1$-dimensional Ising model [12].

The *coherent state* representation [13] provides a direct path integral formulation of the Heisenberg model, the basis of a field theory of Heisenberg antiferromagnets [14]. For a spin s the *coherent states* $|n\rangle$ are labelled by a unit vector n on the Bloch sphere, representing the direction of the spin in the sense $n \cdot S|n\rangle = s|n\rangle$. This is illustrated in Figure 1 (in section 3.2). The explicit form for spin $\frac{1}{2}$ is (up to choice of gauge)

$$|n\rangle = \begin{pmatrix} \cos{(\theta/2)}\, e^{-i\phi/2} \\ \sin{(\theta/2)}\, e^{i\phi/2} \end{pmatrix}. \tag{13}$$

These form an overcomplete set of states for a system of spins S_i:

$$\prod_i \frac{1}{2\pi} \int_{-1}^{1} d\,\cos\theta_i \int_0^{2\pi} d\varphi_i \, \big|\{n_i\}\big\rangle\big\langle\{n_i\}\big| = 1. \tag{14}$$

This resolution of unity can now be inserted between factors at each time step in the TS formula to generate an integral over the paths $n(\tau)$:

$$Z = \int \prod_i \mathcal{D}n_i(\tau)\, e^{-S[\{n_i\}]}. \tag{15}$$

The imaginary part of the action S, the *Wess-Zumino term*, is the Berry phase of the spin system: as the direction of a spin s rotates about a solid angle Ω, it acquires a phase angle $s\Omega$. This term provides the distinction between ferromagnets and antiferromagnets, and between integer and half-integer spins, that is lost in the classical Heisenberg model [14]. This formulation has rarely been used as the basis of a qMC calculation [15]: the phase results in poor convergence for large systems.

We will use an auxiliary-field formulation of the Heisenberg model. This is even less often used in qMC calculations on spins, but sheds light on the application to other models. Applying the HST to the Heisenberg model gives [16]

$$Z = \frac{1}{\mathcal{N}} \int \prod_i \mathcal{D}^3 \Delta_i(\tau) \exp\left(-\int_0^\beta \sum_{ij} \frac{1}{4} \Delta_i(\tau) \cdot J_{ij}^{-1} \Delta_j(\tau)\, d\tau \right) z[\Delta], \tag{16}$$

with $\qquad z[\Delta] = \mathrm{Tr}\, U[\Delta;\beta]$ where $U[\Delta;\tau_1] = \prod_i \mathcal{T} \exp\left(\int_0^{\tau_1} \Delta_i(\tau) \cdot S_{i\tau}\, d\tau \right). \tag{17}$

This requires calculation of spin dynamics in an arbitrary time-dependent magnetic field, which is not integrable (even for spin $\frac{1}{2}$). As time-reversal invariance is broken, the weight is complex. This is discussed in section 3 as our main example of the phase problem.

2.3 The one-band Hubbard model

The Hubbard Hamiltonian, originally proposed to describe correlations between d electrons in transition metals [17,18,19],

$$H = \sum_{ij} \sum_{s=\uparrow}^{\downarrow} t_{ij}\, c_{is}^{\dagger} c_{js} + I \sum_i n_{i\uparrow} n_{i\downarrow} \tag{18}$$

describes a single band of electrons with an on-site repulsion. It interpolates between un-correlated band electrons (described by the hopping matrix elements t_{ij}) and atomic energy levels, as represented by the single parameter I.

This model has been used to describe the metal-insulator transition and itinerant mag-netism, and has recently been studied in the context of high-temperature superconductiv-ity. The phase diagram is largely unknown: despite its application to ferromagnetism in transition metals, the existence of a ferromagnetic state in a physically reasonable part of parameter space is still a matter for debate [20]. The auxiliary-field functional integral method has been used for the statistical mechanics of the model for more than twenty years [21]. The picture that developed was of a transition driven by the disordering of local magnetic moments, formed by the correlations of itinerant electrons. The long-standing controversy [22] has been whether or not the magnetic short-range order in the paramagnetic state is significantly higher than that in an insulator described by the Heisenberg model. The HST is well suited to this question, as it integrates out the elec-tronic degrees of freedom to leave an effective Hamiltonian for the magnetic degrees of freedom. Recent classical MC simulations on iron [23] using a many-spin effective Hamiltonian [24] based on electronic structure calculations show no evidence of signifi-cant short-range order. However, such calculations resort to the static approximation; there has not been a detailed study based on qMC calculations.

The interaction term in the one-band model allows a large number of decompositions, each of which can form the basis of the HST. Common choices are

$$ n_{i\uparrow}n_{i\downarrow} = \tfrac{1}{2}n_i - 2S_{iz}^2 = \tfrac{1}{4}n_i^2 - S_{iz}^2 \tag{19} $$

and

$$ n_{i\uparrow}n_{i\downarrow} = \tfrac{1}{2}n_i - \tfrac{2}{3}\mathbf{S}_i^2 = \tfrac{1}{4}n_i^2 - \tfrac{1}{3}\mathbf{S}_i^2, \tag{20} $$

where

$$ n_i = \sum_{s=\uparrow}^{\downarrow} c_{is}^{\dagger} c_{is} \quad \text{and} \quad \mathbf{S}_i = \tfrac{1}{2}\sum_{st} c_{is}^{\dagger}\,\boldsymbol{\sigma}_{st}\,c_{it}. \tag{21} $$

They rely on identities such as $(n_{i\uparrow})^2 = n_{i\uparrow}$ and $S_{iz}^2 = \tfrac{1}{3}\mathbf{S}_i^2$, which only hold for a single band. The decompositions in Eq. 19 lead to a *scalar* field coupling to the z component of the spin; those in Eq. 20 lead to a *vector* field coupling to the spin: the interaction term in the functional integral becomes respectively

$$ \sum_{\alpha\beta} u_{\alpha\beta}(\tau)c_{\alpha}^{\dagger}(\tau)c_{\beta}(\tau) = -\sum_i \Delta(\tau)\,S_{iz}(\tau) \quad \text{and} \quad \sum_{\alpha\beta} u_{\alpha\beta}(\tau)c_{\alpha}^{\dagger}(\tau)c_{\beta}(\tau) = -\sum_i \boldsymbol{\Delta}(\tau)\cdot\mathbf{S}_i(\tau). \tag{22} $$

The second of the two decompositions in Eq. 19–20 also require an imaginary scalar field coupling to electron number as $iw_i n_i$. Other combinations are possible, and the direction of quantisation may vary, giving a wide choice of decompositions. A further possibility is the discrete HST, which reduces the problem to an Ising model [25].

All of these transformations are exact and must therefore give the correct result if no approximations are made. For MC calculations the choice depends on convenience and rate of convergence. The discrete HST appears to be best on these grounds [26], and very few MC studies have considered vector fields [27]. If approximations are made, then the results do depend on the decomposition. Despite appearances, the scalar decompositions (19) are rotationally invariant *but only if no approximations are made*. The static approx-imation breaks the symmetry. Furthermore, the interaction term in the first of the decompositions in Eqs. 19-20 includes self-interactions of the form $n_{i\uparrow}n_{i\uparrow}$, which give rise to spurious diagrams in a perturbation expansion. While in an exact calculation these will cancel, truncation at finite order will always leave some residue of these terms [28].

The scalar decompositions are only applicable to the one-band Hubbard model: for the degenerate Hubbard model discussed below the interaction cannot be split in this way. The transition metals have five d-bands and so cannot be described by scalar fields. The three-band Hubbard model proposed and simulated for the cuprate superconductors [29, 30] falls under the category of one-band models discussed here, as inter-band interactions between d and p electrons are neglected. An Ising auxiliary field is therefore allowable. Should a more accurate treatment be required of the Hund's rule interactions in an atom, a vector-field method would be needed. One can envisage Hamiltonians for which more complicated fields are needed, for example coupling to a quadrupole moment.

There is a puzzle here that does not seem to have been much discussed. Let us perform a numerical thought experiment to measure finite-temperature critical exponents in the Hubbard model. (This is unlikely to be feasible in the near future.) The HST reduces the Hubbard model to classical statistical mechanics in $d+1$ dimensions. A magnetic transition at finite temperature will be in the classical d-dimensional Heisenberg universality class; there is a crossover from quantum to classical behaviour as the correlation length exceeds the width of the system in the time direction [31]. This poses the question of what would be observed with a scalar or Ising HST. If there is a finite-temperature magnetic transition in the 3D Hubbard model, the effective Hamiltonian of the Ising field must be singular in such a way as to fall into the Heisenberg universality class. A similar problem would arise if there is a superconducting phase. There is evidence for a finite-temperature antiferromagnetic transition in three-dimensional simulations [32], although only small clusters were considered and a mean field theory was used to find the Néel temperature. The question also arises in the world-line method for the Heisenberg model, which appears to give Ising-like exponents in simulations of the two-dimensional spin $\frac{1}{2}$ XY model (with a finite number of time slices) [33], in contradiction to high-temperature expansions. (There is no problem for zero-temperature transitions. Because of the difference in the q and ω dependences of the susceptibility, the effective dimensionality is $d+3$ and the critical behaviour is mean field like [31].) We can draw two conclusions: that it may be better to use auxiliary fields that reflect the order parameter of interest, and that the effective Hamiltonians can be pathological.

One pathology of the effective Hamiltonian is of course the sign problem. With a scalar or Ising auxiliary field the weight in the functional integral is negative for certain configurations. Only for some special cases, such as a half-filled band, is the weight positive-definite. With a vector field (or with an imaginary field iw_i coupling to charge) the breaking of time-reversal invariance implies a complex weight.

2.4 A Hubbard-Heisenberg Model

To discuss the qMC method in more general terms, it is helpful to consider a more general Hamiltonian, a Hubbard-Heisenberg model with N degenerate bands:

$$H = H_{band} - \frac{1}{N}\sum_{ij} J_{ij}\, \mathbf{S}_i\mathbf{S}_j + \frac{1}{4N}\sum_{ij} U_{ij}\, n_i n_j \qquad (23)$$

where
$$H_{band} = -\sum_{ij} \sum_{a=1}^{N} \sum_{s=\uparrow}^{\downarrow} t_{ij} c_{ias}^{\dagger} c_{jas} \qquad (24)$$

and S_i and n_i are total charge and spin operators on site i,

$$S_i = \sum_{a=1}^{N} S_{ia} \; ; \; n_i = \sum_{a=1}^{N} n_{ia} . \qquad (25)$$

We allow for non-zero on-site coupling constants J_{ii} and U_{ii}, which model Hund's rules. The Hamiltonian reduces to a number of more familiar models:

(a) If $J_{ij}=U_{ij}=0$ for $i{\neq}j$, it is an N-band Hubbard model; if $N=1$ and $U_{ii}=J_{ii}=I$, it is Eq. 18.

(b) If $t_{ij}=0$ and there is a constraint of half occupancy, it is a Heisenberg model:

(i)   If $N=1$, it is a spin $\frac{1}{2}$ Heisenberg model.

(ii)   If $J_{ij}=0$ for $i{\neq}j$, it is a van der Waals model with N equally coupled spins on a site.

(iii) If $|U_{ij}| \ll |J_{ij}|$ for $i{\neq}j$, the low-energy structure is a spin $N/2$ Heisenberg model.

Application of the HST to this Hamiltonian gives the functional integral

$$Z \propto Z_{band} \int \prod_i \mathcal{D}^3 \Delta_i(\tau) \mathcal{D}w_i(\tau) \exp\left(-N\beta\left(V_0[\Delta,w] - kT \ln z[\Delta,w]\right)\right) \qquad (26)$$

with
$$V_0[\Delta,w] = \int_0^\beta \left(\sum_{ij} \frac{1}{4} J_{ij}^{-1} \Delta_i(\tau){\cdot}\Delta_j(\tau) + \sum_{ij} \frac{1}{4} U_{ij}^{-1} w_i(\tau) w_j(\tau)\right) d\tau \qquad (27)$$

and
$$z[\Delta,w] = \left\langle \mathcal{T} \exp \int_0^\beta \sum_i \left(\Delta_i(\tau){\cdot}S_{i1}(\tau) + iw_i(\tau){\cdot}n_{i1}(\tau)/2\right) d\tau \right\rangle_{band} . \qquad (28)$$

The operators S_{i1} and n_{i1} refer to a single band. $A(\tau)$ and $\langle A \rangle_{band}$ denote explicit time-dependence and thermal averaging of the operator A with respect to H_{band}. Since the degeneracy N appears only as a factor in the exponent, we only need to compute the response of one band or one spin$\frac{1}{2}$ to the auxiliary field. The Hamiltonian is of a rather restricted form that admits an expansion of this type, in which case the coherent-state representation forms the classical limit [34], but it does contain much of the physics of an interacting many-band system.

3   Dynamics and the Sign Problem

The sign problem is often thought of as a consequence of Fermi statistics: the exchange of two electrons can clearly lead to a minus sign and, in a large system, it is difficult to control whether the number of such interchanges is even or odd .

We shall discuss the problem from a somewhat different viewpoint that does not rest on Fermi statistics — the dynamics of the state in a time-dependent auxiliary field. This has been discussed by a few authors, mostly in the context of the projector MC method [35]. Fahy and Hamann [36] interpret the stochastic auxiliary field in the functional integral representation of the Hubbard model as a diffusion equation on the manifold of Slater determinants. A state Ψ can diffuse to $-\Psi$ on the opposite side of the manifold under this dynamics — hence the sign problem — and the eigenstate of the diffusion operator with the slowest decay is nodeless on the manifold — hence the exponential decay of the average sign with β. Muramatsu *et al* [37] relate the sign problem to a topological invariant: the rotation of the one-electron basis needed to restore the final Slater determinant to the initial state is either proper or improper. The sign problem has also been shown to arise for spins [3] and in a Jahn-Teller system [38]. The general problem is the evolution of a quantum state in a time-dependent field: under what circumstances does the final state have negative overlap with the initial state? As the Heisenberg model provides a simple realisation of the sign problem as a geometrical (or Berry) phase, we concentrate on this case and return at the end to applications to fermion Hamiltonians.

3.1 Geometrical Phases

It is remarkable that a general phenomenon in elementary quantum mechanics escaped widespread recognition until ten years ago. Berry [39] considered the adiabatic evolution of a wave function in a time-dependent field. Suppose a Hamiltonian H depends on parameters $\mathbf{R}$ in some parameter space $\mathcal{F}$, and has eigenstates $\{|n; \mathbf{R}\rangle\}$ for each point $\mathbf{R}$:

$$H(\mathbf{R})\left|n;\mathbf{R}\right\rangle = E_n(\mathbf{R})\left|n;\mathbf{R}\right\rangle . \tag{29}$$

An example might be the Born-Oppenheimer approximation for a molecule, where $\mathbf{R}$ are nuclear coordinates. Prepare a state $|\Psi(0)\rangle = |n; \mathbf{R}(0)\rangle$ at time 0 and take $\mathbf{R}$ round a closed path C in $\mathcal{F}$ in time T. If the state remains non-degenerate, the wave function will evolve nearly adiabatically for a slow parameter variation and the final wave function will differ by a phase from the initial wave function:

$$\left|\Psi(T)\right\rangle = \exp\left(-\frac{i}{\hbar}\int_0^T E_n\left(\mathbf{R}(t)\right) dt\right) \exp\left(i\,\gamma_n(C)\right)\left|\Psi(0)\right\rangle . \tag{30}$$

The first exponent is simply the dynamical phase expected from the time-dependent frequency of the system. The second exponent is the *Berry phase*

$$\gamma_n(C) = \oint \mathbf{A}_n(\mathbf{R})\cdot d\mathbf{R}, \text{ where } \mathbf{A}_n(\mathbf{R}) = i\,\left\langle n; \mathbf{R}\left|\nabla_\mathbf{R}\right|n; \mathbf{R}\right\rangle , \tag{31}$$

a geometrical property of the path in parameter space. If the $\mathbf{R}$ variables are quantised, $\mathbf{A}_n(\mathbf{R})$ becomes a gauge potential; if the system separates into fast and slow variables (such as electronic and nuclear degrees of freedom), this potential appears in the Hamiltonian of the slow variables after elimination of the fast variables. Such a potential will also appear in the Hamiltonian of the auxiliary field after elimination of the electronic or spin degrees of freedom.

Aharonov and Anandan generalised Berry's argument to non-adiabatic evolution by viewing the dynamics in state space rather than parameter space. The wave function lives in *Hilbert space* $\mathcal{H}$. All wave functions in the ray $c|\Psi\rangle \in \mathcal{H}$, where $|\Psi\rangle$ is normalised and c is a non-zero complex number, correspond to the same state in *state space* $\mathcal{P}$. Since points in $\mathcal{P}$ can also be thought of as normalised projection operators $|\Psi\rangle\langle\Psi|$, $\mathcal{P}$ is also called projective Hilbert space. The geometrical phase depends only on the path in state space and the curvature of that space [40] and not on the Hamiltonian, or on the adiabaticity of the evolution, or on whether the path is parametrised by real or imaginary time.

3.2   Dynamics of Paths

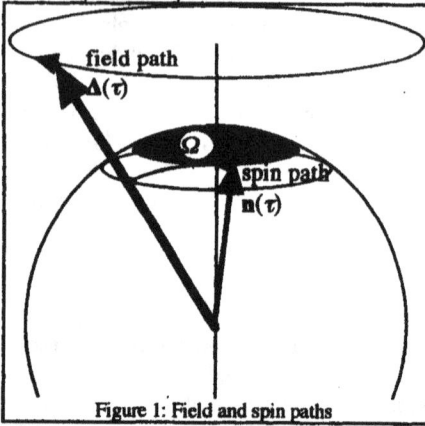

Figure 1: Field and spin paths

In the auxiliary-field formulation of the Heisenberg model, all that is needed is the solution of the equation of motion of a spin $\frac{1}{2}$ in a time-dependent magnetic field Δ. This has been analysed as follows by the present author [41]. Here the parameter space $\mathcal{F}$ is $\mathbb{R}^3$, Hilbert space is $\mathbb{C}^2$, the space of two-component spinors, and state space $\mathcal{P}$ is the sphere S_2. Points on the sphere correspond to the coherent states defined in section 2.2. The evolution operator $U[\Delta;\tau]$ of a *static* field can be resolved into projection operators $P(\mathbf{n})=|\mathbf{n}\rangle\langle\mathbf{n}|$:

$$e^{\Delta\cdot\sigma\tau/2} = e^{\Delta\tau/2}P(\hat{\Delta}) + e^{-\Delta\tau/2}P(-\hat{\Delta}). \qquad (32)$$

In real time the spin would precess around the field; in imaginary time the spin $\mathbf{n}$ relaxes along a great circle towards the field direction $\hat{\Delta}$: $\tan(\alpha/2) \propto e^{-\Delta\tau}$, with α the angle between $\hat{\Delta}$ and $\mathbf{n}$. For a time-dependent field this becomes a pursuit problem on the sphere, with $\mathbf{n}$ relaxing towards the instantaneous field direction. Figure 1 shows an example where the field path is conical: the spin, whose motion is always tangent to the great circle containing the spin and the field, follows a cone of smaller angle, enclosing the cap Ω. There is no solution in closed form for an arbitrary time dependence; however, for any *field path* $\Delta(\tau)$ the evolution operator has two eigenstates, describing closed *spin paths* $\pm\mathbf{n}(\tau)$ enclosing solid angle Ω:

$$|\psi(\beta)\rangle = U[\Delta;\beta]|\psi(0)\rangle = Re^{i\Phi}|\psi(0)\rangle. \qquad (33)$$

As $\det U=1$, we have $\chi[\Delta] = Re^{i\Phi}+R^{-1}e^{-i\Phi}$. Application of the argument of Aharonov and Anandan then shows that the phase is $\Phi=\Omega/2$. This is simply the consequence of rotating a spin $\frac{1}{2}$ particle: rotation by 2π about the equator gives a sign change. The amplitude is

$$R = \exp\left(\int_0^\beta \frac{1}{2}\Delta(\tau)\cdot\mathbf{n}(\tau)d\tau\right) \le \exp\left(\int_0^\beta \frac{1}{2}|\Delta(\tau)|d\tau\right) \qquad (34)$$

where the inequality is an equality if the spin follows the field adiabatically. The derivation is simpler than in real time, as the dynamical phase becomes an amplitude factor while the geometrical phase remains a phase. Although the field path is discontinuous and its enclosed solid angle cannot be defined, the spin path is continuous (but non-differentiable) and does enclose a well-defined solid angle. There is a close connection to the coherent-state functional integral of Eq. 15, but since the space of paths in $\mathbb{R}^3$ is much larger than that of paths in S_2 a family of field paths maps onto a single spin path.

The main purpose of this study is to gain insight into the sign problem. The weight is clearly complex in this case. Since the conjugate path also appears in the functional integral, the imaginary part cancels and the main concern is with the sign of the real part of the weight. For smooth field paths at low temperatures the spin follows the field nearly adiabatically, and the phase becomes a Berry phase (or the Wess-Zumino phase of the spin path). An expansion of the auxiliary-field effective action at low temperature does show a gauge potential responsible for the phase [42,43]. The spin will describe a random walk on the sphere whose typical enclosed area will increase with β in such a way that the average sign $\langle$ sgn cos $\Phi \rangle$ decreases. At high temperatures the spin can only respond to the average field and the static approximation is recovered (as expected).

It is possible to consider the functional integral for the Hubbard model in a similar way: in a static field $\{\Delta_i, w_i\}$, the electron state will relax towards the ground state of the electron system in that field. Let us call the manifold of such states the ground state manifold $\mathcal{D}$. We can also, in the spirit of density-functional theory, label these states by the spin and charge density. In a smoothly varying field at low temperatures a state in $\mathcal{D}$ will pursue the field and will develop a phase in an analogous fashion to the spin system. The important low-energy excitations are often due to spin degrees of freedom and this may be an adequate description, although it may be necessary to consider dynamics in the larger space of Slater determinants.

4 The Classical Effective Hamiltonian

We wish to understand the quantum statistical mechanics of an electron or spin system in d dimensions. The qMC method represents this as classical statistical mechanics on a $d+1$-dimensional slab, with a finite extent in the imaginary time direction for finite temperature. It would be of interest to describe this as a classical Hamiltonian on the original d-dimensional lattice, renormalised in some way by quantum fluctuations. For a particle in a potential, one can formally integrate the finite-frequency components out of the partition function to leave an effective action dependent only on the mean position. Feynman and Kleinert developed a variational approximation to this action [44], which was later applied to many-body systems [45]. This yields a classical effective Hamiltonian from which the statistical mechanics can be derived. We look at the possibility of such a classical Hamiltonian for the auxiliary fields in the Heisenberg and Hubbard model. The sign problem, absent in the bosonic path integrals in these references, leads to an unusual structure in the classical Hamiltonian. The non-positive-definite nature of the weight appears to preclude a variational approximation along the lines of Feynman and Kleinert, but the classical Hamiltonian can be expanded in powers of $1/N$. We first look at the simplest approximation to this Hamiltonian, the static approximation, and then at the nature of the exact effective Hamiltonian.

4.1 The Static Approximation

An approximation frequently made, particularly in the context of itinerant magnetism, is the *static approximation* (SA). Here the auxiliary field is constrained to be time-independent, and the functional integral becomes an ordinary integral over the $u_{\alpha\beta}$:

$$Z \approx Z_{SA} = \frac{1}{\mathcal{N}} \int \prod_{\alpha\beta} d\,u_{\alpha\beta} e^{-\beta V_{SA}(u)} = \frac{1}{\mathcal{N}} \int \prod_{\alpha\beta} d\,u_{\alpha\beta} e^{-\beta V_{0SA}(u)} z_{SA}(u) , \qquad (35)$$

with

$$V_{0SA}(u) = -\sum_{\alpha\beta\gamma\delta} \frac{1}{2} u_{\alpha\beta} v_{\alpha\beta,\gamma\delta}^{-1} u_{\gamma\delta} \qquad (36)$$

and

$$z_{SA}(u) = \text{Tr} \exp \sum_{\alpha\beta} \left(-t_{\alpha\beta} - u_{\alpha\beta} + \mu\delta_{\alpha\beta} \right) c^\dagger_\alpha c_\beta .$$ (37)

This yields a problem in classical statistical mechanics with an effective Hamiltonian that can be calculated by electronic structure techniques. The ground state is a single Slater determinant, a variational ground state of the original Hamiltonian (4) that (depending on the choice of auxiliary fields) may be the Hartree-Fock state. This is an upper bound for the true ground state energy. The SA is however correct in the high-temperature limit. One consequence is an underestimate of the heat capacity, which can even become negative, as will now be illustrated for a two-site spin $\frac{1}{2}$ Heisenberg Hamiltonian:

$$H = -J_0 S_1^2 - J_0 S_2^2 - 2J_1 S_1 \cdot S_2.$$ (38)

The partition function in the SA is

$$Z_{SA} = \frac{\int d^3\Delta_1 d^3\Delta_2 \exp\left(-\frac{1}{4}\beta J_{ij}^{-1}\Delta_i \cdot \Delta_j\right) 2\cosh\left(\frac{1}{2}\beta \Delta_1\right) 2\cosh\left(\frac{1}{2}\beta \Delta_2\right)}{\int d^3\Delta_1 d^3\Delta_2 \exp\left(-\frac{1}{4}\beta J_{ij}^{-1}\Delta_i \cdot \Delta_j\right)}.$$ (39)

This converges if $J_0 > |J_1|$, so that the matrix J_{ij} is positive-definite. Manipulation of Gaussian integrals gives

$$Z_{SA} = 2e^{\beta J_0/2}\left(e^{\beta J_1/2} + e^{-\beta J_1/2} + \frac{\beta}{2J_1}\left[(J_0 + J_1)^2 e^{\beta J_1/2} - (J_0 - J_1)^2 e^{-\beta J_1/2} \right] \right).$$ (40)

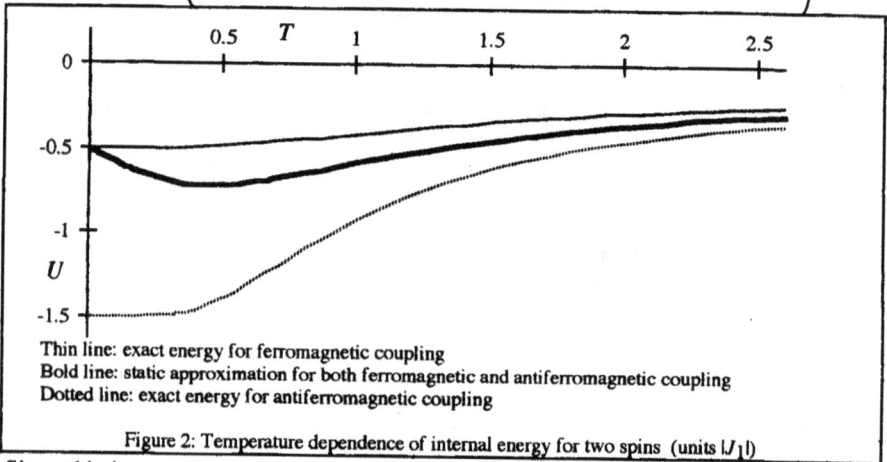

Thin line: exact energy for ferromagnetic coupling
Bold line: static approximation for both ferromagnetic and antiferromagnetic coupling
Dotted line: exact energy for antiferromagnetic coupling

Figure 2: Temperature dependence of internal energy for two spins (units $|J_1|$)

Since this is an entire function of both J_0 and J_1 for finite β, we henceforth set $J_0 = 0$. Figure 2 shows the temperature dependence of the internal energy U_{SA} (defined as $-\partial \ln Z_{SA}/\partial \beta$). This tends to the correct high-temperature limit of 0. For both ferromagnetic and antiferromagnetic coupling the SA gives the ground state energy as $-\frac{1}{2}|J_1|$. This is correct for ferromagnetic coupling: triplet states such as $|\uparrow\uparrow\rangle$ are in the ground state manifold $\mathcal{D}$. For antiferromagnetic coupling the SA does not find the correct singlet ground state energy $-\frac{3}{2}|J_1|$, since the singlet state is not in $\mathcal{D}$. Thus the SA misses the quantum distinction between ferromagnetism and antiferromagnetism. More alarmingly, the $T \to 0$

limit of the heat capacity (as calculated by differentiation of Z_{SA}, but not from a calcula-tion of $\langle H \rangle$) is $-k_B$. This negative heat capacity has also been noted in other cases [46].

4.2   Exact Classical Effective Hamiltonian

We wish to find a Hamiltonian $V_{eff}(a)$, a function of a the time-averaged auxiliary field, that will give the exact partition function and expectation values. This is of course harder than solving the full quantum Hamiltonian in the first place but may be instructive. For the general form of the partition function we can define V_{eff} by integrating the finite-frequency components out of the functional integral 8:

$$e^{-\beta V_{eff}(\bar{u})} = \frac{1}{\mathcal{N}} \int \prod_{\alpha\beta} \delta\left(\bar{u}_{\alpha\beta} - \int_0^\beta u_{\alpha\beta}(\tau)\,d\tau \right) \mathcal{D}u_{\alpha\beta}(\tau) e^{-\beta M[u]}. \tag{41}$$

This renormalisation group transformation collapses the time direction to a point, leaving a renormalised spatial Hamiltonian. By using the integral representation of the delta func-tion and partially undoing the HST, we can derive an expression for V_{eff} [47]. It can be written as a diagrammatic expansion in which fermion propagators represent the electrons in the static field $\bar{u}$ and interaction lines are only summed over non-zero Matsubara fre-quencies. In the N-band model of section 2.4 the effective Hamiltonian is expanded as

$$V_{eff}\left(\bar{\Delta}, \bar{w}; T, N\right) = V_{SA}\left(\bar{\Delta}, \bar{w}; T\right) + \frac{1}{N} V_{RPA}\left(\bar{\Delta}, \bar{w}; T\right) + O(N^{-2}), \tag{42}$$

with the SA as the leading term and the RPA´ [21] as the first correction [48]. (The RPA´ keeps zero-frequency modes to all orders but finite-frequency modes only quadratically.) Thus work on itinerant ferromagnetism can use a single-band Hubbard model, which has not been shown to have a ferromagnetic phase for physically plausible parameters. Such work uses the SA, which, as the $N=\infty$ limit, can therefore exhibit ferromagnetism.

We have seen that the integrand in Eq. 41 is not positive-definite: the paths can have complex weight. We can ask what effect this has on the V_{eff}. This can be evaluated in closed form for a few cases such as the van der Waals spin system (to be published). For a single spin s with Hamiltonian $-J\mathbf{S} \cdot \mathbf{S}$, the effective classical Hamiltonian is given by

$$e^{-\beta V_{eff}(\bar{\Delta})} = -e^{\beta Js(s+1)} \sum_{m=-s}^{s} \frac{2JkT}{\Delta} \frac{\partial}{\partial \Delta} e^{-\beta\left(\bar{\Delta}-2Jm\right)^2 / 4J}. \tag{43}$$

The paths of negative weight have an important consequence: this expression is not positive-definite at low temperatures, and V_{eff} is complex. We should not be too shocked that a classical Hamiltonian which we have contrived to give the correct quantum me-chanics must be singular in some way. Figure 3 shows the effective Hamiltonian for spin $\frac{1}{2}$. For $k_B T < J/2$ the weight $\exp(-\beta V_{eff})$ is negative for small values of Δ, as shown by the non-zero imaginary part and the singularities corresponding to quantum numbers of the spin.

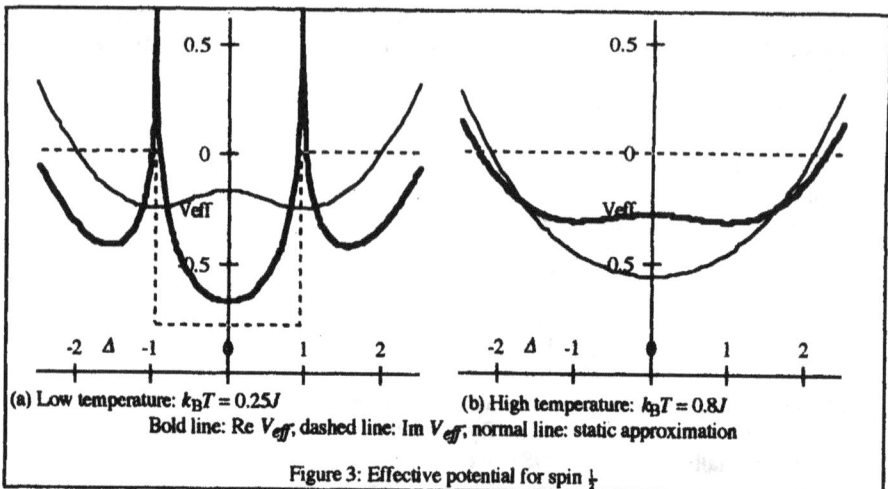

(a) Low temperature: $k_B T = 0.25J$ (b) High temperature: $k_B T = 0.8J$

Bold line: Re V_{eff}; dashed line: Im V_{eff}; normal line: static approximation

Figure 3: Effective potential for spin $\frac{1}{2}$

The reader may dismiss the analysis as mere recreational mathematics, since the auxiliary field and V_{eff} have little physical content. Nevertheless, auxiliary-field correlation functions are closely related to those of the spins (or, more generally, of the operators to which the fields couple). The absence of a positive distribution for the auxiliary field might suggest that spin correlations cannot be described by a positive distribution. This is the case for two antiferromagnetically coupled spins, where certain correlations in the singlet ground state are incompatible with the existence of a positive-definite distribution for the spin components [49]. At high temperature, where the density matrix has an admixture of triplet states, a positive distribution may exist. Similarly, the Wigner function for the joint distribution of the components of a single spin [50] is not positive-definite.

5 Discussion

This chapter has examined the auxiliary-field qMC method with special reference to the origin and meaning of the sign problem. It is easiest to interpret in systems with vector order parameters, where the components of the order parameter do not commute. In that case the sign problem appears as a geometrical phase as the time-dependent auxiliary field transports the electron state. This is similar to the phases responsible for topological terms in low-dimensional Heisenberg models, which suggests that the sign problem may contain important information. One can speculate that the MC algorithm is attempting to ask impermissible questions about the distribution of non-commuting variables. At high temperatures the electrons only respond to the average field and the static approximation is recovered. The sign problem occurs even with scalar fields coupling to, say, up and down spin densities. Although the commutator $[n_{is}(\tau), n_{jt}(\tau')]$ now vanishes at equal times, it is non-zero for $\tau \neq \tau'$ and the sign problem can still be considered rather less directly as the transport of a state round a manifold.

If we are not interested in the possible physical significance of the sign problem, we are still left with the problem of how to live with it. The work described here does not provide any answers on how to circumvent the problem and even suggests it may get worse before it gets better, as more complex systems are studied in the future. However, a better understanding of the oscillatory nature of the integrand in the functional integral may eventually suggest means of reordering the integral, grouping paths according to their phase or developing a combinatorial analysis of the paths.

References

1. R. P. Feynman and A. R. Hibbs, *Quantum Mechanics and Path Integrals* (McGraw Hill, New York, 1965)
2. J. Hubbard, *Phys Rev Lett* **3** 77-8 (1959)
3. W. von der Linden, *Phys Rep* **220** 53-162 (1992) gives a recent review
4. M. Suzuki, *Commun Math Phys* **51** 183-90 (1976)
5. J. E. Hirsch, R. Sugar, D. Scalapino and R. Blankenbecler, *Phys Rev* **B26** 5033-55 (1982)
6. J. W. Negele and H. Orland, *Quantum Many Particle Systems* (Addison Wesley, Redwood City, 1988)
7. D. J. Scalapino and R. L. Sugar, *Phys Rev Lett* **46** 612-4 (1981)
8. R. Blankenbecler, D. J. Scalapino and R. L. Sugar, *Phys Rev* **D24** 2278-86 (1981)
9. P. L. Silvestrelli, S. Baroni and R. Car, *Phys Rev Lett* **71** 1148-51 (1993)
10. E. Y. Loh Jr., J. E. Gubernatis, R. T. Scalettar, S. R. White, D. J. Scalapino and R. L. Sugar, *Phys Rev* **B41** 9301-7 (1990)
11. S. Sorella, in *The Hubbard Model – Recent Results* 73-112, ed. M. Rasetti (World Scientific, Singapore, 1991)
12. M. Suzuki, *J Statist Phys* **43** 883-909 (1986)
13. W-M. Zhang, D. H. Feng and R. Gilmore, *Rev Mod Phys* **62** 867-927 (1990)
14. E. Fradkin, *Field Theories of Condensed Matter Systems* (Addison Wesley, Redwood City, 1991); E. Fradkin and M. Stone, *Phys Rev Lett* **38** 7215-8 (1988)
15. H. Takano, in *Quantum Monte Carlo Methods in Equilibrium and Nonequilibrium Systems* 144-52, ed. M. Suzuki (Springer, Berlin, 1987)
16. S. Leibler and H. Orland, *Ann Phys (NY)* **132** 277-91 (1981)
17. M. C. Gutzwiller, *Phys Rev Lett* **10** 159-62 (1963)
18. J. Hubbard, *Proc Roy Soc London* **A276** 238-57 (1963)
19. T. Moriya, ed, *Electron Correlations and Magnetism in Narrow-Band Systems* (Springer, Berlin, 1981), references therein and citations thereof
20. M. W. Long, in *The Hubbard Model – Recent Results* 1-19, ed. M. Rasetti (World Scientific, Singapore, 1991)
21. W. E. Evenson, J. R. Schrieffer and S. Q. Wang, *J Appl Phys* **41** 1199-204 (1970)
22. V. Heine (chair) *et al*, Panel discussion, *Inst Phys Conf Ser* **55** 669-87 (1981)
23. K. S. Chana, J. H. Samson, M-U. Luchini and V. Heine, *J Phys: Condens Matter* **3** 6455-71 (1991)
24. M. U. Luchini and V. Heine, *J Phys: Condens Matter* **1** 8961-77 (1989)
25. J. E. Hirsch, *Phys Rev* **B28** 4059-61 (1983)
26. G. M. Buendia, *Phys Rev* **B33** 3519-21 (1986)
27. K. S. Chana, PhD Thesis, Loughborough University of Technology (1993)

28. C. A. Macêdo and M. D. Coutinho-Filho, *Phys Rev* **B43** 13515-24 (1991)
29. S. J. Sciutto, U. Marina Bettolo Marconi and R. Medina, *Physica* **A171** 139-58 (1991)
30. G. Dopf, A. Muramatsu and W. Hanke, *Europhys Lett* **17** 559-64 (1992)
31. J. A. Hertz, *Phys Rev* **B14** 1165-84 (1976)
32. J. E. Hirsch, *Phys Rev* **B35** 1851-9 (1987)
33. H. de Raedt, B. de Raedt and A. Lagendijk, *Z Phys* **B57** 209-20 (1984)
34. P. Coleman, *J Magn Magn Mater* **47-8** 323-8 (1985)
35. S. Sorella, S. Baroni, R. Car and M Parrinello, *Europhys Lett* **8** 663-8 (1989)
36. S. Fahy and D. R. Hamann, *Phys Rev* **B43** 765-79 (1991)
37. A. Muramatsu, G. Zumbach and X. Zotos, *Int J Mod Phys* **C3** 185-93 (1992)
38. P. de Vries, PhD thesis, Universiteit van Amsterdam (1991)
39. M. V. Berry, *Proc Roy Soc Lond.* **A392** 45-57 (1984)
40. Y. Aharonov and J. Anandan, *Phys Rev Lett* **58** 1593-6 (1987)
41. J. H. Samson, *Phys Rev* **B47** 3408-11 (1993)
42. A. Angelucci and G. Jug, *Int J Mod Phys* **B3** 1069-83 (1989)
43. A. Angelucci, *Phys Rev* **B44** 6849-57 (1991)
44. R. P. Feynman and H. Kleinert, *Phys Rev* **A34** 5080-4 (1986)
45. A. Cuccoli, V. Tognetti, P. Verrucchi and R. Vaia, *Phys Rev* **A45** 8418-29 (1992)
46. J. H. Samson, *J Physique* **45** 1675-80 (1984)
47. J. H. Samson, *Phys Rev* **B30** 1437-47 (1984)
48. J. H. Samson, *J Magn Magn Mater* **54-7** 983-4 (1986)
49. J. S. Bell, *Physics* **1** 195-200 (1964)
50. C. Chandler, L. Cohen, C. Lee, M. Scully and K. Wódkiewicz, *Found Phys* **22** 867-78 (1992)

QUANTUM SIMULATIONS OF THE DEGENERATE SINGLE-IMPURITY ANDERSON MODEL

J. Bonča and J. E. Gubernatis
Theoretical Division and Center for Nonlinear Studies,
Los Alamos National Laboratory, Los Alamos, New Mexico 87545

Abstract

We summarize results of quantum Monte Carlo simulations of the degenerate, single-impurity, Anderson model. Using Maximum Entropy methods, we performed the analytic continuation of the imaginary-time Green's functions produced by these simulations to obtain their real-frequency, single-particle, spectral densities for degeneracies of $N = 2$, 4, and 6. In several cases, we split the degeneracies to mimic crystal electric field and external magnetic field effects. All simulations were done on a cluster of workstations using a simple message-passing model. This model is also briefly described.

1 INTRODUCTION

Although the single-impurity (spin-degenerate) Anderson model [1] was first proposed 30 years ago as a model for the properties of dilute magnetic alloys, theoretical and numerical work on the model remains very active because it is one of the simplest paradigms for a system of strongly interacting electrons. Over the years, considerable progress has been made in understanding the properties of the model by several significant advances in analytic and numerical techniques. These techniques have sought to calculate various static and dynamic correlation functions to reveal the relevance of the model for such many-body phenomena as the Kondo effect, mixed valence fluctuations, and local magnetic moment formation that are observed in dilute magnetic alloys. However, in spite of numerous experimental and theoretical works in the field of dilute magnetic alloys, some disagreement still exists between theory and experiment, and even among different experimental groups [2, 3].

Incorporating orbital degeneracies into the original Anderson model brings the model closer to physical systems. For example, a degeneracy of $N = 6$ matches the degeneracy of a Ce impurity in a host with cubic symmetry and strong spin-orbit splitting. The generally accepted belief is that this single impurity model reproduces the main spectral features in Ce or Yb heavy fermion compounds [2], but recent studies by Joyce et al. [3] show features near $\omega = 0$ that do not appear to scale with the Kondo temperature T_K or display the appropriate temperature dependence. These findings remain a puzzle.

While the main features of the spectral density function of the Anderson model, such as the position of the broad side peaks and the existence of a sharp resonance close to the Fermi energy, are likely well reproduced by different analytical [5, 6] and numerical methods [7, 8, 9, 10, 11]. Relative spectral weights and their temperature dependence, however, often seem to be dependent on the underlying approximation. Therefore, there is a need for a method which calculates the spectral density function of the degenerate Anderson model at arbitrary interaction strength U, hybridization Γ, degeneracy N, and temperature T. In many respects, the quantum Monte Carlo method fulfills this need.

In this paper, we summarize our results [4] for spectral densities of the orbital-degenerate single-impurity Anderson model obtained by means of the quantum Monte Carlo method proposed by Hirsch and Fye [12], which we extended to higher degeneracies. This algorithm is one of the nicest quantum Monte Carlo algorithms: It can embed the impurity in an infinite medium, is stable at low temperatures without the need of special programming considerations [13], and even lacks the sign problem [14] that plagues most quantum Monte Carlo algorithms in the absence of particle-hole symmetry. As a direct result of the method, we obtain the impurity part of the many-body, imaginary-time Green's function and then perform the analytic continuation, using the maximum entropy method [15], to obtain the impurity contribution to the spectral function. Well-known drawbacks of the method are its scaling of the computation time by the cube of the inverse temperature and the square of the degeneracy. As we increased the degeneracy of the Anderson model, we discovered it was necessary to scale the computation time by the degeneracy in order to keep producing statistically independent measurements. Using parallel computing, we were able to reduce the consequences of these drawbacks.

The paper is organized as follows: In Section II, we briefly discuss the model and some of the relevant sum rules appropriate to it. In Section III, we summarize the quantum Monte Carlo, maximum entropy, and our parallelization methods. In Section IV, we present results. We conclude with summary comments in Section V.

2   Model and Sum Rules

We treated the following form of the degenerate Anderson model [6]

$$
\begin{aligned}
H &= H_0 + H_1 \\
H_0 &= \sum_{km} \epsilon_k n_{km} + \sum_{km} V_{km}(c_{km}^\dagger f_m + f_m^\dagger c_{km}) + \sum_m \epsilon_m n_m, \\
H_1 &= \frac{1}{2} \sum_{m,m'} U_{mm'} n_m n_{m'},
\end{aligned}
\tag{1}
$$

where $c_{km}^\dagger$ creates a state in the conduction band with the energy ϵ_k in the channel m, $f_m^\dagger$ creates an orbital state m at the impurity site with the unrenormalized energy ϵ_m, and n_{km} and n_m are the numbers operators for the conduction band and orbitals at the impurity site. V_{km} represents hybridization between the conduction band and the localized impurity states. We assume that the conduction band is infinitely wide and structureless; therefore, V_{km} is neither energy nor channel dependent. This assumption leads to the simple relation for the impurity level half-width $\Gamma = \pi N(0)V$, where $N(0)$ is the energy density of states per spin at the Fermi energy. The symmetric matrix $U_{mm'}$, with the additional condition $U_{mm} = 0$, represents the Coulomb repulsion between two electrons occupying different orbitals at the impurity site. Furthermore, we associate the channel index m with the magnetic quantum number $m \equiv m_j$ since we want to model systems with strong spin-orbit coupling, such as Ce impurities in a metal. In particular, the low-lying multiplet in Ce has a total angular momentum $j = 5/2$ and therefore a degeneracy $N = 2j + 1 = 6$, which represents the highest degeneracy reached in our calculations.

In the special case, when $\epsilon_m = \epsilon_f$ does not depend upon m and $U_{mm'} = U$ does not depend on m and m', the Hamiltonian (2) has particle-hole symmetry when $\epsilon_f = -(N-1)U/2$. In this case, the parameter space is limited to the values of U and Γ. In the asymmetric case, where we have an additional parameter ϵ_f, the particle-hole transformation preserves H if ϵ_f is replaced by $-[\epsilon_f + (N-1)U]$. Thus, it is sufficient to study a limited parameters space where $\epsilon_f > -(N-1)U/2$. In other cases, we split the ϵ_m into two groups, ϵ_1 and ϵ_2, to mimic the crystal-filed splitting of the orbitally degenerate impurity state. For $N = 6$, these groups had a degeneracy 2 and 4. The double spin degeneracy case [16] corresponds to $N = 4$ with two groups of 2. The flexibility in assigning different values to ϵ_m can also be used to simulate the effects of an external magnetic field.

We will be mainly concerned with the computation of the single-particle spectral density associated with the impurity state. Several important features of this function are known quite generally [17]. The imaginary-time Green's function $G(\tau)$, which we will obtain using quantum Monte Carlo simulation procedures, is directly connected to the spectral function $A(\omega)$ through the following relation

$$G(\tau) = \int_{-\infty}^{+\infty} d\omega \frac{e^{-\tau\omega}}{1 + e^{-\beta\omega}} A(\omega), \qquad (2)$$

where β is the inverse temperature. In the case of the particle-hole symmetry, the Green's function obeys the relation $G(\tau) = G(\beta - \tau)$ and therefore $A(\omega)$ is an even function of frequency. Furthermore, $A(\omega)$ obeys

$$\int_{-\infty}^{+\infty} d\omega A(\omega) = 1, \qquad (3)$$

$$A(\omega) \geq 0 \qquad (4)$$

These properties allow us to interpret $A(\omega)$ as a probability density.

3   Methods

3.1   Quantum Monte Carlo

The Monte Carlo method we used was originally developed to treat the single-impurity, spin-degenerate (N=2) Anderson model [12] and was later generalized to treat the doubly spin-degenerate Anderson model [16]. Here, we present the method for arbitrary degeneracy.

Dividing the imaginary-time scale into L discrete time intervals $\Delta\tau = \beta/L$ allows us to write a path-integral formulation for the partition function as

$$Z = \mathrm{Tr}\, e^{-\beta H} = \mathrm{Tr}\prod_{l=1}^{L} e^{-\Delta\tau H} \simeq \mathrm{Tr}\prod_{l=1}^{L} e^{-\Delta\tau H_0}e^{-\Delta\tau H_1}, \tag{5}$$

where we used a Suzuki-Trotter approximation to separate the exponents since H_0 and H_1 (2) do not commute. Next, we transform the electron-electron interaction part of the Hamiltonian H_1 into an non-interacting one by introducing discrete Hubbard-Stratonovich variables [12]. In the case of general degeneracy, this transformation at time-step l is

$$\exp\left(-\Delta\tau U_{mm'}n_m n_{m'}\right) = \frac{1}{2}\exp\left[-\Delta\tau U_{mm'}(n_m + n_{m'})/2\right]$$
$$\sum_{S_{mm'}^l = \pm 1} \exp\left[S_{mm'}^l J_{mm'}(n_m - n_{m'})\right], \tag{6}$$

where $\cosh(J_{mm'}) \equiv \exp(\Delta\tau U_{mm'}/2)$ and auxiliary fields $S_{mm'}^l$ form an antisymmetric matrix. The Hubbard-Stratonvich transformation enables us to take the trace over fermionic degrees of freedom exactly for a fixed configuration of the auxiliary fields $\{S\} \equiv \{S_{mm'}^1, S_{mm'}^2, ..., S_{mm'}^L\}$. The partition function (5) becomes

$$Z = \mathrm{Tr}_S \prod_{m=1}^{N} \det O_m(\{S\}), \tag{7}$$

where O_m is a matrix of dimension $(N_k+1)L \times (N_k+1)L$, with N_k being the number of k vectors of the conduction electrons. The matrix elements of O_m are

$$\begin{aligned}(O_m)_{l,l} &= 1,\\ (O_m)_{l,l-1} &= e^{-\Delta\tau K}e^{V_m^{l-1}}(1-2\delta_{l,1}),\\ (O_m)_{l,l'} &= 0 \quad \text{otherwise,}\end{aligned} \tag{8}$$

where K represents the noninteracting part of the Hamiltonian in Eq. (2). The potential V_m^l, which couples to the orbital degrees of freedom, is a diagonal matrix and depends on the auxiliary fields which act only at the impurity site,

$$V_m^l = \sum_{m'} S_{mm'}^l J_{mm'}|f\rangle\langle f|. \tag{9}$$

The desired one-particle Green's function is the inverse of the O matrix: $G_m = O_m^{-1}$ [18]. Using Dyson's equation we can connect different Green's functions corresponding to different potentials produced by different configurations of Hubbard-Stratonovich fields

$$G'_m = G_m + (G_m - I)(e^{V'_m - V_m} - I)G'_m. \tag{10}$$

We start the Monte Carlo calculation with the Green's function where all auxiliary fields are zero. We do this calculation exactly for an infinite, structureless conduction band. Then, for each time slice, we generate an arbitrary configuration of auxiliary fields $\{S\}$ and calculate the interacting Green's function G' using the Dyson equation (10). With this Green's function, we then begin the Monte Carlo steps in which we flip (change the sign) each component of the auxiliary fields $S^l_{mm'}$ while conserving the asymmetry of $S^l_{mm'}$. In this process, we use the Metropolis algorithm to determine whether we accept the flip. If we accept, we update the Green's function for the new configuration of auxiliary fields using (10) and need only to consider its components among impurity orbitals since the interaction couples only these states.

The method produces the Green's function as a function of τ as its natural product. For a given configuration of Hubbard-Stratonovich fields, this is the exact Green's function, within the systematic error caused by the Trotter approximation, because the Hubbard-Stratonovich transformation converts the interacting problem into an non-interacting problem which is, of course, solvable. This conversion also means that various thermodynamic averages are directly computable from the Green's function by use of Wick's theorem. Averaging these quantities over many Hubbard-Stratonovich configurations restores the interactions.

3.2 Maximum entropy

Using Eq. (2), we seek to determine the spectral function at a large number of discrete frequencies $A(\omega_i)$, called the image, from the Green's function $G(\tau)$ calculated at smaller number of discrete imaginary-time values, called the data. There are several difficulties associated with this problem: the Green's function is almost insensitive to changes in $A(\omega_i)$ at large frequencies due to the exponentially small kernel. This insensitivity makes the problem extremely ill-posed. With $G(\tau)$ being determined by a Monte Carlo procedure, variations in the data (noise) are a fact. Furthermore, the number of data is smaller than the desired number of image points $A(\omega_i)$; thus we cannot solve the problem exactly. There is also a practical limit to the number of $G(\tau)$ values that can be produced. Increasing their number by decreasing $\Delta\tau$ for a fixed β leads to a point where the difference between successive values of $G(\tau)$ becomes smaller than the accuracy of the calculation, and hence the production of new information ceases.

Bayesian statistical inference, with the principle of maximum entropy, yields procedures that enable the analytic continuation and represent a major improvement over a constrained least-squares method [20]. The approach is based on probability theory and relies on the specification of probability and conditional probability functions connected by Bayes's Theorem. The principle of maximum entropy uses the *a priori* knowledge that the spectral density is non-negative and is normalizable (satisfies a sum rule), and it enters the process by specifying the prior probability of the image, namely

$$\Pr[A] \propto e^{-\alpha S} \tag{11}$$

where

$$S = \sum_i [A_i - m_i - A_i \ln(A_i/m_i)] . \tag{12}$$

The function m_i is called the model and it sets the zero of the entropy S. This approach is quite different than the constrained least-squares method which would use the non-negativity and sum rule as constraints on the solution and then one would be forced to determine the associated Lagrange multipliers in an *ad hoc* manner.

Our results are calculated from [19, 20]

$$\langle A \rangle = \int d\alpha \, \Pr[\alpha|G, m] A(\alpha), \tag{13}$$

where the conditional probability function $\Pr[\alpha|G, m]$ is found by using Bayes's Theorem. Details are given elsewhere [20], but the main ingredient, besides (11), is the choice of the likelihood function

$$\Pr[G|A] \propto e^{-\chi^2/2}, \tag{14}$$

where χ^2 is the least-squares function

$$\chi^2 = \frac{1}{2} \sum_{ik} (G_i - \sum_j K_{ij} A_j) \, C_{ik}^{-1} (G_k - \sum_j K_{kj} A_j), \tag{15}$$

with K_{ij} being the kernel from the Eq. (2) and C_{ij} the covariance matrix [20] for the different τ components of $G(\eta)$.

The choice (14) of the likelihood function assumes that the data are statistically independent and Gaussian-distributed. These assumptions, which are implicit in a least-squares procedure, are not naturally satisfied. Promoting the consistency of the data with them was achieved by using large bins to reduce the correlations between binned measurements and a large number of bins to generate the Gaussian behavior. The large number of binned measurements also reduces the statistical error associated with the measurements. Because the covariance matrix is not diagonal and its eigenvalues span 4 to 6 orders of magnitude for this problem, the standard root-mean-square error estimate is meaningless. The number of bins calculated was found empirically to be the number needed so the results did not change if this number was increased.

3.3 Parallelization

The details of the Hirsch-Fye algorithm afford several different strategies for parallelization, all of which, however, require distributing parts of the calculations over all the processors and passing data (messages) on a regular basis from processor to processor. For some strategies, message passing is frequent and only a small amount of floating point operations are done after the messages are passed. To avoid communication overhead, we decided to exploit the inherently parallel nature of almost all Monte Carlo calculations. Since our code can run on a single processor, we gave each of n processors a copy and a separate random number seed, had it run the code, and then collected data from all processors. To see some of the issues involved in this strategy, we will first consider the general aspects of doing a Monte Carlo simulation [21].

In Monte Carlo, we want thermodynamic average such as

$$\langle A \rangle = \frac{\int dx\, A(x)\rho(x)}{Z} \qquad (16)$$

where

$$Z = \int dx\, \rho(x) \qquad (17)$$

is the partition function (a normalization constant). The basic property of a Monte Carlo method is to replace the thermal average by a sample average

$$\langle A \rangle \approx \bar{A} \qquad (18)$$

where

$$\bar{A} = \frac{1}{M} \sum_{i=1}^{M} A_i \qquad (19)$$

and M is the number of measurements A_i of A. If M is large enough and the A_i are *statistically independent* estimates of A, then error estimates are taken to be $\pm m\sigma/\sqrt{M}$ where m is a small integer, usually equal to 1 and

$$\sigma^2 = \frac{1}{M-1} \sum_{i=1}^{M} (A_i - \bar{A})^2 \qquad (20)$$

With our parallelization strategy, the basic Monte Carlo relation (19) becomes

$$\bar{A} = \frac{1}{M} [\underbrace{\sum_{i=1}^{M/n} A_i}_{\text{proc 1}} + \underbrace{\sum_{i=M/n+1}^{2M/n} A_i}_{\text{proc 2}} + \cdots + \underbrace{\sum_{i=(n-1)M/n+1}^{M} A_i}_{\text{proc n}}] \qquad (21)$$

however, proper error estimation precludes collecting from each processor only the sum of its measurements. To be able to test for correlations between successive measurements, one needs to collect the measurements from a given processor sequentially and combine these measurements with those from the other processors into a single file (or buffer). Then, after all node processes are finished, proper error estimates are attempted. If successive measurements from a given processor are correlated and the measurements from the processors are interspersed, then in the combined set of measurements the distance between decorrelated measurements is increased. By grouping measurements by processors, one can test if successive measurements are correlated, and if they are, then one can easily "rebin" successive pairs, triplets, etc. to produce statistical independence.

All the calculations reported here were done on the LANL/IBM workstation cluster which has 16 IBM RISC/6000-560 workstations, each with 128 MB memory, connected ethernet and FDDI networks. For our message passing software, we used PVM [22] because it has the functionality we needed and it is free and easy to use.

Because of cluster usage, the time each processor takes to do M/n measurements can vary widely. To *load balance*, we start the simulation on each node in an infinite loop and whenever a node completes a measurement, the measurement is sent to the host process which writes the result sequentially as a record in a direct-access file and records in a index array which processor wrote to that record. When the host records the desired number of measurements, it terminates all node processes and then uses the index to rewrite the data into another file in proper processor sequence. The number of measurements made varies from node to node but each node does the best it can under the constraints of the time sharing system.

With this overall strategy virtually all inter-processor communication is eliminated. Our code runs 3/2 to 3 times faster on the cluster than on one processor of a Cray-YMP computer, depending on whether the inner loop vector lengths are short (high temperatures) or long (low temperatures). For long loop lengths, the code runs at 250 Mflops sustained on one processor of Cray Y-MP computer. On the cluster, the code has run as long as three days.

4   Results

In this section, we present results for the spectral functions obtained by quantum Monte Carlo simulations of the degenerate Anderson model. We used an energy scale in which the full width of the resonance equals unity, that is, $2\Gamma = 1$. Relying on previous works [20], we assumed that $\Delta\tau = 0.125$ yields a systematic error which is smaller than the statistical error for $U \lesssim 4$ and $\Gamma \lesssim 0.5$. We reached temperatures as low as $T = 1/16$, in which case, for the assumed $\Delta\tau$, the number of time slices was $L = \beta/\Delta\tau = 128$.

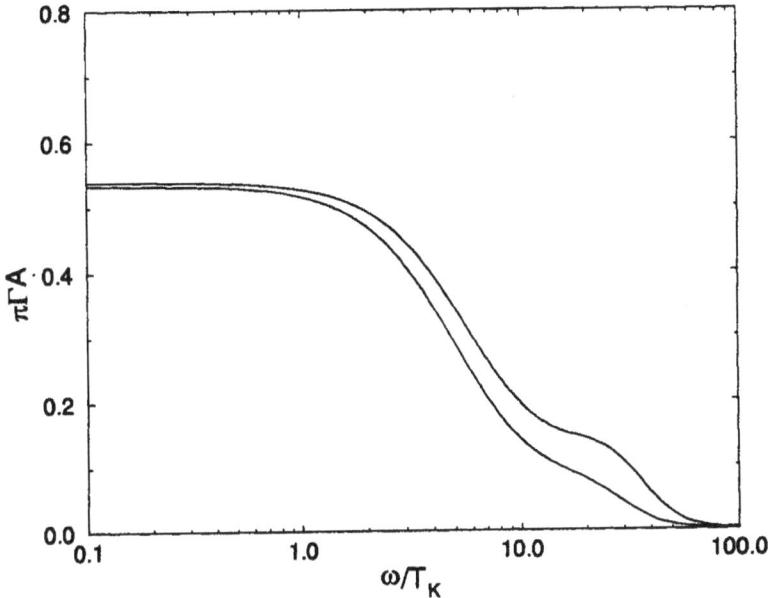

Figure 1: The spectral densities times $\pi\Gamma A$ as functions of ω/T_K in the symmetric case calculated at $T/T_K = 2.97$ and $U = 4$ for $N = 4$ and 6.

4.1 Symmetric case

One striking feature of the Anderson impurity model is the universal behavior of various physical properties if the temperature is scaled by the Kondo temperature. For the case of $N = 2$, it has also been shown [20] that the spectral functions as functions of ω/T_K, when calculated at fixed T/T_K, follow a universal curve in the low frequency regime. In Fig. 1, we present spectral functions of the systems $N = 4$, $U = 4$ with $T_K = 0.108$, and for $N = 6$, $U = 4$ with $T_K = 0.164$ calculated at fixed ratio $T/T_K = 2.98$. As seen from Fig. 1, the behavior in the low frequency region $\omega/T_K \lesssim 10$ is universal within the statistical error.

4.2 Asymmetric case

Since the asymmetric case has an additional parameter ϵ_f, completely covering parameter space becomes difficult. To simplify things, we limited our calculations to $N = 4$, $U = 4$ and $\epsilon_f > -(N-1)U/2$. In Figs. 2(a-d), we plot the spectral functions for $N = 4$, $U = 4$ and $\epsilon_f = -4.5$, -3, -2, and -1. For the given N and U, the value of ϵ_f in the symmetric case would be $\epsilon_f = -6$. As ϵ_f increases, the Kondo peak moves from its position at $\omega = 0$ in the symmetric case to $\omega > 0$. A rough estimate

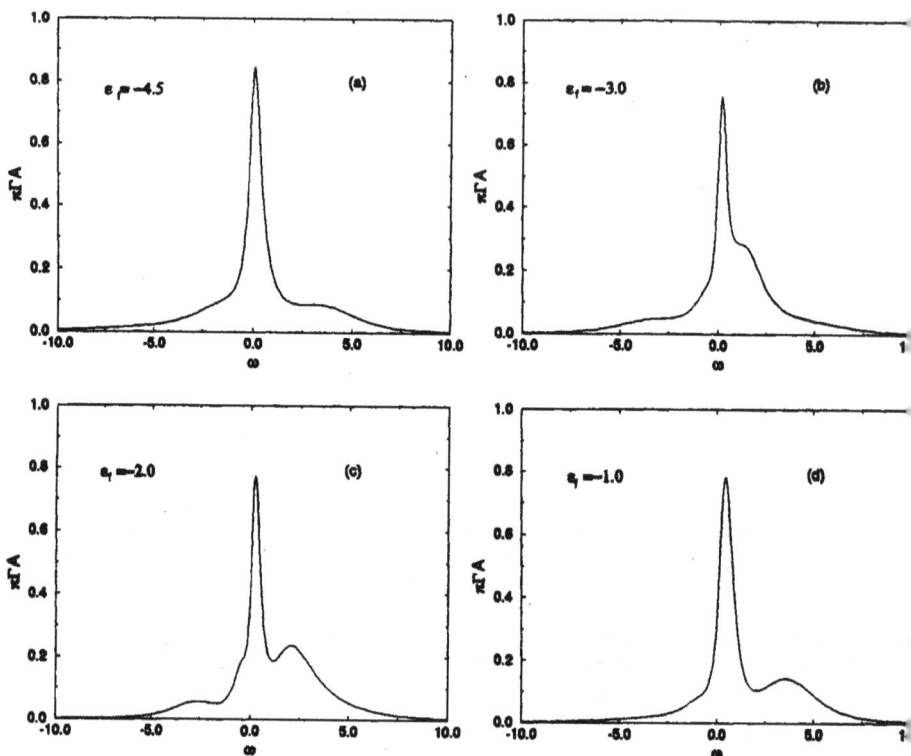

Figure 2: The spectral densities $\pi \Gamma A$ in the asymmetric case as functions of the frequency ω calculated for $N = 4$, $U = 4$, $\Gamma = 0.5$, $T = 1/8$, and different impurity energies: (a) $\epsilon_f = -4.5$, (b) $\epsilon_f = -3.0$, (c) $\epsilon_f = -2.0$, and (d) $\epsilon_f = -1.0$.

of this shift, valid for $\langle n_f \rangle \gtrsim 1$, is $\Delta\omega = T_K(N - 2\langle n_f \rangle)$. Fig. 2(b) we have the singly-occupied state f^1 being the state with minimum energy; therefore, the peaks in PES ($\omega < 0$) and BIS ($\omega > 0$) spectra represent transitions $f^1 \rightarrow f^0$ and $f^1 \rightarrow f^2$. On the PES side, we observe a small peak located approximately at $\omega \sim \Delta_{01} = 3$. The peak on the BIS side is not well developed because its position at $\omega \sim \Delta_{21} = 1$ is close to the origin. Thus, we observe only a shoulder. The choice of parameters in Fig. 2(c) is particularly significant for understanding the effects of higher degeneracy. The lowest energy impurity state is f^1; however, at $\epsilon_f = -2$, the energy differences between f^1 and the next lowest states f^0 and f^2 are equal, that is, $\Delta_{01} = \Delta_{21} = 2$. One could, therefore, naively expect the spectral function to be symmetric around the origin. Contrary to this expectation, the difference between peaks is significant and results in a different phase space for transitions $f^1 \rightarrow f^0$ and $f^1 \rightarrow f^2$.

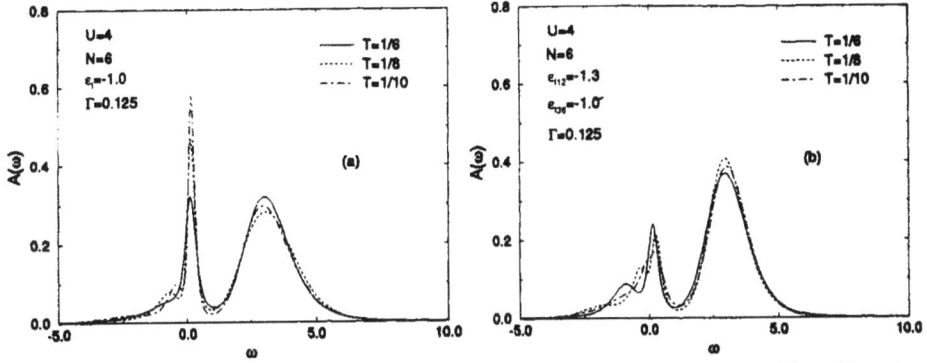

Figure 3: Spectral functions of the asymmetric six-fold degenerate Anderson model at $U = 4$, $\Gamma = 0.125$, and three different temperatures $T = 1/6$, $1/8$, and $1/10$ a) without the crystal field and $\epsilon_f = -1$ and b) with the crystal field and $\epsilon_{f_{12}} = -1.3$ and $\epsilon_{f_{36}} = -1.0$.

4.3 Crystal-Field Case

A large spin-orbit coupling in a cerium atom splits its fourteen $4f$-states into a six-fold degenerate ground state with total angular momentum $j = 5/2$ and a eight-fold degenerate excited state with $j = 7/2$ that is widely separated from the ground state. The introduction of cubic crystal fields further splits the $j = 5/2$ multiplet into the doublet Γ_7 and the quadruplet Γ_8. In Fig. 3a, we first present spectral properties of the asymmetric $N = 6$ Anderson model in the absence of the crystal field splitting, meaning that $\epsilon_m = \epsilon_f$ for $m = 1, \ldots, 6$. In contrast to the previous cases, here we chose a smaller hybridization, $\Gamma = 0.125$, to model our calculations closer to the real physical systems. However, since the Kondo temperature decreases exponentially with decreasing Γ, we were unable to reach temperatures close to the Kondo temperature. Nevertheless, we still observed a peak close to $\omega = 0$ (Fig. 3a) that is increasing with decreasing temperature. In Fig. 3b, we present the spectral density for the same system but now in the presence of the crystal field splitting. The lowest lying multiplet is a doublet with energy $\epsilon_{f_{12}} = \epsilon_f + \Delta_c$. We chose $\epsilon_f = -1.0$ and the crystal field splitting $\Delta_c = -0.3$. The energy of the quadruplet is $\epsilon_{f_{36}} = \epsilon_f$. We observed an unusual temperature dependence of the $\omega = 0$ peak. In contrast to the case with no splitting, the central peak decreases with decreasing temperature. We speculate that this behavior is due to existence of two different low-energy scales in the system. At higher temperatures, a small crystal field splitting is "washed out" by temperature fluctuations; therefore, the six-fold degeneracy is restored. As temperature is lowered, the splitting becomes important, which in our case decreases the degeneracy from six to two. Existence of two different effective degeneracies then leads to the existence of two different Kondo temperatures.

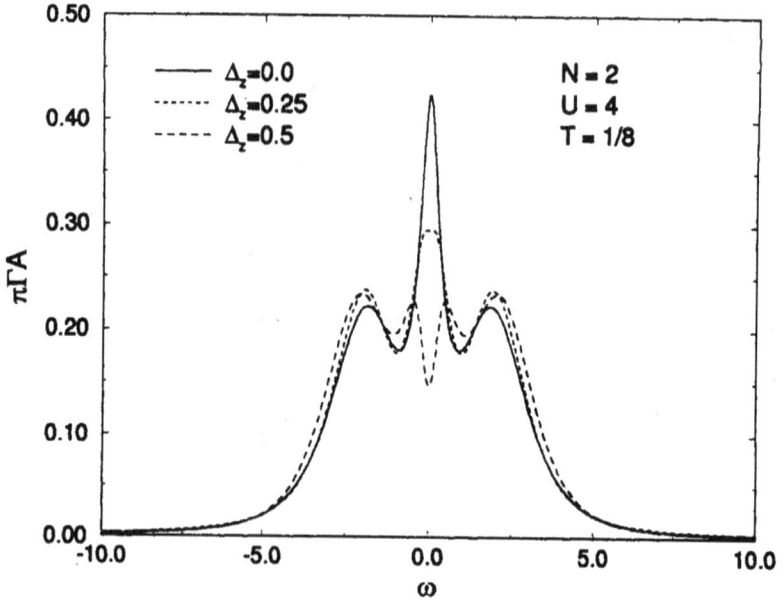

Figure 4: Spectral functions $A(\omega)$ of the Anderson model in an external field for the particle-hole symmetric case. Functions are presented for three different values of the Zeeman energy $\Delta_s = 0.0$, 0.25, and 0.5 with $N = 2$, $U = 4$, and $T = 1/8$.

4.4 External Magnetic Field

In the case of an external magnetic field, we were mainly interested in the effect of a static magnetic field on the low energy spectral properties of the $N = 2$ Anderson model. To preserve the particle-hole symmetry, we choose the on-site energies of spin up and down states on the impurity site to be $\epsilon_\uparrow = -(U + \Delta_Z)/2$ and $\epsilon_\downarrow = -(U - \Delta_Z)/2$, where Δ_Z is the Zeeman splitting. In Fig. 4, we show the spectral functions for three different values of Zeeman splitting, $\Delta_Z = 0.0$, 0.25, and 0.5. As expected, a Zeeman splitting of $\Delta_Z = 0.25$, which is smaller than the hybridization $\Gamma = 0.5$, only decreases the Kondo resonance. When Δ_c becomes comparable with the hybridization, i.e., $\Delta_Z = \Gamma = 0.5$, the Kondo resonance splits into two peaks separated by Δ_Z.

5 Conclusions

By using parallel computing, in an embarassingly simple but effective way, we performed a series of simulations of the degenerate, single-impurity Anderson model. Then, by using methods of Bayesian statistical inference, coupled with the principle of maximum entropy, we extracted from the imaginary-time Green's function the single-particle spectral densities of the model. We studied the behavior of the spectral densities as a function of temperature and degeneracy for a variety of model parameters. With $N = 6$ we have a situation that begins to model the actual degeneracy of a Ce impurity. By splitting the degeneracy of the impurity level, we were able to give addition "realism" to the model. In this paper we summarized previous work [4]. We direct the reader to these papers for more extensive details.

6 Acknowledgments

This work was supported by the U.S. Department of Energy. We acknowledge helpful discussions with A. Arko, M. Jarrell, and J.W. Wilkins. We also thank S.W. Hodson and K.H. Winkler for making their workstation cluster available.

References

1. P.W. Anderson, Phys. Rev. **124**, 41 (1961).
2. J.W. Allen, Adv. Phys. **35**, 275 (1986).
3. J.J. Joyce *et al.*, Phys. Rev. Lett. **68**, 236 (1992).
4. J. Bonča and J.E. Gubernatis, Phys. Rev. B **47**, 13137 (1993); Phys. Rev. B, submitted.
5. A.M. Tsvelik and P.B. Wiegmann, Adv. Phys. **32**, 453 (1983).
6. N.E. Bickers, D.L. Cox, and J.W. Wilkins, Phys. Rev. B **36**, 2036 (1987); N.E. Bickers, Rev. Mod. Phys. **59**, 845 (1987); O. Gunnarsson and K. Schönhammer, Phys. Rev. B **28**, 4315 (1983); N. Read and D.M. Newns, J. Phys. C. **16**, 3273 (1982).
7. H.R. Krishna-murty, J.W. Wilkins, and K.G. Wilson, Phys. Rev. B **21**, 1003 (1980); 1044 (1980).
8. L.N. Oliveira and J.W. Wilkins, Phys. Rev. B **24**, 4863 (1981).
9. H.O. Frota and L.N. Oliveira, Phys. Rev. B **33**, 7871 (1986).
10. O. Sakai, Y. Shimizu, and T. Kasuya, J. Phys. Soc. Jpn. **58**, 3666 (1989).
11. T.A. Costi and A.C. Hewson, Physica B **163**, 179 (1990); Physica C **185-189**, 2649 (1991); Phil. Mag. B **65**, 1165 (1992); J. Magn. Magn. Mat. **108**, 129 (1992).
12. J.E. Hirsch and R.M. Fye, Phys. Rev. Lett. **56**, 2521 (1986).
13. E.Y. Loh, Jr. and J.E. Gubernatis, in *Electronic Phase Transitions*, edited by W. Hanke and Yu.V. Kopaev (North-Holland, Amsterdam, 1991), chap. 3.
14. The lack of a sign problem is an empirical finding true for the current model and is not a result based on any known symmetry. A sign problem is sometimes found in simulations of the Kondo single-impurity model (R.M. Fye, private communication) and $d = \infty$ lattice models (M. Jarrell, private communication).

15. R.N. Silver, D.S. Sivia, J.E. Gubernatis, and M. Jarrell, Phys. Rev. Lett. **65, 496** (1990)
16. H.Q. Lin and J.E. Hirsch, Phys. Rev. B **37**, 1864 (1988).
17. G.D. Mahan, *Many-Particle Physics* (Plenum, New York, 1983), chap. 3.
18. R. Blackenbecler, D.J. Scalapino and R.L. Sugar, Phys. Rev. D **24**, 2278 (1981).
19. R.K. Bryan, Eur. Biophys. J. **18**, 165 (1990).
20. J.E. Gubernatis, R.N. Silver, and M. Jarrell, Phys. Rev. B **44**, 6011 (1991).
21. M.H. Kalos and P.A. Whitlock, *Monte Carlo Methods* (Wiley, New York, 1986).
22. A. Beguelin, *et al.*, "PVM User's Guide, ver. 2.3," Oak Ridge National Laboratory Report ORNL/TM-12187.

QUANTUM MONTE CARLO SIMULATION
BY AUXILIARY FIELDS

S. Sorella

International School for Advanced Studies

Via Beirut 2-4 34013 Trieste, Italy

e-mail: sorella@sissa.it

Abstract

The ground state properties of a many body hamiltonian H can be calculated by a recently developed technique, known as " the Projection Quantum Monte Carlo method". Instead of simulating the many body system at finite temperature, one attempts to reach the zero temperature limit by applying to a trial wavefunction, suitably chosen, a many body propagator e^{-Ht}. For large enough t -t playing a role of an effective inverse temperature- the ground state component of the trial state is filtered out usually much faster than the corresponding finite temperature method. The Hubbard Stratonovich transformation allows to formulate the many body propagation of the trial state as that of sampling a distribution. This distribution is constructed by propagating the trial wavefunction under the influence of a one-body time dependent external field. However this distribution may be not positive definite, especially for large t, and serious problems occur when the average sign of the distribution is too small. We introduce a symmetric Hubbard Stratonovich transformation that in principle may solve this problem with large enough computer time. We also discuss how to apply the "Projection Quantum Monte Carlo" technique to the infinite U Hubbard model in a simple way.

1 Introduction

The many-body problem of interacting quantum particles has been a subject of interest for many years. Unfortunately very few models which take correlations into consideration are exactly solvable. Therefore several approximate techniques have been used to treat correlation for systems of physical interest. Among this methods it is worth mentioning the Configuration Interaction CI (which is limited to very few electrons) , the many body perturbation theory (Hartree, Hartree-Fock) and the Local Density Approximation -LDA-. However these methods have two important drawbacks: either they are not systematically convergent to the exact solution, even if an infinite computation time were available, or they require a prohibitive amount of computer time. As an example in the full CI method the computation costs grows exponentially as the number of electrons in the molecule increases. Current approximation methods have costs ranging typically from the third power to the seventh power of the number of electrons.

Recently there has been progress,which looks very promising, in the simulation of quantum systems, using the so called Quantum Monte Carlo method (QMC). The goal of this method[1, 2] is to obtain the exact ground state properties of a many body system by numerically solving the Schrödinger equation. The basic idea is the observation that the imaginary time evolution $|\psi(t)>= e^{-Ht}|\psi_T>$ of any initial state $|\psi_T>$ is proportional after long time to the exact ground state of the hamiltonian H, provided $|\psi_T>$ has non zero overlap with the true ground state and the ground state is not degenerate. $|\psi(t)>$ can be obtained as a solution of the time dependent Schrödinger equation at imaginary times. If $|\psi_T>$ is positive definite, its propagation at imaginary times can be viewed as a diffusion process that can be simulated stochastically as done in the Diffusion Monte Carlo and the Green Function Monte Carlo (DMC and GFMC). The method works very well for bosons, however it faces serious problems when dealing with fermions. In such case in fact, because of the antisymmetry of the fermionic wavefunctions, $|\psi(t)>$ is no longer positive definite and a straightforward application of the diffusion ideas is no longer possible. This is the well known "fermion sign problem" which in this case has a very clear physical meaning. In fact the propagated wavefunction $|\psi(t)>$ can be decomposed into two positive definite wavefunctions $|\psi(t)>= |\psi^+(t)> -|\psi^-(t)>$. Each of them has a non vanishing component with the more stable boson ground state. Thus the imaginary time evolution becomes numerically intractable since both the positive definite components are attracted exponentially by the much more stable boson ground state, until their fermion component becomes numerically undetectable.

In the present paper we review the "Projection Quantum Monte Carlo technique" (PQMC), first introduced by Koonin[3] and then developed later by several authors[4, 5, 6]. The imaginary time evolution e^{-Ht} is numerically tractable when the Hamiltonian contains only one body operators and no interaction terms. Thus instead of sampling the imaginary time evolution with a stochastic diffusion scheme one retrieves a very old-fashioned technique: the Hubbard-Stratonovich transformation (HST). The HST is basically a method that allows to transform a two-body operator into a superposition of one-body operators. In this way the HST transforms the many body problem into a functional integral over a fluctuating external potential. We calculate the functional integral using standard Monte Carlo technique such as the thermal bath[7] and the Metropolis algorithms. Some problems arise since it is not guaranteed that one can define a positive statistical weight in the Monte Carlo simulation. Although some preliminary results[4] erroneously indicated that the PQMC were not affected by the "fermion sign problem", it is now clear instead [8, 9, 10] that an instability at low temperature occurs, very similar to the Diffusion Monte Carlo one. I will show in this paper that the nature of the low temperature instability of the PQMC is not due to the fermion character of the ground state wavefunction, anyway preserved by the HST, but it has a completely different meaning. This instability is more properly related to an analytical difficulty rather than to a physical limitation.

The reason of the difficulty in this case comes from the discretization of the imaginary time evolution of the many body propagator, a necessary step to neglect the commutator between the one body and the two body term in H for short time evolution $\Delta\tau$. I will show that, using a more symmetric Hubbard Stratonovich transformation the fermion sign problem disappears in the limit of small discretization time $\Delta\tau$. This symmetrization however leads to a great enhancement in computational time although it allows to work with a positive definite distribution.

The paper is organized as follows. In section (2) we introduce the HST, summarizing the basic steps for the calculation of ground state properties. In section (3) we apply the HST to a specific and well known model -the Hubbard model- and we describe the details of the standard PQMC algorithm as well as its simple generalization to the infinite U Hubbard model. Finally in section (4) we show that a simple symmetrization of the HST allows one to define a positive definite weight and solve, at least theoretically, the famous "fermion sign problem".

2   The Hubbard-Stratonovich transformation

As we have already mentioned in the introduction the ground state $|\psi_G>$ of a many body system can be written in the following formal way :

$$|\psi_G> = \lim_{t\to\infty} e^{-Ht} |\psi_T> \qquad (1)$$

where for simplicity we assume a vanishing ground state energy for H, so that the previous limit is finite and well definite. Hereafter we will use the symbol $|\psi_T>$ to indicate the trial wave function. By expanding the R.H.S. of Eq. (1) in terms of the eigenstates of H, this equation is easily verified provided the trial wavefunction has a non vanishing component of the exact ground state ($<\psi_T|\psi_G> \neq 0$). The convergence in Eq. (1) is of the following exponential form:

$$e^{-E_{gap}t}$$

where $E_{gap} = E_1 - E_0$ is the gap between the first excited energy and the ground state energy. The problem of determining the ground state eigen-function of an Hamiltonian H is equivalent to that of solving Eq. (1).

It is worth noting that if the ground state is degenerate Eq. (1) gives one of the possible combination of states depending on the particular choice of the trial wavefunction:

$$\psi(t) \to \sum_i <\psi_T|\psi_G^i> \psi_G^i$$

where we indicate with $\{\psi_G^i\}$ the set of all the possible degenerate ground states.

The evaluation of the propagator e^{-Ht} is numerically tractable when the Hamiltonian H contains only one body operators. In this case in fact the time dependent evolution of a simple Slater determinant remains a Slater determinant through such time propagation, yielding a big simplification in the calculation. However the hamiltonians of physical interest usually contain a two-body interaction which does not allow to apply the previous simple strategy. In order to overcome the latter difficulty a very simple relation allows to write a two-body propagator $e^{\frac{\alpha^2}{2}O^2}$ by means of an integral containing only the one-body propagator O:

$$
e^{\frac{\alpha^2}{2}O^2} = \frac{\int_{-\infty}^{+\infty} d\sigma\, e^{-\alpha\sigma O - \frac{1}{2}\sigma^2}}{\int_{-\infty}^{+\infty} d\sigma\, e^{-\frac{1}{2}\sigma^2}}.
\tag{2}
$$

The previous transformation (2) can be applied only for negative definite two-body interaction contained in the hamiltonian H unless considering imaginary α. [1]

Now the problem of calculating e^{-Ht} has almost been solved. A general hamiltonian is usually the sum of a kinetic operator H_0 and a two body term V. The non-commutativity between V and H_0 does not allow to replace the full propagator naively as: $e^{-(H_0+V)t} = e^{-Vt}e^{-H_0t}$. On the other hand for infinitesimal evolution $\Delta\tau$ the effect of the commutator can be neglected up to $O(\Delta\tau^2)$

$$
e^{-H\Delta\tau} = e^{-V\Delta\tau}\, e^{-\Delta\tau H_0} + O(\Delta\tau^2)
\tag{3}
$$

and for a better accuracy:

$$
e^{-H\Delta\tau} = e^{-\frac{\Delta\tau}{2}H_0}\, e^{-V\Delta\tau}\, e^{-\frac{\Delta\tau}{2}H_0} + O(\Delta\tau^3)
\tag{4}
$$

Although in practice it is very convenient to use the latter more accurate expression, for sake of simplicity we will use in the following the more simple Eq. (3), since our results can be easily extended to the more accurate expression (4). Finally the full propagator e^{-Ht} can be evaluated by a repeated application of the previous equation (3):

$$
e^{-Ht} = (e^{-H\Delta\tau})^L \quad \text{where } \Delta\tau = \frac{t}{L}
\tag{5}
$$

Hence, using Eqs. (3) and (5) and having in mind the HST in Eq. (2) the full many body propagator can be in general expressed by means of a superposition of one-body propagators. When expression (3) is substituted in (5), an error in the evaluation of the ground state of H will appear in (1), due to discretization in time. In this sense it is useful to remark that formula (1) still provides the exact ground state $|\psi_G'>$ of the Hamiltonian

$$
H' = -\frac{1}{\Delta\tau}\ln\left(e^{-\Delta\tau V}\, e^{-\Delta\tau H_0}\right)
$$

$|\psi_G'>$ differing by $O(\Delta\tau)$ from the ground state of H. Analogously, using the more accurate Eq. (4), we get an error in the ground state by $O(\Delta\tau^2)$. Thus the $\Delta\tau$ error in the resulting path integral is well controlled and depends only on the commutator $[H_0, V]$, as expected.

---

[1]After many years of experience one can certainly state that it is better to work with a complex HST rather than for instance to implement unphysical shifts to make negative the two-body coulomb potential[11].

3 Hubbard Stratonovich transformation for the Hubbard model

Let us now consider in details the application of this method to the Hubbard model. Consider a crystal of N_a lattice sites with a total of $N \leq N_a$ electrons. It is supposed that the electrons can hop between localized states of neighboring lattice sites, and that each site is capable of accommodating two electrons of opposite spins, with an interaction energy $U > 0$. The Hamiltonian to consider is then:

$$H = H_0 + V$$

$$\text{where}:$$

$$H_0 = -\sum_{<i,j>,\alpha} c_{i\alpha}^{+} c_{j\alpha} \quad \text{and} \quad V = -\frac{U}{2}\sum_i m_i^2 \tag{6}$$

$c_{i\alpha}^{+}$, $c_{i\alpha}$ are the creation and annihilation operators for an electron of spin α at the ith lattice site, the sum $\sum_{<i,j>}$ is restricted to nearest neighbor sites and the local site magnetization $m_i = n_{\uparrow} - n_{\downarrow}$, $n_\alpha = c_{i\alpha}^{\dagger} c_{i\alpha}$ being the local density operator of given spin α. For reason of convenience we have not chosen here the standard notations for the model. The term $-\frac{U}{2}m^2$ substitutes the more conventional $Un_{\uparrow}n_{\downarrow}$ for a direct application of the HST. Anyway this change add only irrelevant constant terms to the hamiltonian, without affecting its ground state. There are many reason why the "Projection Monte Carlo" is very efficient for this model:

- It is a model defined on a discrete lattice and we do not have to deal with an ultraviolet regularization of the HST fields.

- At half filling or negative U there is no sign problem[7], as a consequence of the reflection positivity[12]. No other QMC method is able to make use of this property of the Hubbard model so far.

- In the small U limit the HST fields do not fluctuate and the statistical method becomes exact.

We introduce Hubbard-stratonovich fields σ_r to decouple the interaction term, having in mind the basic relation (2):

$$e^{-\Delta_\tau V} = \pi^{-\frac{N_a}{2}} \left(\prod_r \int_{-\infty}^{\infty} d\sigma_r \right) e^{\sum_r -\frac{\sigma_r^2}{2} + \lambda m_r \sigma_r} \tag{7}$$

where $\lambda = \pm\sqrt{U \Delta_\tau}$. Similar transformation is possible with a discrete field s which takes only the two values $s = \pm 1$:

$$e^{-\Delta_\tau V} = 2^{-N_a} \sum_{s_r = \pm 1} e^{\sum_r \lambda s_r m_r} \tag{8}$$

with $\cosh \lambda = e^{\frac{U\Delta\tau}{2}}$ [7]. The latter transformation is by far more convenient for numerical applications of the HST. In the following however we use the original continuous formulation since the discrete version may be seen as a numerical evaluation of the integral (7) with only two special points per variable σ_r. This is also why the discrete version of the HST is so much convenient. Finally by labeling the discrete imaginary times $t_l = \Delta\tau l$ with an index $l = 1,2,\dots L$ we can write the full propagator in (5) with compact notations [7, 4]:

$$e^{-H\tau} = \int [d\mu_\sigma]_{(\tau,0)} U_\sigma(\tau,0) \tag{9}$$

where :

$$[d\mu_\sigma]_{(\tau,0)} = \prod_{t=\tau,\tau-\Delta\tau,\dots,\Delta\tau} d\mu_\sigma(t) \tag{10}$$

$$d\mu_\sigma(t) = \prod_r \frac{d\sigma_{r,t} e^{-\frac{\sigma_{r,t}^2}{2}}}{\sqrt{\pi}}$$

$$U_\sigma(t,t') = e^{\lambda V_t} e^{-\Delta\tau H_0} e^{\lambda V_{t-\Delta\tau}} e^{-\Delta\tau H_0} \dots e^{\lambda V_{t'+\Delta\tau}} e^{-\Delta\tau H_0} \tag{11}$$

$$V_t = \sum_r \sigma_{r,t}(n_{r,\uparrow} - n_{r,\downarrow})$$

The propagator $U_\sigma(t,t')$ is defined for $t > t'$. It is easy to extend its definition for $t' > t$ by $U_\sigma(t,t') = U_\sigma^\dagger(t',t)$, i.e. with a backward propagation after taking the hermitian conjugate of $U_\sigma(t',t)$ as defined in (11). Thus U_σ satisfies the usual rules for a propagator $U_\sigma(t,t') = U_\sigma^\dagger(t',t)$ and $U_\sigma(t,t')U_\sigma(t',t'') = U_\sigma(t,t'')$ if $t > t' > t''$.

The many body operator (9) is written in this way as an integral of product of single particle operators with a considerable advantage from the practical and numerical point of view. In this way the ground state wavefunction is given by:

$$|\psi_G> = \lim_{t\to\infty} \int [d\mu_\sigma]_{(t,0)} U_\sigma(t,0) |\psi_T> \tag{12}$$

Although the last expression can appear quite cumbersome it is basically, for finite L, a multidimensional integral over the LN_a variables σ. There are many different equivalent ways to get the same transformation, which correspond, for example, to all the possible change of variables in such multidimensional integrals. In principle we can get all the details of the ground state wavefunction $|\psi_G>$ using (12). However the information contained in ψ_G is prohibitive for available computers because the memory to handle a many body wavefunction is exponentially growing with N_a. For this reason it is useful to define a pseudo partition function:

$$Z = <\psi_T|e^{-H\tau}|\psi_T> = \int [d\mu_\sigma]_{(0,\tau)} <\psi_T|U_\sigma(\tau,0)|\psi_T> \tag{13}$$

at an effective temperature $kT = \frac{1}{\tau}$. At the given temperature the true partition function Z of the quantum electron system described by the Hamiltonian (6) can be obtained with a particular choice of the trial function ψ_T (and surely academic):

$$\psi_T = \sum_{all\ eigenstates of H} \tilde{\psi}_i$$

However in order to reach the $t \to \infty$ using (1) it is clearly more convenient to make a good choice of ψ_T. For instance $e^{-Ht}|\psi_T>$ conserves all the symmetries of the trial state. Thus a wavefunction $|\psi_T>$ with the same symmetry property of the ground state filters out at the very beginning all the unwanted excited states of H with different symmetry properties.

The task of our statistical simulation is not the evaluation of this pseudo-partition function (13), but the evaluation of its logarithmic derivative with respect to suitable parameters. As an example the ground state energy, using Eq. (1) and the definition of Z in Eq. (13), can be written as $E_G = -\frac{1}{L}\frac{d}{d\Delta\tau}\ln Z$, yielding::

$$E_G = \lim_{L \to \infty} \frac{1}{L} \frac{\int [d\mu_\sigma]_{(\tau,0)} - \frac{d}{d\Delta\tau} < \psi_T \,|\, U_\sigma(\tau,0) \,|\, \psi_T >}{\int [d\mu_\sigma]_{(\tau,0)} < \psi_T \,|\, U_\sigma(\tau,0) |\, \psi_T >} \tag{14}$$

An estimation of E_G in expression (14) requires the numerical calculation of two multi-dimensional integrals over $L \times N_a$ $\sigma_{r,t}$ variables, with large enough L. For this reason we are forced to use a statistical method.

3.1 Monte Carlo sampling

We consider the statistical system with σ-degrees of freedom interacting with a weight:

$$w(\sigma) = < \psi_T \,|\, U_\sigma(\tau,0) \,|\, \psi_T > \tag{15}$$

From this definition the ground state energy in Eq. (14), (and the expectation value of any operator), represents the statistical thermal average

$$E_G = \frac{\int [d\mu_\sigma]_{(\tau,0)}\, w(\sigma) E_{E_G}((\sigma)}{\int [d\mu_\sigma]_{(\tau,0)}\, w(\sigma)} \tag{16}$$

of a correspondent classical operator:

$$E_{E_G}(\sigma) = -\frac{1}{L}\frac{d}{d\Delta\tau}\ln < \psi_T \,|\, U_\sigma(\tau,0) \,|\, \psi_T > . \tag{17}$$

Hence the quantum many body problem is simplified to the simulation of a classical system. Note also that the gaussian factors are implicit in the definition of $[d\mu_\sigma]_{(\tau,0)}$ in (10).

In order to apply the standard monte carlo technique we start with an initial random configuration of the field σ. Then one can generate a new configuration σ_{new}

by changing only one of the spin, say $\sigma_{r,t} = \sigma_{old}$ at a given time and at a given site. We accept the new configuration if the ratio

$$R = w(\sigma_{new})/w(\sigma_{old})\, e^{\frac{\sigma_{old}^2 - \sigma_{new}^2}{2}} > z \qquad (18)$$

,where z is a pseudo-random number uniformly distributed in the interval $(0, 1)$. By iterating this Markov process, named the Metropolis algorithm, the stationary distribution converges to the desired one and the evaluation of the classical average (17) is possible. An alternative method is available when the discrete HST is used. In this case the gaussian factors drops out in Eq. (18) and we have only two possibility for the spin at a given time and site. Thus if we assume that the distribution is already at equilibrium the new configuration has to be accepted with a probability

$$\frac{R}{1+R} > z \qquad (19)$$

in order to remain at equilibrium. This is the so called thermal bath algorithm. As it is easy to see by comparing Eqs. (18) and (19) the thermal bath algorithm has a smaller acceptance ratio compared with the usual Metropolis one but for the way it is defined it may converge faster to the equilibrium distribution. In practice both the algorithms work well for the given weight although for small U, where the acceptance is quite large anyway the thermal bath is more efficient. The MC step is applied to any of the independent variables σ by scanning first the lattice sites in some defined order and then the Trotter imaginary slices from the final $t = \tau$ to the initial $t = 0$ time. A single *Monte Carlo sweep* is obtained after the MC step (Eq. 18 or Eq. 19) is applied to all the variables of the functional integral (13).

The potential V_t in Eq. (12) , as well as H_0 in Eq. (6), can be split into two commuting parts with different spin component:

$$V_t = V_t^\uparrow + V_t^\downarrow \quad \text{with} \quad V_t^\uparrow = \sum \sigma_{r,t} n_r^\uparrow \quad \text{and} \quad V_t^\downarrow = -\sum \sigma_{r,t} n_r^\downarrow$$
$$H_0 = H_0^\uparrow + H_0^\downarrow. \quad \text{with} \quad H_0^\alpha = -\sum_{\langle i,j\rangle} c_{i\alpha}^+ c_{j\alpha}$$

Thus the one body propagator can be divided into to independent part $U_\sigma(t,0) = U_\sigma^\uparrow(t,0)\, U_\sigma^\downarrow(t,0)$ where U_σ^α is given in Eq. (11) by replacing H_0 and V_σ with the corresponding H_0^α and V_t^α. This simplification is important because we can formally write the weight in a product of two independent terms:

$$w_\sigma = w_\sigma^\uparrow w_\sigma^\downarrow \qquad (20)$$

provided the trial wavefunction is a Slater determinant with definite spin orbitals, i.e. $|\psi_T> = |\psi_T^\uparrow> \otimes |\psi_T^\downarrow>$. In the following we give an efficient way to calculate the ratio R of the weights, and due to the previous simplification we focus our attention only to one spin component, by omitting the spin index for the sake of simplicity.

3.2 Imaginary time propagation and updating of the weight

In order to evaluate the weight (15) we first need to know how to calculate the imaginary time propagation of a Slater determinant under a time dependent potential. We first define the matrix elements of the kinetic term propagation:

$$G^t_{r,r'} = (e^{-H_0 t})_{r,r'} \tag{21}$$

The effect of the $e^{-H_0 t}$ on a Slater determinant $|\psi>$, is to propagate each of its orbitals $\phi_n(r)$ by the matrix -vector multiplication $\phi_n(r) \to \sum_{r'} G^t_{r,r'} \phi_n(r')$. The potential term instead is diagonal in space and $e^{\lambda V_t}|\psi>$ leads to the multiplication of the orbitals by $\phi_n(r) \to \eta_{r,t} \phi_n(r)$, where

$$\eta_{r,t} = e^{\alpha \lambda \sigma_{r,t}} \quad \text{with} \quad \eta_{r,t=0} = 1 \tag{22}$$

where the $\alpha = 1$ or -1 for the up and down spin propagation, respectively. We can identify therefore the propagation of the full Slater determinant with the one of the orbitals at times $t = l\Delta\tau$ with $l = 1, \ldots, L$. In the application of the method it is important to follow both the forward propagated orbitals $\psi^>_{n,t}$ of $U_\sigma(t,0)|\psi_T>$ and the back propagated ones $\psi^<_{n,t}$ of $<\psi_T|U(\tau,t)$:

$$\psi^>_{n,t}(r) = \eta_{r,t} \phi^>_{n,t}(r) \tag{23}$$
$$\psi^<_{n,t}(r) = \phi^<_{n,t}(r) \tag{24}$$

where we have introduced for later convenience the orbitals $\phi^>_{n,t}$ and $\phi^<_{n,t}$ that can be iteratively calculated by:

$$\phi^>_{n,t+\Delta\tau}(r) = \sum_{r'} G^{\Delta\tau}_{r,r'} \eta_{r',t} \phi^>_{n,t}(r') \tag{25}$$

$$\phi^<_{n,t-\Delta\tau}(r) = \sum_{r'} G^{\Delta\tau}_{r,r'} \eta_{r',t} \phi^<_{n,t}(r') \tag{26}$$

with the initial conditions $\phi^>_{n,0} = \phi^<_{n,\tau} = \phi^T_n$, where ϕ^T_n are the orbitals of the trial state $|\psi_T>$.

It is worth mentioning that, in order to be consistent with the more accurate time discretization (4) U_σ slightly changes in (11) and it is enough to replace::

$$\psi^>_{n,t}(r) = \sum_{r'} G^{\frac{\Delta\tau}{2}}_{r,r'} \eta_{r',t} \phi^>_{n,t}(r') \tag{27}$$

$$\psi^<_{n,t}(r) = \sum_{r'} G^{-\frac{\Delta\tau}{2}}_{r,r'} \phi^<_{n,t}(r') \tag{28}$$

and change the initial condition of $\phi^>_{n,0}(r) = \sum_{r'} G^{-\frac{\Delta\tau}{2}}_{r,r'} \phi^T_n(r')$ and $\phi^<_{n,\tau}(r) = \sum_{r'} G^{\frac{\Delta\tau}{2}}_{r,r'} \phi^T_n(r')$.

The overlap matrix $A_{n,m} = \sum_r \psi^<_{m,t}(r)\psi^>_{n,t}(r)$ between the orbitals of the wavefunctions $U_\sigma(t,0)|\psi_T >$ and $< \psi_T|U(\tau,t)$, is clearly independent of t and can be written directly in terms of the iterated orbitals $\phi^>$ and $\phi^<$ (see Eqs. 25-28) :

$$A_{n,m} = \sum_r \phi^<_{n,t}(r)\eta_{r,t}\phi^>_{m,t}(r). \tag{29}$$

Finally the weight w_σ is the determinant of the matrix $A_{n,m}$.

The straightforward application of the Monte Carlo scheme as described in section 3.1 would be very time consuming if a full determinant [2] has to be computed for each evaluation of the ratio R. We note in fact that if only one spin is changed at the time t the matrix A is modified in the following way:

$$A'_{n,m} = A_{n,m} + \delta_{r,t}\ \phi^<_{n,t}(r)\phi^>_{m,t}(r) \tag{30}$$

where $\delta_{r,t} = \eta^{new}_{r,t} - \eta^{old}_{r,t}$ The new determinant det A' is then very easily related to the old matrix A. To this end it is useful to write $(A'/A)_{n,m} = \delta_{n,m} + \delta_{r,t}\bar\phi^<_{n,t}(r)\phi^>_{m,t}(r)$, where the barred propagated orbitals are defined by:

$$\bar\phi^<_{n,t}(r) = \sum_m A^{-1}_{n,m}\phi^<_{m,t}(r) \qquad \bar\phi^>_{n,t}(r) = \sum_m \phi^>_{m,t}(r)A^{-1}_{m,n}. \tag{31}$$

The ratio of determinants $P = \frac{\det A'}{\det A}$ is then the determinant of the previous matrix A'/A. This can be easily evaluated since $v_n = \bar\phi_{n,t}(r)$ is the unique right-eigenvector of A'/A with eigenvalue different from one, and thus equal to P:

$$P = 1 + \delta_r \sum_n \bar\phi^<_{n,t}(r)\phi^>_{n,t}(r) \tag{32}$$

Note in fact that all the vectors orthogonal to $\phi^>_{m,t}$ are right eigenvectors of A'/A with eigenvalue one.

In order to evaluate the ratio of determinants P it is convenient to update the inverse matrix A^{-1} with the useful algorithm :

$$A'^{-1}_{n,m} = A^{-1}_{n,m} + g_{r,t}\ \bar\phi^<_{n,t}(r)\bar\phi^>_{m,t}(r) \tag{33}$$

where $g_{r,t} = -\delta_{r,t}/P$. Eq. (33) can be readily verified using the above definitions (30,31,32).

The updating of the inverse matrix has to be done each time the Monte Carlo move is accepted. Note that the ratio of the weight is $R = P_\uparrow P_\downarrow$ for the discrete HST (19) and $R = P_\uparrow P_\downarrow e^{\frac{\sigma^2_{old}-\sigma^2_{new}}{2}}$ for the continuous one (18).

Using expressions (32-33) we can compute the ratio of determinants or we can update the inverse matrix using only $O(N^2)$ instead of the much more heavy $O(N^3)$ needed to evaluate a full determinant. This is the analogous of the Blankenbecker at al. algorithm[13], applied to the 'Projection Monte Carlo method". The total time required to make one single PQMC sweep takes N^2LN_a floating point operations, i.e. the computational time is only linear in the number of lattice sites. This is particularly useful for electronic structure calculations where the number of electrons is much smaller than the lattice size.

---

[2]indeed two, one for each spin component

3.3 Numerical stability

The imaginary time propagation is not unitary and the orthonormality conditions, initially satisfied by the orbitals, are not preserved during such propagation. Therefore after repeating many times the iterations (25,26) an orthogonal basis set $\phi_i(r)$ $i = 1, \ldots, N$ will no longer remain orthogonal. This circumstance can produce a numerical instability of the algorithm. In fact suppose the orbitals are independently propagated through an imaginary time one-body propagator e^{-Ht}. The fermionic ground state can be considered as an excited state (with the right symmetry) of a many-body hamiltonian; its true ground state being a boson-symmetric wavefunction. When the orbitals freely propagate, after long time, they are spontaneously led to the bosonic ground state of H. In this way the numerical information of the fermionic state is gradually lost until the Slater determinant exactly vanishes when the N orbitals of the Slater determinant are no longer linear independent within the given computer precision. Therefore, in order to have a stable propagation, we have to rewrite, any -say-l_o steps, the Slater determinant which is at the time t:

$$|\psi(t)> = \det \phi_{m,t}(r_j)$$

in terms of an orthogonal basis set. This is always possible with a transformation

$$\phi_{m,t} = \sum_n U_{m,n} \, \tilde{\phi}_{n,t}$$

where the matrix $U_{m,n}$ is chosen in such a way that $< \tilde{\phi}_m | \tilde{\phi}_n > = \delta_{m,n}$. The matrix $U_{m,n}$ is not univocally determined by the previous condition. A convenient choice is to use the Gram- Schmitd orthogonalization scheme because, in this case, $U_{m,n}$ is a triangular matrix and its numerical implementation is very efficient.

Then $|\psi(t)>$ can be written as

$$|\psi(t)> = \det \left[\sum_n U_{m,n} \, \tilde{\phi}_{n,t}(r_j) \right] = \det(U) \det \left[\phi_{n,t}(r_j) \right]$$

where the latter equality simply follows by expanding the determinant of the product of two square matrices: $U_{m,n}$ and $\phi_{n,t}(r_j)$. Hence we easily get that the Slater determinant can be written, up to a constant, by means of orthogonal orbitals. Such a constant affects only the norm of the propagated wavefunction which can be independently updated. After that one can proceed as before using Eqs. (25, 26) until the numerical stability will require another orthogonalization. Numerically such a strategy is very useful and we can propagate for long time any function without any numerical problem. The updating procedure for the matrix A^{-1} in Eq. (33) can be then applied only between two consecutive orthogonalization.

3.4   Ground state expectation values of operators

The ground state expectation value of arbitrary operators O can be calculated using the fundamental relation (1) in a way analogous to what done for the ground state energy (14):

$$< \psi_G|O|\psi_G> = \frac{<\psi_T|e^{-H(\tau-t)}Oe^{-Ht}|\psi_T>}{<\psi_T|e^{-H\tau}|\psi_T>} + O(e^{-\frac{E_{gap}\tau}{2}}) \qquad (34)$$

with $t \sim \frac{\tau}{2}$. Notice that it is not necessary to insert the operator O in the middle of the imaginary time interval in Eq(34). For large enough τ it is sufficient to divide the total time interval into two approximately equal parts. As a consequence the statistical error in the evaluation of (35) can be reduced by using several estimators of O differing for the way in which τ has been divided. A convenient choice is to average the estimator (35) over the interval $(\frac{3}{8}\tau < t < \frac{5}{8}\tau)$.

The many body propagators in Eq. (34) are calculated using the functional expression derived in (11). Hence to each operator O we associate a classical function (called estimator) $E_O^t(\sigma)$

$$E_O^t(\sigma) = \frac{<\psi_T|U_\sigma(\tau,t)OU_\sigma(t,0)|\psi_T>}{<\psi_T|U_\sigma(\tau,0)|\psi_T>} \qquad (35)$$

The expectation value of the operator O is then obtained from the statistical average of the estimator $E_O^t(\sigma)$ over the equilibrium distribution (15) that can be sampled by Monte Carlo.

Let us now consider the equal time Green's function at the two sites r_j and r_k, needed for evaluating single-particle correlation functions, like momentum distribution, kinetic energy etc. In this case the corresponding operator O reads:

$$O = c_{j\alpha} c_{k\alpha}^\dagger$$

and Eq. (35) can be easily written in the following way:

$$E_O^t(\sigma) = \frac{<c_{\alpha j}^\dagger U_\sigma(t,\tau)\psi_T|c_{\alpha k}^\dagger U_\sigma(t,0)\psi_T>}{<U_\sigma(t,\tau)\psi_T|U_\sigma(t,0)\psi_T>} \qquad (36)$$

Therefore if ψ_T is a N–state single Slater determinant, Eq. (36) involves the scalar products of two $N+1$- determinants, whose orbitals are given by Eqs. (23,24) and the ones localized at sites j and k. Then by using that the scalar product of two Slater determinants is just the determinant of the corresponding overlap matrix we get:

$$E_O(\sigma) = \frac{\det \underline{A}(j,k)}{\det \underline{A}} \qquad (37)$$

where $\underline{A}_\alpha(j,k)$ is a $(M+1)\times(M+1)$ matrix indexed by the single particle wavefunction components:

$$A_{m,n}(j,k) = \begin{pmatrix} \delta_{j,k} & \psi_{n,t}^>(r_j) \\ \psi_{m,t}^<(r_k) & A_{m,n} \end{pmatrix} \qquad (38)$$

and $A_{m,n}$ is the overlap matrix $N \times N$:

$$A_{m,n} = <\psi_{m,t}^< | \psi_{n,t}^> >$$

By letting the indices j and k assume all the possible values, we obtain N_a^2 matrices of the form (38). Therefore the full calculation of the two points Green's function requires the evaluation of $2N_a^2$ determinants of order $(N+1)$. This is clearly too much expensive for an efficient calculation of single particle correlation functions. In fact the problem can be greatly simplified by noticing that the N_a^2 matrices differ one from another simply by the exchange of one row and one column. It is convenient to formulate all the calculation in terms of the quantities:

$$g_{j,k} = \sum_{m,n} \psi_{m,t}^>(r_j) A_{m,n}^{-1} \psi_{n,t}^<(r_k) \tag{39}$$

It is a well known property that a determinant remains unchanged if one adds to a column any linear combination of the other ones. Hence we may add to the first column of the matrix $\underline{A}$ in (38) a linear combination of the other columns, in order to make vanishing all the elements of the first column but the one in the first row:

$$\det \underline{A}_{n,m}(j,k) = \det \begin{pmatrix} \delta_{j,k} - \sum_n c_n \psi_{n,t}^>(r_j) & \psi_{n,t}^>(r_j) \\ \psi_{m,t}^<(r_k) - \sum_n A_{m,n} c_n & A_{n,m} \end{pmatrix} \tag{40}$$

where in order to have:

$$\psi_{m,t}^<(r_k) - \sum_n A_{m,n} c_n = 0 \tag{41}$$

one must take:

$$c_m = \sum_n A_{m,n}^{-1} \psi_{n,t}^<(r_k). \tag{42}$$

Substituting (42) in the R.H.S. of (40) one obtains:

$$\det \underline{A}_{n,m}(j,k) = \begin{pmatrix} \delta_{j,k} - g(j,k) & \psi_{n,t}^>(r_j) \\ 0 & \\ 0 & A_{n,m} \\ 0 & \\ \vdots & \end{pmatrix} = (\delta_{j,k} - g_{j,k}) \det A \tag{43}$$

In Eq. (37) the factor $\det A$ cancels out with the denominator, yielding a simple expression for the estimator in terms of $g_{j,k}$ only:

$$E_{c_j c_k^\dagger}(\sigma) = \delta_{j,k} - g_{j,k}$$

The computation (39) of the matrix g requires the inversion of a $N \times N$ matrix A amounting N^3 operations, a change of basis $A^{-1}\psi^<$ ($N^2 N_a$ operations) and remaining

N multiplications for each couple of lattice sites (N_a^2) for which the matrix g is defined. In this way all the matrix g can be updated with less than $O(N_a^3)$ operations.

For higher order correlation functions, by using standard properties of determinants as before, it is easy to verify that Wick's theorem applies and that all the estimators $E_O(\sigma)$ can be expressed in terms of the matrix g as well, e.g.:

$$< c_i^\dagger c_j c_k^\dagger c_l > = < c_i^\dagger c_j > < c_k^\dagger c_l > + < c_i^\dagger c_l > < c_j c_k^\dagger >$$

Here the brackets means the quantum expectation value over a fixed configuration of σ fields and each pair average in the RHS is calculated by replacing $< c_i^\dagger c_j >$ with $g_{j,i}$ and obviously $< c_j c_i^\dagger >$ with $\delta_{j,i} - g_{j,i}$.

3.5 Generalization to infinite U

We note here that the basic relations (32,33) that we need to apply the algorithm are not defined for $U \to \infty$, since $\lambda \propto \sqrt{U}$. However if we consider the discrete hubbard Stratonovich transformation we can replace the terms $\eta_{r,t}$ in Eqs. (32,33) with the finite one:

$$\eta_{r,t} = e^{\lambda(s_r - 1)} \tag{44}$$

This amounts to make some irrelevant constant shift in the finite U partition function. The same of course can be done for the spin down particles with $\eta_{r,t} = e^{-\lambda(s_r + 1)}$. In this way $\eta = 0$ or 1 for $U \to \infty$ is finite and the updating scheme in Eqs. (32,33) is unambiguously defined. In other words, for $U \to \infty$ the HST in Eqs. (8) translates the projector over non doubly occupied sites (LHS) as a sum of simpler one-body projector operators defined by the values of the fields s_r. This of course up to irrelevant constants that have been adsorbed in the new definition of η (44). The limit $U \to \infty$ of the Eqs. (32,33) is well defined and leads to a $U = \infty$ algorithm very similar to the one described in ([14], which was derived using the Grassmann algebra approach directly to the $U = \infty$ case.

3.6 The algorithm

We assume in the following that the total number of Trotter slices L is a multiple of l_0, the number of Trotter slices between to consecutive orthogonalizations: $L = p \times l_0$. The calculation of correlation functions described in Sec.(3.4) can be done inside the loop (7.) after calculating the orbitals (23,24) (or 27 and 28 for a better Trotter accuracy), using the already known orbitals $\phi^>$ and $\phi^<$. Since the operations to update the Green's function is quite time consuming it is convenient to perform only few measurements per sweep. For the infinite U algorithm the following single spin-flip update is not ergodic and it is better to move clusters of spins toghether as it is done in [14].

Algorithm for a PQMC sweep

1. Initialize the orbitals $\phi^>_{n,0}$ and $\phi^<_{n,\tau}$ as described in (25,26).

2. Store the advanced orbitals $\phi^>_{n,t}$ for $t = \Delta\tau, \ldots, L\Delta\tau$ by applying Eq. (25) and by orthonormalizing them each l_0 steps (see Sec. 3.3).

4. For $k = p, p-1, \ldots 1$;

    5. orthogonalize $\phi^<_{n,t}$ for $t = t_k = k l_0 \Delta\tau$

    6. compute A^{-1} by scratch using the definition of A (29) with $\phi^>$ and $\phi^<$ at times t_k;

    7. for $j = k l_0, k l_0 - 1, \ldots, (k-1) l_0 + 1$;

        8. for each site $i = 1, \ldots, N_a$;

            9. calculate R using (32) and generate a random number z;

            10. if condition (18) (or 19) is fulfilled update the matrix A^{-1} with Eqs. (33,31) [3] and the fields $\sigma_{r,t}$ at the time $t_j = \Delta\tau j$ and site i;

        8. continue;

        11. back propagate $\phi^<_{n,t}$ with Eq. (26) for $t = t_j$;

    7. continue;

1. continue.

4 Positiveness of the statistical weight

Up to now we have implicitly assumed that the weight (15) is positive definite , so that we can apply Monte Carlo to evaluate statistical averages like (14). Unfortunately the weight is positive definite only for particular cases in the Hubbard model [9]. The standard way to solve this problem is to apply the Monte Carlo algorithm to the absolute value of the weight $|w(\sigma)|$, and take into account the sign s_σ of w in the average of the estimator (16). Thus we obtain the usual umbrella average, a ratio of two classical averages over the positive weight $|w|$:

$$\frac{< \psi_T | e^{-H(\tau-t)} O | e^{-Ht} | \psi_T >}{< \psi_T | e^{-H\tau} | \psi_T >} = \frac{< E^t_O(\sigma) s_\sigma >_{|w|}}{< s_\sigma >_{|w|}}. \qquad (45)$$

The quantity appearing in the denominator of the RHS of Eq. (45) is known as the average sign $S(\tau)$ of the weight w and reads more explicitly:

$$S(\tau) = \frac{\int [d\mu_\sigma]_{(\tau,0)} < \psi_T | U_\sigma(\tau,0) | \psi_T >}{\int [d\mu_\sigma]_{(\tau,0)} | < \psi_T | U_\sigma(\tau,0) | \psi_T >|} \qquad (46)$$

---

[3] In order to speed up the algorithm instead of $\phi^<$ we can back propagate in (11.) $\bar\phi^<$ with (26) and optionally $\bar\phi^>$ with the inverse propagation of (25). In the most important inner loop (8.) the matrix update (33) is much faster since (31) has not to be computed.

As usual in Monte Carlo one divides the MC time in several bins of length at least larger than the autocorrelation time of the estimator. Then the variance of the average quantity over the bins times the inverse square root of the number p of independent bins gives a good estimate of the statistical error in the total average. An analogous scheme can be applied even with the umbrella average (45). One gets estimates of the physical quantity over the bins E_i together with an estimate of the average sign s_i. The total average in this case is $E = \frac{\sum_i s_i E_i}{\sum_i s_i}$. Thus if the bins are large enough that the partial sign averages are positive [4] the previous sum can be considered a weighted average over a distribution of weights s_i. An estimate of the statistical error is proportional in this case to the weighted standard deviation of E_i:

$$\delta E = \sqrt{\frac{\sum_i s_i (E_i - E)^2}{p \sum_i s_i}}. \qquad (47)$$

Although there are no relevant complications to generalize the standard QMC sampling to Eq. (45), serious problems occur when the quantity $S(\tau)$ becomes very small. In fact the physical quantity in Eq. (45) is obtained after cancellation of positive and negative terms, spoiling the accuracy of the statistical calculation. In the following we will show that by choosing a more symmetric type of Hubbard Stratonovich decoupling in (2,9,11) we can in principle solve the so called "fermion sign problem".

4.1   Symmetric Hubbard-stratonovich transformation

In the usual path integral expression for Z (13) the sign of $\lambda \propto \sqrt{\Delta\tau}$ in the HST (7) is arbitrary and the basic expression in Eq (7) is valid for both determinations of the sign of λ. The resulting propagator U_σ contains many unphysical terms since it is an analytic function of $\sqrt{\Delta\tau}$ and not $\Delta\tau$ itself as the physical propagator $e^{-H\Delta\tau}$. By symmetrizing the Hubbard Stratonovich transformation , we can remove these difficulties and obtain a better behaved partition function. In fact the basic Hubbard Stratonovich transformation can be symmetrized easily by writing the formal identity:

$$e^{-\Delta\tau V} = \int d\mu_\sigma(t) \cosh \lambda V_t \qquad (48)$$

with the definitions given in Eq. (12).

The new partition function looks very similar to the previous one (13) and is in fact obtained by replacing $e^{\lambda V_t} \rightarrow \cosh(\lambda V_t)$ in the propagator U_σ of Eqs. (9). The new path integral formulation is equivalent to take the previous expression for Z (13) and, for each given field $\sigma_{r,t}$, make a partial summation of the configurations, obtained by taking all possible 2^L sign change of λ, on each imaginary time slice.

---

[4] the average sign is always a positive quantity for long enough runs

In the following we will show that this partial summation of configurations will much alleviate the sign problem, known to be a serious difficulty in Quantum Monte Carlo simulation of fermion systems. In particular we will show that *in the limit of $\Delta\tau \to 0$ the sign problem does not exist in the symmetrized path integral formulation.*

We consider a complete basis $|\phi_i>$ in the Hilbert space at fixed number of spin up and spin down electrons. Each state ψ in this Hilbert space can be written

$$\psi = \sum_i \phi_i |\phi_i > \tag{49}$$

At a given time t the propagated wavefunction $\psi_t = U_\sigma(t,0)|\psi_T >$ is distributed in the Euclidean Hilbert space according to a given probability distribution, determined by the random Hubbard Stratonovich fields:

$$P(\psi,t) = \int [d\mu_\sigma]_{(t,0)} \, \delta(\psi - \psi_t) \tag{50}$$

Notice that the initial distribution is given by:

$$P(\psi, t = 0) = \delta(\psi - \psi_T). \tag{51}$$

Using the definition of the symmetrized propagation we can easily derive the Markov process that this distribution satisfies:

$$P(\psi, t+\Delta\tau) = \int d\mu_\sigma(t+\Delta\tau) \int [d\mu_\sigma]_{(t,0)} \, \delta(\psi - \cosh(\lambda V_{t+\Delta\tau})e^{-H_0\Delta\tau}\psi_t) \tag{52}$$

In the previous integrand we can make the substitution:

$$\psi = \cosh(\lambda V_{t+\Delta\tau})e^{-H_0\Delta\tau}\psi' \tag{53}$$

and by factoring out the jacobian from the delta function

$$\frac{\partial \psi'}{\partial \psi} = \det^{-1}\left[\cosh(\lambda V_{t+\Delta\tau})e^{-H_0\Delta\tau}\right] \tag{54}$$

we get:

$$P(\psi, t+\Delta\tau) = \int d\mu_\sigma(t+\Delta\tau) \, |\frac{\partial \psi'}{\partial \psi}| \, P(e^{H_0\Delta\tau}\cosh(\lambda V_{t+\Delta\tau})^{-1}\psi, t) \tag{55}$$

Then by expanding to first order in $\Delta\tau$ both Eq. (54) and Eq. (55) we find:

$$P(\psi, t+\Delta\tau) = \int d\mu_\sigma(t+\Delta\tau)\left\{(1 + a_\sigma\Delta\tau)P(\psi,t) + \Delta\tau\sum_i \frac{\partial}{\partial \phi_i}P(\psi,t)\dot\phi_i\right\}$$

$$a_\sigma = tr\left[H_0 - \frac{U}{2}\sum_{r,r'}\sigma_{r,t+\Delta\tau}\sigma_{r',t+\Delta\tau}(n_{\uparrow,r} - n_{\downarrow,r})(n_{\uparrow,r'} - n_{\downarrow,r'})\right]$$

$$\dot\phi_i = \sum_j\left[H_0^{i,j}\phi_j - \frac{U}{2}\sum_{r,r'}\sigma_{r,t+\Delta\tau}\sigma_{r',t+\Delta\tau}\left[(n_{\uparrow,r} - n_{\downarrow,r})(n_{\uparrow,r'} - n_{\downarrow,r'})\right]^{i,j}\right]\phi_j$$

$$\tag{56}$$

Finally by taking the integral over $\sigma_{r,t+\Delta\tau}$, using that $\int d\mu_\sigma(t)\sigma_{r,t}\sigma_{r',t} = \delta_{r,r'}$, we get for $\Delta\tau \to 0$:

$$\frac{\partial}{\partial t}P(\psi,t) = tr(H)P(\psi,t) + \sum_{i,j}H^{i,j}\phi_j\frac{\partial}{\partial\phi_i}P(\psi,t) \tag{57}$$

The solution of the previous first order differential equation is known exactly and depends only on the many body hamiltonian matrix in the chosen basis $\{\phi_i\}$:

$$P(\psi,t) = P(e^{Ht}\psi,0)\det(e^{Ht}) \tag{58}$$

The previous solution satisfies $\int [d\psi]P[\psi,t] = 1$, as it is easy to verify, and by the initial condition (51) is non zero only for $\psi = e^{-Ht}|\psi_T>$, i.e.:

$$P(\psi,t) = \delta(\psi - e^{-Ht}|\psi_T>). \tag{59}$$

The average sign $S(\tau)$ is defined in Eq. (46) Using Eq. (50) it is easy to show that $S(\tau)$ can be expressed in terms of $P(\psi,\tau)$ only:

$$S(\tau) = \frac{\int [d\psi] < \psi_T|\psi > P(\psi,\tau)}{\int [d\psi]| < \psi_T|\psi > |P(\psi,\tau)} \tag{60}$$

Thus, substituting the solution (59) in the previous equation, it follows that $S(\tau) = 1$ independent of τ.[5] Reaching the $\tau \to \infty$ limit does not present any particular problem in this approach since $S(\tau)$ is constant and equal to its maximum value.

This shows that our symmetrized Hubbard Stratonovich transformation does not suffer the sign problem in the limit of $\Delta\tau \to 0$. Of course the symmetrization procedure causes an exponential increase of computational time $\sim 2^L$. On the other hand the statistical fluctuations are much improved because $P(\psi,t)$ is sharply peaked around ψ_G for $t \to \infty$ and , in principle, only a single configuration $\{\sigma_{r,t}\}$ might be enough to sample the ground state. This is also an interesting self-averaging property satisfied by the symmetric HST and *not* by the conventional one.

5   Conclusion

We have reviewed here a numerical method to simulate fermion systems by means of auxiliary fields. The interested reader can find the details of the algorithm in section (3). We have derived how to apply the functional integral formulation in the infinite U limit, where it becomes more or less similar to the infinite U algorithm in [14]. Usual Monte Carlo simulations face enormous difficulties with the so called "fermion sign problem". In the HST formulation we have shown that, although a "sign problem" exists, it is not related to the difficulty to sample the fermion wavefunction, but to a more subtle analytical property of the HST. The basic reason why the HST suffers the sign problem is because the Trotter-Suzuki discretization of the imaginary

[5]The many body propagator is positive definite and $< \psi_T|e^{-Ht}|\psi_T >> 0$

time introduces unphysical terms in the limit $\Delta\tau \to 0$. A theoretical proof that it is possible to define a symmetric HST without generating the mentioned unphysical terms is described in details in the last section. From the practical point of view this kind of transformation may have useful application to perform simulations at lower temperatures or larger U in the Hubbard model. For the time being, a preliminary study[15] of the two-dimensional electron gas interacting with the long range Coulomb potential suggests that a big improvement in computational time may be obtained by implementing the symmetrized HST (48). Of course it is not necessary to symmetrize each of the L imaginary time slices. A good compromise is to choose a reasonable time of symmetrization to increase the average sign up to a feasible value, such to obtain meaningful statistical averages. The practical application of the symmetrized HST to the Hubbard model is under way.

I gratefully acknowledge S. Silvestrelli, S. Baroni and R. Car for useful discussions about their preliminary results[15].

References

1. Kalos M. H., in *Monte Carlo Method in Quantum Physics*, edited by M. H. Kalos NATO ASI Series, (D. Reidel Publ. Co., Dordrecth, 1984) p.19.
2. Ceperley D.M., in *Recent Progress in Many-Body Theories*, edited by J. G. Zabolitzky, M. de Liano, M. Fortes and J. M. Clark (Springer Verlag, Berlin, 1981), p. 262.
3. G. Sugiyama and S. E. Koonin, Ann. Phys. **168**, 1 (1986).
4. S. Sorella, S. Baroni, R. Car and M. Parrinello, Europhys. Lett. **8**, 663 (1989).
5. S.R. White, D. J. Scalapino, R.L. Sugar, E.Y. Loh, J.E. Gubernatis and R.T. Scalettar, Phys. Rev. B **40**, 506 (1989).
6. M. Imada and Y. Hatsugai, J. Phys. Soc. Jpn. **58**, 3752 (1989).
7. J.E. Hirsch, Phys. Rev. **B 31**, 4403 (1985).
8. E. Y. Loh, J. E. Gubernatis, R. T. Scalettar, S. R. White, D. J. Scalapino, and R. J. Sugar, Phys. Rev. B **41**, 9301 (1990).
9. S.Sorella Int. J. Mod. Phys. B **5** 937 (1991).
10. S. Fahy and D.R. Hamman, Phys. Rev. B **43**, 765 (1991).
11. White S. R., Wilkins J. and Wilson K., Phys. Rev. Lett. **56**, 4 (1986).
12. E.H. Lieb, Phys. Rev. Lett. **62**, 1201 (1989).
13. Blankenbecker R., Scalapino D.J., and Sugar R.L., Phys. Rev. D **24** 2278 (1981).
14. X. Y. Zhang, Elihu Abrahams and G. Kotliar, Phys. Rev. Lett. **66**, 1236 (1991).
15. S. Silvestrelli, S. Baroni and R. Car, in preparation.

GROUND-STATE PROJECTION USING AUXILIARY FIELDS

S. Fahy
Department of Physics
University College, Cork, Ireland

Abstract

This paper discusses the method of projecting out the ground-state of an interacting fermion system using propagation of a trial state in imaginary-time, with particle interactions represented exactly by the auxiliary-field (Hubbard-Stratonovitch) approach. The sampling of fields using the hybrid molecular-dynamics-Monte-Carlo method is discussed in this context. An understanding of sign problems which occur in the auxiliary-field ground-state projection method in terms of a diffusion-like equation is reviewed and an approximate ground-state projection method which arises naturally from this understanding is presented.

1   Introduction

One of the fundamental problems in electronic structure theory is that of adequately treating correlation in strongly interacting systems. Even if one is interested purely in ground-state properties (energy, particle density, correlation functions, etc.), leaving entirely aside the dynamical properties, one is faced in general with technical and conceptual problems which have defied any all-embracing solution. A number of techniques have been developed to address this problem, each with its own strengths and technical difficulties.

In this paper, I will discuss one such method, the auxiliary-field or Hubbard-Stratonovitch method of projecting the ground-state from a trial state using the method of "imaginary-time" evolution. Sugiyama and Koonin [1] first used the auxiliary-field approach in this context. It has much in common, both technically and conceptually, with the use of auxiliary-fields in calculating finite temperature grand-canonical averages [2], where the inverse of the temperature plays the role of imaginary time, though there are some important differences.

The auxiliary-field approach can be said in a certain sense to map the problem of finding the ground-state of the interacting quantum system into a problem of sampling averages over the auxiliary-fields subject to a classical potential. The mapping is in principle exact, and can yield exact quantum ground-state correlation functions as long as the classical sampling can be achieved exactly. This promise of "exactness" is the appeal of the method, and the reason that considerable effort has been put into its implementation.

However, in practice, technical problems intervene. The "fermion sign problem", which has been the focus of much attention and effort [3-6], makes it difficult to obtain reliable estimates of quantum observables from the classical sampling of the fields. In this paper I will review one recent approach to understanding the sign problem as it occurs in the auxiliary field ground-state projection approach and an approximate method, which we have called the "positive projection approximation", for projecting the quantum ground state which arises in a fairly natural way from this understanding and which has given satisfactory results in a number of lattice models. It is worth noting, however, that even in the absence of any sign problems, the classical sampling of the fields is far from trivial, with many of the same problems of slow dynamics (critical-slowing-down) that arise in the simulation of purely classical systems. Slow sampling dynamics can be as much of an impediment to the practical implementation of the auxiliary-field approach as the sign problems [7-9].

2   The Auxiliary-Field Representation

The imaginary-time projection approach [1] is based on the idea that if we formally solve the time-dependent Schrödinger equation,

$$-i\hbar\frac{\partial|\Psi>}{\partial t} = H|\Psi>,$$ (1)

in imaginary time $\tau = it/\hbar$, we find that the weight of excited states of the Hamiltonian dies exponentially with increasing τ:

$$|\Psi(\tau)> = \exp[-\tau H]|\Psi(0)> = \sum_i \exp[-\tau E_i]|\Psi_i><\Psi_i|\Psi(0)>$$

$$\rightarrow \exp[-\tau E_0]|\Psi_0><\Psi_0|\Psi(0)>, \quad \text{as } \tau \rightarrow \infty,$$

leaving only the ground-state of the system at large imaginary times. This obviously assumes that the initial, trial state $|\Psi(0)>$ is not orthogonal to the ground state $|\Psi_0>$ and that the product $\tau(E_1 - E_0)$, where E_1 is the first excited state, is large enough to ensure that the ground-state dominates the sum. Moreover, if this concept is to be useful as a way of calculating ground-state properties, we must have a way of representing the imaginary time propagator $\exp[-\tau H]$ which can be explicitly evaluated in practice.

To do so, we need to make explicit use of the form of the Hamiltonian H. In the Green's Function or Diffusion Monte Carlo approach [10] (for continuous, not lattice, systems), one makes use of the specific ∇^2 form of the kinetic energy operator to make the imaginary-time Schrödinger equation look like the diffusion equation. The realization of the propagator is then achieved by a random walk process in the coordinates of the particles.

In the auxiliary-field approach [2,11], we will rely on a different aspect of the form of the many-body Hamiltonian, namely, the fact that two-body operators (the interaction terms in the Hamiltonian) can be written as the square of one-body operators. Thus, we will assume the Hamiltonian can be written in the form

$$H = H_0 - \frac{1}{2}\sum_{i=1}^{n}(H_i)^2 , \tag{2}$$

where the H_i are all one-body operators. The term H_0 gives the non-interacting part of the Hamiltonian, and the other terms give the interactions. (It turns out that the choice of the set of H_i is not unique, and this freedom allows many different auxiliary-field representations of the same interacting system.)

In order to see how this helps us to evaluate the propagator, let us first look at the case of a non-interacting system. Here the explicit evaluation of the many-particle propagator is trivial once we have the propagator for a single-particle. The non-interacting Hamiltonian consists of one-body terms only. If the initial state $|\Psi(0)>= |\psi_1(0),\ldots,\psi_N(0)>$ is a Slater determinant (i.e., an anti-symmetric product) of single particles states $\psi_i(0)$, $i = 1, N$, then the propagated state is again a Slater determinant, $|\Psi(\tau)>= |\psi_1(\tau),\ldots,\psi_N(\tau)>$, where each of the single-particle states has propagated independently:

$$|\psi_i(\tau) >= \exp[-\tau H]|\psi_i(0) > .$$

This is nothing other than a mathematical statement of the physical fact that the motion of non-interacting particles is independent of each other. From a computational point of view, it means that if we know how to propagate single-particle states (a relatively trivial operation), we can construct the many-particle propagator.

Moreover, this same result holds even if the Hamiltonian propagating the states is time-dependent, as long as it always has only one-body terms, corresponding to a non-interacting system. Thus, if H is non-interacting but depends on τ, we can define a propagator $U(\tau,0)$ which satisfies

$$\frac{dU(\tau,0)}{d\tau} = -H\,U(\tau,0) .$$

Then $|\Psi(\tau)>= |\psi_1(\tau),\ldots,\psi_N(\tau)>$, satisfies the time-dependent Schrödinger equation when each $|\psi_i(\tau)>= U(\tau,0)|\psi_i(0)>$ propagates independently. (An analogous result [2] exists for density matrices of non-interacting systems in the finite-temperature, grand canonical approach.)

If we divide the interval from 0 to τ into m discrete intervals (so-called time-slices) of length $\Delta\tau$, and assume that the Hamiltonian is constant on each interval, the propagator U can be written as an ordered product of simple exponentials on each of the intervals:

$$U(\tau,0) = \exp[-\Delta\tau H(m)]\exp[-\Delta\tau H(m-1)]\ldots\exp[-\Delta\tau H(1)] ,$$

where $H(l)$ is the Hamiltonian on the lth time interval.

In order to take advantage of this simple form of the propagators for a non-interacting system in the case of an interacting Hamiltonian of the form given above in Eq. 2, we make use of the Hubbard-Stratonovitch transformation [12], based on the following simple mathematical operator identity:

$$\exp[\frac{1}{2}\Delta\tau O^2] = \int_{-\infty}^{+\infty} \exp(-xO)\, \frac{e^{-x^2/(2\Delta\tau)}}{\sqrt{2\pi\Delta\tau}}\, dx\ ,$$

where O is an operator. Here we have replaced a propagator involving O^2 on the l.h.s. with an integral over propagators involving just O on the r.h.s.. We can think of the variable x on the r.h.s. as a gaussian random field (the auxiliary-field) which couples to the system through the operator O.

We can now use this relation in evaluating the propagator for the interacting Hamiltonian of the form given above. We break the total propagation interval τ into m time-slices of length $\Delta\tau$ (the Trotter-Suzuki breakup [13]). Neglecting terms of order $\Delta\tau^2$, we now use the Hubbard-Stratonovitch transformation on each $(H_i)^2$ on each time-slice to get:

$$\exp[-\tau H] \approx \int dG[\mathbf{x}]\, U_{\mathbf{x}}(\tau,0)\ ,$$

where the propagator,

$$U_{\mathbf{x}}(\tau,0) = \prod_{l=1}^{m}\left[\exp(-\Delta\tau H_0)\prod_{i=1}^{n}\exp(-x_{il}H_i)\right]$$

is determined by the random auxiliary-fields $\mathbf{x} \equiv \{x_{il}\}$ with gaussian weights

$$dG[\mathbf{x}] = \prod_{l=1}^{m}\prod_{i=1}^{n}\frac{e^{-x_{il}^2/(2\Delta\tau)}}{\sqrt{2\pi\Delta\tau}}dx_{il}\ .$$

Thus, we have replaced all interacting propagators with non-interacting propagators which are τ-dependent. The price to be paid is that on each time-slice, and for each interaction term $(H_i)^2$, we have a gaussian random field which must be integrated over to reconstruct the interacting propagator. Also note that to obtain an exact representation of the interacting system, we must take the limit of a very large number of time-slices, $m \to \infty$.

3 Sampling of Fields

In the practical implementation of the ground-state projection method, we do the appropriate integration over the fields by Monte Carlo sampling methods [1]. Thus, if we wish to evaluate the ground state expectation of an observable A, we calculate

$$< A > = \frac{\int < \Psi_t | U_\mathbf{x}(\tau, \tau/2) A U_\mathbf{x}(\tau/2, 0) | \Psi_t > dG[\mathbf{x}]}{\int < \Psi_t | U_\mathbf{x}(\tau, 0) | \Psi_t > dG[\mathbf{x}]} ,$$

where Ψ_t is a trial state (a Slater determinant) and τ is sufficiently large that all excited states are projected out in an imaginary-time interval $\tau/2$. The magnitude of the matrix element $< \Psi_t | U_\mathbf{x}(\tau, 0) | \Psi_t >$ varies enormously for different choices of fields $\mathbf{x}$ and so importance sampling of the fields is used, including the magnitude of the matrix element along with the gaussians in the importance sampling scheme:

$$d\mu[\mathbf{x}] = | < \Psi_t | U_\mathbf{x}(\tau, 0) | \Psi_t > | \, dG[\mathbf{x}] . \tag{3}$$

Note that the matrix element is not always positive, giving rise to ill-conditioned statistical estimators when the size of the positive and negative contributions to the normalization integral are almost equal — the fermion sign problem, which we will discuss in detail below.

This sampling process can be viewed in terms of classical statistical mechanics [14] as sampling from a Boltzmann distribution of the fields $\mathbf{x}$ at temperature $= 1$, with a potential energy defined as

$$V(\mathbf{x}) = -\log | < \Psi_t | U_\mathbf{x}(\tau, 0) | \Psi_t > | + \frac{\mathbf{x}^2}{2\Delta\tau} . \tag{4}$$

Fields $\mathbf{x}$ which have very high "potential energy" $V(\mathbf{x})$ make very little contribution to the integral and are sampled very rarely. Fields which have low energy make the dominant contribution to the integral and are sampled very often in the importance sampling scheme.

In almost all importance sampling schemes, the fields are sampled using a random walk process, each field being generated from the previous field by a "step" which is determined by some set of rules. The efficiency of the sampling scheme is dependent on how efficiently the rules for making steps - the "simulation-time dynamics" - allow the process to explore all important regions of the fields.

One scheme for importance sampling of the fields, the hybrid molecular-dynamics Monte Carlo (MD/MC) method [15,16], is based in a very direct way on the analogy to the Boltzmann distribution and its appearance as the distribution in the physical process of Brownian motion. The method is closely related to the Langevin approach [16] for sampling the Boltzmann distribution, but has the advantage that longer steps can be taken in the random walk using an approximate integration of the classical equations of motion without introducing any error into the steady-state distribution sampled. The longer steps may be thought of as reducing the correlation between successive fields selected in the random walk, thereby improving the efficiency of the sampling.

One may think of the field values **x** as defining the coordinates of particles moving in the potential $V(\mathbf{x})$ as defined in Eq. 4. The particles are stopped at regular intervals t_0 (similar to collisions in Brownian motion, except occuring at regular intervals) and have their velocities reset to values chosen at random from a Maxwell distribution for temperature = 1. If the trajectories of the particles between stops and starts are determined by Newton's equations in the potential $V(\mathbf{x})$ (each particle having unit mass), then the steady-state distribution of positions **x** is the Boltzmann distribution $\exp[-V(\mathbf{x})]$. Thus the scheme achieves importance sampling of the fields with weighting given by Eq. 3.

Moreover, even if the integration of Newton's equations is not exact (but is exactly reversible and exactly preserves phase-space volumes — the simple "leap-frog" integration method [15] satisfies this condition); it can be shown that the steady state distribution remains $\exp[-V(\mathbf{x})]$ if we introduce a Monte Carlo accept/reject decision (as in the Metropolis Method) at the end of each attempted step. If the total classical energy associated with the fields in the molecular-dynamics, $H = K + V$, where K is the classical kinetic energy associated with the motion of the fields in simulation-time, changes by an amount δH due to inaccurate numerical integration of Newton's equations, we accept the move with probability

$$P = \min[1, \exp(-\delta H)] .$$

If the MD integration is exact, moves are always accepted and one has a pure MD sampling scheme. As the MD integration becomes less accurate, the Monte Carlo accept/reject process compensates exactly for any errors. In practice, there is a balance between keeping the acceptance rate for moves high enough and spending too much effort integrating the MD equations of motion accurately.

An inexact integration of Newton's equations proves to be not only computationally convenient but strictly necessary for correct integration in cases where positive and negative matrix elements $< \Psi_t|U_\mathbf{x}(\tau, 0)|\Psi_t >$ can occur for different values of the fields **x**. The potential $V(\mathbf{x})$ has infinite barriers where the matrix element changes sign, so that if the integration were exact the sampling method would never step from regions of positive sign to regions of negative sign. Such steps are necessary if the detailed balance of integration weight between the different regions is to be satisfied. In practice, with a reasonable numerical integration step size, the hybrid MD/MC sampling moves from positive to negative sign in a time typically less than one hundred MD/MC steps.

Even in the absence of infinite barriers associated with the change of sign of the matrix elements, the more general character of the landscape defined by the potential $V(\mathbf{x})$ affects the simulation-time dynamics in a very important way [7-9,17]. In the MD/MC method, moves are made within contiguous highly probable regions of the space of fields — the valleys of the potential $V(\mathbf{x})$ — and moves over regions with high (but finite) values of the potential — the mountain peaks and passes — may be difficult and rare. Understanding the general features of this landscape in any specific model is a very important step towards developing better rules for governing the random walk to ensure efficient sampling of all important regions of the auxiliary-fields, and even for devising better forms of the Hubbard-Stratonovitch transformation itself [7-9].

4 Sign Problems and the Positive Projection Approximation

Even when detailed balance can be achieved between regions of $\mathbf{x}$ where the sign,

$$s(\mathbf{x}) \;=\; \frac{<\Psi_t|U_{\mathbf{x}}(\tau,0)|\Psi_t>}{|<\Psi_t|U_{\mathbf{x}}(\tau,0)|\Psi_t>|} \; ,$$

of the matrix element is positive and the regions where it is negative, there remains a fundamental numerical difficulty in calculating observables using the auxiliary-field approach when negative matrix elements are possible. Although there are important exceptions (see below), it is usually the case that the matrix element $<\Psi_t|U_{\mathbf{x}}(\tau,0)|\Psi_t>$ can be either positive or negative, depending on the choice of fields $\mathbf{x}$. In that case, the average sign,

$$<s> \;=\; \frac{\int <\Psi_t|U_{\mathbf{x}}(\tau,0)|\Psi_t> dG[\mathbf{x}]}{\int |<\Psi_t|U_{\mathbf{x}}(\tau,0)|\Psi_t>|dG[\mathbf{x}]} \; ,$$

will usually decay exponentially to zero as τ increases [3-6]. Any Monte Carlo estimate of $<s>$ is then ill-conditioned, the fluctuations in the estimate overwhelming the average value. The same is true of any other observable, which is calculated in the importance sampling method as,

$$<A> \;=\; \frac{1}{<s>} \int \frac{<\Psi_t|U_{\mathbf{x}}(\tau,\tau/2)AU_{\mathbf{x}}(\tau/2,0)|\Psi_t>}{<\Psi_t|U_{\mathbf{x}}(\tau,0)|\Psi_t>} s(\mathbf{x})d\mu[\mathbf{x}] \; ,$$

where $d\mu[\mathbf{x}]$ specifies the weighting used in the importance sampling Monte Carlo integration, as in Eq. 3.

In this equation, the correct value of $<A>$ appears to depend crucially (in the large τ limit) on a delicate cancellation of contributions from the positive and negative regions of $s(\mathbf{x})$. It was then quite remarkable when Sorella *et al.* [3] found that numerically accurate results could be obtained for many observables when the sign $s(\mathbf{x})$ is neglected in the equation (i.e., setting $s(\mathbf{x}) = 1$).

In order to understand the surprising success of this approximation and to introduce the positive-projection approximation [4,18], which we believe to be in general more accurate than the simple neglect of the sign $s(\mathbf{x})$, it is useful to look again at the representation of the propagated many-body state at imaginary time τ:

$$|\Psi(\tau)> \ = \ \exp[-\tau H]|\Psi_t> \ \approx \ \int dG[\mathbf{x}] \ U_{\mathbf{x}}(\tau,0)|\Psi_t> \ \approx \ \int dG[\mathbf{x}] \ |\Psi_{\mathbf{x}}> \ ,$$

where $|\Psi_{\mathbf{x}}> \ = U_{\mathbf{x}}(\tau,0)|\Psi_t>$ is a Slater determinant [1], dependent on the fields $\mathbf{x}$ on the interval $[0,\tau]$ and the initial trial state $|\Psi_t>$. Here it is clear that $|\Psi(\tau)>$ is being represented as an integral linear combination (or, in practice, a finite sum) of Slater determinants [4]. (Note that the determinants involved in the sum are *not* linearly independent.)

One can then transform this integral representation of $|\Psi(\tau)>$ to an integral over the set of all normalized Slater determinants $|\psi>$ with a positive weight function f:

$$|\Psi(\tau)> \ = \ \int d\psi f(\psi,\tau)|\psi> \ ,$$

where $|\psi> \ = |\Psi_{\mathbf{x}}> /||\Psi_{\mathbf{x}}>|$ [4]. Here we are explicitly writing the propagated state as a linear combination of a distribution f of normalized Slater determinants $|\psi>$. Formally, we have replaced the integral over the set of all auxiliary-fields $\mathbf{x}$ with an integral over the intrinsic coordinates ψ which label the set of all normalized Slater determinants by means of a coordinate transformation. (The set of normalized Slater determinants is a manifold, but not a linear subspace, in the Hilbert space of many-body states – see Ref. 4 for details.) The positive weight function $f(\psi,\tau)$ then includes both the Jacobian of the coordinate transformation from $\mathbf{x}$ to ψ and the magnitude $||\Psi_{\mathbf{x}}>|$ of the propagated state.

It can be shown [4] that the distribution f obeys a diffusion-like equation,

$$\frac{\partial f}{\partial \tau} \ = \ \frac{1}{2}\mathbf{D}(\psi)f - [\nabla_\psi V_1(\psi)] \cdot \nabla_\psi f + V_2(\psi)f \ .$$

The differential diffusion operator $\mathbf{D}$, the drift potential V_1, and the branching potential V_2 can all be written explicitly in terms of the original Hamiltonian and the Hubbard-Stratonovitch transformation used. (It should be emphasized that V_1 and V_2 have nothing to do with the effective classical potential V discussed above in connection with the MD/MC sampling of fields.) Formally solving this equation of motion for f, with the initial condition $f(\psi,0) = \delta(\psi - \psi_t)$, gives the evolution of the true many-body wavefunction $|\Psi(\tau)>$.

Note that, although $|\Psi(\tau)>$ is uniquely determined by the original interacting many-body Hamiltonian, f is not. There is more than one possible Hubbard-Stratonovitch transform which (when all auxiliary fields are integrated out) gives the same physical Hamiltonian. These different Hubbard-Stratonovitch transformations will give rise to different $\mathbf{D}$'s and V's in the diffusion equation, leading to different functions $f(\psi,\tau)$ but the same $|\Psi(\tau)>$. This is not too surprising when we consider that the auxliiary field approach represents the propagated state as a linear combination of an arbitrary collection of Slater determinants. This is a highly over-complete representation, since no conditions of linear independence are imposed on the Slater determinants used and the number of possible Slater determinants is much larger than the dimension of the many-body Hilbert space — indeed the number of determinants is infinite. This will prove to be an important point in understanding the generic sign problems associated with the auxiliary-field approach.

To understand the exponential decay of the average sign $< s >$, it is helpful to consider [4] the expansion of the solution of the evolution equation for f in terms of eigenfunctions f_i and eigenvalues ϵ_i of its right hand side, as follows:

$$f(\psi, \tau) = \sum_i c_i \exp[-\epsilon_i \tau] f_i(\psi) .$$

The evolution equation is invariant under inversion of the Hilbert space (and the set of Slater determinants), $\psi \to -\psi$, and so the eigenfunctions f_i are either odd $(f(\psi) = -f(-\psi))$ or even $(f(\psi) = f(-\psi))$ with respect to this inversion. Moreover, from the general properties of diffusion-like equations (in particular, their property that functions which are initially positive always remain positive), it follows that the dominant eigenfunction f^+ at large τ must be positive everywhere, and so must be even under inversion. But an even function contributes nothing to the physical state $\Psi(\tau)$ since $\int d\psi f^+(\psi)|\psi > \equiv 0$, the contributions of ψ and $-\psi$ exactly cancelling. Thus, while $f^+(\psi)$ dominates the magnitude of the integrand $f(\psi, \tau)$ at large τ, the dominant odd eigenfunction $f^-(\psi)$ determines all physical integrals. The rate of exponential decay of the sign $< s >$ at large values of τ is then equal to the difference between the eigenvalues associated with the dominant odd and even eigenfunctions.

4.1   The Positive Projection Approximation

Although the detailed form of the potentials V_1 and V_2 in terms of the H_i in the Hubbard-Stratonovitch transformation are quite complicated, the overall effect of these "confining potentials" can be understood qualitatively as simply causing the distribution $f(\psi, \tau)$ to be peaked near the mean-field solutions of the many-body Hamiltonian for large values of τ. (This is especially true when the interaction terms H_i^2 are weak.) Thus, V_1 and V_2 have minima near the mean-field solutions, so that drift and branching tends to enhance the magnitude of the solutions $f^+(\psi)$ and $f^-(\psi)$ near the mean-field solutions. Note that ψ_m, the mean-field solution, and $-\psi_m$ are considered as distinct points in the set of normalized Slater determinants and usually have barriers of the potentials V_1 and V_2 separating them. The dominant physically relevant (i.e. odd) eigenfunction f^- will have a node between ψ_m and $-\psi_m$. Intuitively, we expect the node of this function to be where V_1 and V_2 are large and all long-lived odd and even eigenfunctions f_i are small in magnitude. Thus, we expect the nodes of f^- to be well removed from the mean field solutions and we expect f^- and f^+ to be very similar in form close to the mean field solutions, just as in the solution of a standard double-well diffusion problem.

This view of the auxiliary-field approach allows us to understand why the approach of neglecting the sign $s(\mathbf{x})$, as proposed by Sorella *et al.* [3], can make sense in many cases and give accurate results for many physical quantities. In the language we have been using here, ignoring the sign $s(\mathbf{x}) = s(\psi_{\mathbf{x}})$ in calculating physical quantities is equivalent to setting $f^-(\psi) = f^+(\psi)s(\psi)$. This is a reasonably good approximation when the node of f^- occurs in a region where f^+ is small in magnitude.

If we were to solve the diffusion equation with the boundary condition $f(\psi) = 0$ at the nodal surface of f^-, the dominant eigenfunction f^* would be identical to f^- and ground-state physical observables calculated using $|\Psi^* > = \int d\psi f^*(\psi)|\psi >$ would be exact. (We can think of this nodal condition as setting $V_2(\psi) = \infty$ on one side of the node.) If the node is placed approximately at the true nodal surface of f^-, f^* is approximately equal to f^- and $|\Psi^* >$ is a good approximation to the ground state wavefunction. A practical way of imposing a node in f is to demand that the projection $< \psi(\tau)|\Phi_c >$ onto a chosen state $\Phi_c >$ is positive for all $\tau > 0$ for any allowed set of auxiliary fields $\mathbf{x}$. If Φ_c is chosen with some physical insight (for example, an appropriate choice is often $\Phi_c = \psi_m$, the mean-field solution, or a linear combination of symmetry-related mean-field solutions [18]), then the node will be forced to lie where V_1 and V_2 are large and $|\Psi^* >$ will be a close approximation to the ground state. We have called this approach the "positive projection approximation".

There is a formal similarity between the positive projection approximation and the fixed-node diffusion Monte Carlo method [10] applied to continuous coordinate fermion problems. However, the relation to the physical problem is entirely different. This can be seen when one realizes that the diffusion in the standard diffusion Monte Carlo approach arises from the kinetic energy operator ∇^2, whereas the diffusion arises in the present problem from the interaction terms H_i^2 in the many-body Hamiltonian. Moreover, the fixed-node condition is applied in diffusion Monte Carlo to prevent collapse at large times into the boson ground state; in the auxiliary-field projection method anti-symmetry is automatically enforced and the collapse that is prevented by positive projection is into a function f^+ which has mathematical and numerical significance, but no physical relevance. In typical cases, the splitting between the eigenvalues associated with f^+ and f^- is much smaller than the difference of energies between the boson and fermion ground states [3-6,19].

We have introduced the diffusion equation for f above principally for conceptual reasons. Practical implementation [18,20] of the positive projection approach uses the auxiliary fields $\mathbf{x}$ in the functional integral form, rather than working directly with the diffusion equation in ψ. The nodal condition is then enforced by imposing the condition $< \psi(\tau_i)|\Phi_c > \geq 0$ at each time-slice τ_i. (We note that these constitute constraints on the fields $\mathbf{x}$ which are non-local in time, making a "global" update scheme for sampling the auxiliary-fields, like hybrid MD/MC, much more efficient than one where the fields on each time-slice are updated sequentially.) The constraint is applied in the Monte Carlo accept/reject part of each MD/MC move simply by rejecting moves where the fields $\mathbf{x}$ cause a violation of the constraint $<\psi(\tau_i)|\Phi_c> \geq 0$ on any time-slice τ_i. The constraint condition does not enter into the MD part of the sampling step. Thus, in situations where the MD moves do not give rise too often to violations of the positive projection constraints, the additional cost of the method over the unconstrained auxiliary-field method is small.

The positive projection approximation has been used [18,20] for a number of models where the sign problem precludes obtaining exact results from the auxiliary-field approach. With the positive projection constraint, the average sign saturates in the large-τ limit at values typically between 0.5 and 1. The comparison of positive projection results with exact numerical diagonalization for Hubbard models (with and without interaction between electrons on different sites) with a small number of sites has shown agreement for the total energy and kinetic energy, occupation of single-particle states, and pair-correlation functions within a few percent.

In certain special cases, no sign problem occurs. A well-known example, first pointed out by Hirsch [21] in the grand canonical approach, is the standard half-filled Hubbard model [22] (with on-site repulsive interactions only) on bipartite lattices. In the ground-state projection method, a similar lack of negative signs is found if the initial trial state is chosen with particle-hole symmetry. In this case exact particle-hole symmetry (i.e. the spin-up hole dynamics identical to the spin-down particle dynamics) causes the spin-up overlap $<\psi_\uparrow(\tau)|\Psi_{t,\uparrow}>$ to have the same sign as the spin-down overlap $<\psi_\downarrow(\tau)|\Psi_{t,\downarrow}>$ always. Thus, no overall sign changes occur. In the language we have been using to describe the auxiliary-field projection approach, particle-hole symmetry causes the functions f^+ and f^- to have degenerate eigenvalues because diffusion from ψ_t to $-\psi_t$ is impossible.

This lack of any negative signs has also been empirically observed [23] for the "three-field" approach, introduced in Ref. 20, when applied to half-filled one-dimensional chains with nearest-neighbour hopping but with *arbitrary intersite interactions* when the number of sites is twice an odd number. In this case, spin-up hole dynamics and spin-down particle dynamics are not identical and the spin-up and spin-down overlaps are both always positive. It is likely that this behaviour is again indicative of some fundamental symmetry in the problem, though it is not clear what that symmetry might be or if it has any important physical consequences.

From the analysis above it is clear that cases where the sign in the auxiliary-field projection approach does not decay exponentially with increasing τ are exceptional. They arise through a special degeneracy of the odd and even eigenfunctions $f^-(\psi)$ and $f^+(\psi)$ of the abstract diffusion problem on the set of all normalized Slater determinants to which the auxiliary-field projection method is mathematically equivalent. The fundamental symmetry between ψ and $-\psi$, along with the overcompleteness of the representation of the propagated state $\Psi(\tau)$, dictates that exponential decay must be the expected behaviour for an arbitrary problem. The positive projection approach breaks the $\psi \to -\psi$ symmetry of the problem and so generically gives a non-zero sign in the large-τ limit. It is thus a useful tool in cases where the sign problem precludes the application of the unconstrained auxiliary-field approach. The price to be paid is that the positive projection approach is approximate and requires some insight into the physical problem in its application if sensible results are to be obtained.

References

1. G. Sugiyama and S. E. Koonin, Annals of Phys. **168**, 1 (1986).
2. R. Blankenbeckler, D. J. Scalapino, and R. L. Sugar, Phys. Rev. D **24**, 2278 (1981).
3. S. Sorella, E. Tosatti, S. Baroni, R. Car, and M. Parrinello, Int. J. Mod. Phys. B **1**, 993 (1988).
4. S. Fahy and D. R. Hamann, Phys. Rev. B **43**, 765 (1991).
5. E. Y. Loh, J. E. Gubernatis, R. T. Scalettar, S. R. White, D. J. Scalapino, and R. L. Sugar, Phys. Rev. B **41**, 9301 (1990); G. G. Batrouni and R. T. Scaletter, Phys. Rev. B **42**, 2282 (1990).
6. N. Furukawa and M. Imada, J. Phys. Soc. Japan **60**, 810 (1991).
7. R. T. Scalettar, R. M. Noak, and R. R. P. Singh, Phys. Rev. B **44**, 10502 (1991).
8. L. Chen and A.-M. S. Tremblay, (unpublished).
9. S. Fahy, in *Computer Simulations in Condensed Matter Physics VI*, ed. D. P. Landau, K.K. Mon, and H. B. Schüttler (Springer Verlag, Heidelberg, Berlin, 1993).
10. J. B. Anderson, J. Chem. Phys. **65**, 4121 (1976).
11. J. W. Negele and H. Orland, *Quantum Many-Particle Systems*, Chapt. 7 (Addison-Wesley, Redwood City, California, 1988).
12. R. L. Stratonovitch, Dokl. Akad. Nauk. SSSR **115**, 1097 (1957) [Sov. Phys. Dokl. **2**, 416 (1957)]; J. Hubbard, Phys. Rev. Lett. **3**, 77 (1959).
13. H. F. Trotter, Proc. Am. Math. Soc. **10**, 545 (1959); M. Suzuki, Commun. Math. Phys. **51**, 183 (1976).
14. N. Metropolis, A. W. Rosenbluth, M. N. Rosenbluth, A. H. Teller, and E. Teller, J. Chem. Phys. **21**, 1087 (1953).
15. S. Duane, A. D. Kennedy, B. J. Pendleton, and D. Roweth, Phys. Lett. B **195**, 216 (1987), and references therein.
16. R. T. Scalettar, D. J. Scalapino, and R. L. Sugar, Phys. Rev. B **34**, 7911 (1986); S. R. White and J. W. Wilkins, Phys. Rev. B **37**, 5024 (1989);
17. S. Fahy and D. R. Hamann, Phys. Rev. Lett. **69**, 761 (1992).
18. S. Fahy and D. R. Hamann, Phys. Rev. Lett. **65**, 3437 (1990).
19. D. R. Hamann and S. Fahy, Phys. Rev. B **41**, 11352 (1990).
20. D. R. Hamann and S. Fahy, Phys. Rev. B **47**, 1717 (1993).
21. J. E. Hirsch, Phys. Rev. Lett. **51**, 1900 (1983); Phys. Rev. B **31**, 4403 (1985).

22. J. Hubbard, Proc. Roy. Soc. London, Ser. A **276**, 238 (1963).
23. S. Fahy, (unpublished).

FERMION SIMULATIONS OF CORRELATED SYSTEMS

Masatoshi Imada
Institute for Solid State Physics, University of Tokyo,
Roppongi 7-22-1, Minato-ku, Tokyo 106, JAPAN
e-mail: imada@tansei.cc.u-tokyo.ac.jp

Abstract

Recently developed simulation algorithms for strongly correlated lattice fermions are reviewed. The simulation methods which we describe are based on the path-integral formalism. Basic algorithms for auxiliary-field methods as well as prescriptions for reducing the minus-sign problem are presented. As an important application, ground state properties of the Hubbard model are discussed in the light of recent simulation results. Clarified so far, important features of antiferromagnetic transitions, metal-insulator transitions and pairing mechanisms in strongly correlated systems are summarized. Recent applications of the fermion Monte Carlo method to other fermion models with degenerate orbitals such as the d-p model are also reviewed. Heavy fermion systems, exhibiting a variety of crossovers such as the temperature crossover between the intersite antiferromagnetic correlation and the local singlet formation, are also discussed. Pairing mechanisms examined numerically in a variety of extended lattice models are briefly sketched.

1 Introduction

Quantum simulation methods have been developed in a variety of fields such as QCD, quantum chemistry and condensed-matter physics. In all of these fields, an important algorithmic problem is how to treat the fermion degrees of freedom. In this paper, we discuss recent trends in fermion simulations in condensed-matter physics[1,2]. Applications of fermion algorithms in this field range from electrons in a continuum space to tight-binding lattice fermions. The problem of electrons in the continuum space has been studied using, for example, the Green's function Monte Carlo method[3] and the path-integral Monte Carlo method[4]. In this paper, we only describe recent applications of fermion algorithms to lattice fermion models because they have been developed very rapidly in this decade.

In purely one-dimensional lattices, it is possible to transform fermion systems to a boson representation. The transformation has been successfully applied to the Hubbard model[5,6] as well as to the t-J model[7,8]. In two or more dimensions, however, such a transformation does not exist. Then, the notorious sign problem is one of the bottlenecks when one wishes to obtain physical properties in low-energy scale for large lattices. Several algorithmic improvements have been proposed to circumvent the difficulty and their application has yielded some useful information about lattice fermion systems. The auxiliary-field approach may be viewed as one of such devices. The basic points of the auxiliary-field method are briefly summarized in §2. Other prescriptions for the sign problem are described in §3. These methodological developments have made it possible to investigate ground-state and low-temperature properties of the Hubbard model. In fact, a variety of basic properties of the metal-insulator transition and antiferromagnetic transition in the Hubbard model have been clarified. These recent developments are reviewed in §4. In §5, simulations done for other strongly correlated systems such as the d-p model and heavy fermion models are discussed. Several extended models for investigating pairing mechanisms are also introduced and recent studies are summarized. In §6, the present status of fermion simulations and future problems are discussed.

2   Auxiliary-Field Method

To describe the auxiliary-field algorithm, we take the Hubbard model defined by the Hamiltonian

$$
\begin{aligned}
\mathcal{H} &= \mathcal{H}_0 + \mathcal{H}_1, \\
\mathcal{H}_0 &= -\sum_{\substack{\langle ij \rangle \\ \sigma}} t_{ij}(c_{i\sigma}^\dagger c_{j\sigma} + \text{H.c}), \\
\mathcal{H}_1 &= U\sum_i n_{i\uparrow}n_{i\downarrow},
\end{aligned}
\tag{2.1}
$$

where the creation (annihilation) operator for fermions with spin σ at the i-th site is denoted by $c_{i\sigma}^\dagger(c_{i\sigma})$. The transfer term $\mathcal{H}_0$ allows us to treat any kind of band structure; here the summation over $\langle ij \rangle$ may be taken over the nearest-neighbor pairs, the second-nearest-neighbor pairs and so on. The on-site Coulomb interaction is given in the second term $\mathcal{H}_1$. To calculate thermodynamic quantities, the differential equation for the density matrix $\rho(\beta) = e^{-\beta\mathcal{H}}$ given by

$$
\frac{\partial \rho(\beta)}{\partial \beta} = -\mathcal{H}\rho(\beta)
\tag{2.2}
$$

is solved following the path-integral formalism. The symbol β denotes the inverse temperature $1/T$, where the Boltzmann constant k_B is taken unity. We introduce the matrix element of the density matrix

$$
\rho(\varphi_i, \varphi_j; \beta) = \langle \varphi_i | e^{-\beta\mathcal{H}} | \varphi_j \rangle,
\tag{2.3}
$$

where $|\varphi_i\rangle$ is a many-body wave function. Using $\rho(\varphi_i, \varphi_j; \beta)$, we obtain thermodynamic quantities at a finite temperature T from the partition function

$$
Z = \sum_{\{\varphi_i\}} \rho(\varphi_i, \varphi_i; \beta),
\tag{2.4}
$$

where we take the summation over an orthonormalized complete set of $\{\varphi_i\}$. To obtain ground-state properties, one does not need to take the summation over $\{\varphi_i\}$, as we discuss below. The density matrix is formally rewritten as

$$\rho(\varphi_i, \varphi_j; \beta) = \langle\varphi_i|e^{-\beta\mathcal{H}}|\varphi_j\rangle = \langle\varphi_i|(e^{-\tau\mathcal{H}})^L|\varphi_j\rangle, \qquad (2.5)$$

where $\tau L = \beta$. Taking τ sufficiently small, we can apply the Suzuki-Trotter formula to each slice of the path integral as[9,10]

$$e^{-\tau\mathcal{H}} = e^{-\mathcal{H}_0\tau/2}e^{-\mathcal{H}_1\tau}e^{-\mathcal{H}_0\tau/2} + O(\tau^3). \qquad (2.6)$$

The discrete Stratonovich-Hubbard transformation introduced by Hirsch[11] in the form

$$e^{-\alpha n_{i\uparrow}n_{i\downarrow}} = \frac{1}{2}\sum_{s=\pm1}\exp[2as(n_{i\uparrow} - n_{i\downarrow}) - \frac{\alpha}{2}(n_{i\uparrow} + n_{i\downarrow})] \qquad (2.7)$$

is applied to $\exp[-\tau\mathcal{H}_1]$ in Eq.(2.6) to transform the interacting system into a sum of non-interacting systems under the imaginary-time-dependent field s, which is called the auxiliary variable or the Stratonovich variable. Here, a is defined by $a = \tanh^{-1}\sqrt{\tanh(\alpha/4)}$. The diagonal component of the density matrix $\rho(\varphi, \varphi; \beta)$ is then given as

$$\rho(\varphi, \varphi; \beta) = \sum_{\{s\}}W(\{s\}; \beta; \varphi), \qquad (2.8)$$

$$W(\{s\}; \beta; \varphi) = W_\uparrow(\{s\}; \beta; \varphi_\uparrow)W_\downarrow(\{s\}; \beta; \varphi_\downarrow), \qquad (2.9)$$

$$W_\sigma(\{s\}; \beta; \varphi_\sigma) = \langle\varphi_\sigma|\prod_{l=1}^{L}\Omega_{\sigma l}(s_1(l), \cdots, s_N(l))|\varphi_\sigma\rangle, \qquad (2.10)$$

where $\Omega_{\sigma l}(s_1(l), \cdots, s_N(l)) \equiv w_0 w_{1\sigma}(s_1(l), \cdots, s_N(l)) w_0$ and $w_0 = e^{-\tau\mathcal{H}_0/2}$. Here $s_i(l)$ represents the auxiliary variable at the i-th site in the l-th imaginary-time slice introduced in Eq.(2.7). We describe a single-particle representation of $|\varphi_\sigma\rangle$ as $|\Phi_\sigma\rangle$. The interaction part is transformed using

$$w_{1\sigma}(s_1(l), \cdots, s_N(l)) \equiv \prod_{i=1}^{N_s}(\frac{1}{2}\exp[2\sigma a_U s_i(l)n_{i\sigma} - \frac{\tau U}{2}n_{i\sigma}]), \qquad (2.11)$$

where $a_U = \tanh^{-1}\sqrt{\tanh(\tau U/4)}$ and N is the number of lattice sites, while σ takes the value $+1$ for $\uparrow$ and -1 for $\downarrow$. The sum over $\{s\}$ in Eq.(2.9) describes the summation of all the $N \times L$ auxiliary Ising variables $s_i(l)$. Because Ω is given from a non-interacting system, $|\varphi_\sigma\rangle$ may be represented using single-particle basis.

We first describe the method[7,12] for the canonical ensemble at $T = 0$. The ground-state properties in the canonical ensemble (i.e. fixed particle number) are obtained from $\rho(\varphi, \varphi; \beta)$ by taking β sufficiently large if $|\varphi\rangle$ is not orthogonal to the ground state. This is because $e^{-\beta\mathcal{H}}|\varphi\rangle$ converges to the ground state irrespective of the choice of $|\varphi\rangle$. For a given configuration of s, the equal-time single-particle Green's function at the l-th slice is given by

$$(G_\sigma(l))_{jk} = |R_\sigma(l)\rangle g_\sigma(l) \langle L_\sigma(l)|, \qquad (2.12)$$

$$|R_\sigma(l)\rangle = \Omega_{\sigma,l+1}\Omega_{\sigma,l+2}\cdots\Omega_{\sigma,L}|\Phi_\sigma\rangle, \qquad (2.13)$$

$$\langle L_\sigma(l)| = \langle \Phi_\sigma|\Omega_{\sigma 1},\Omega_{\sigma 2}\cdots\Omega_{\sigma l} \qquad (2.14)$$

and

$$g_\sigma(l) = (\langle L_\sigma(l)|R_\sigma(l)\rangle)^{-1}. \qquad (2.15)$$

The imaginary-time-dependent Green's function is obtained similarly. In the matrix representation, since $|\Phi_\sigma\rangle$ is given by an $N \times M$ matrix for M particles in an N-site system, Ω and G are $N \times N$ matrices while g is an $M \times M$ matrix. By taking β large, Green's function in the ground state is obtained from Eq.(2.13) in the region $1 \ll l \ll L$, where $e^{l\tau\mathcal{H}}|\phi\rangle$ and $e^{-(L-l)\tau\mathcal{H}}|\varphi\rangle$ converge well to the ground state. From Green's functions, thermodynamic quantities for a given set of s are obtained using Wick's theorem, because the system is already transformed to a noninteracting system after the Stratonovich-Hubbard transformation. Physical quantities of the original Hamiltonian $\mathcal{H}$ are obtained after the summation over s. In simulations such as the Monte Carlo algorithm, the summation is replaced with a sampling average. Usually, the importance sampling procedure described below is used for efficient sampling.

To calculate $|R_\sigma(l)\rangle$ and $\langle L_\sigma(l)|$, we need the orthogonalization procedure[7], because direct successive multiplications of Ω lead to poor accuracy of the result due to the rounding error of computers.

To update samples in the Monte Carlo steps, repetitions of a local updating procedure are efficient. In one of these procedures, we calculate the ratio of $\mathcal{W}$ to $\mathcal{W}'$, where $\mathcal{W}'$ is a new trial configuration obtained from the auxiliary variables $\{s_i(l)\}$ of $\mathcal{W}$ replaced with $s_i'(l)$ only at the i-th site on the l-th slice while other auxiliary variables are kept unchanged. The ratio $\mathcal{W}'/\mathcal{W}$ is obtained from[7]

$$\mathcal{W}'/\mathcal{W} = \prod_\sigma |I + \Delta_\sigma(G_\sigma(l))| = \prod_\sigma |1 + \delta_\sigma(G_\sigma(l))_{ii}|, \qquad (2.16)$$

where the $N \times N$ matrix Δ_σ is defined by

$$\Delta_{\sigma jj'} = \begin{cases} \delta_\uparrow = \exp[2a_U(s_i'(l) - s_i(l))] - 1 & \text{for } j = j' = i \text{ and } \sigma = \uparrow \\ \delta_\downarrow = \exp[-2a_U(s_i'(l) - s_i(l))] - 1 & \text{for } j = j' = i \text{ and } \sigma = \downarrow \\ 0 & \text{otherwise.} \end{cases}$$

If a new configuration is accepted, the updated Green's functions are obtained from[7]

$$(G_\sigma'(l))_{jk} = [(I + \Delta_\sigma)b_1]_{jj}(b_1^{-1})_{jj}\left[(G_\sigma(l))_{jk} - \frac{(G_\sigma(l))_{ji}\delta_\sigma(G_\sigma(l))_{ik}}{1 + (G_\sigma(l))_{ii}\delta_\sigma}\right]. \qquad (2.17)$$

The number of operations needed for updating one auxiliary variable is proportional to N^2.

At finite temperatures, the grand-canonical-ensemble method is useful to calculate physical properties efficiently[13–15]. In this case, the partition function Z defined in Eq.(2.4) is given by

$$Z = \sum_{\{s\}} \tilde{W}, \tag{2.18}$$

where

$$\tilde{W} = \tilde{W}_\uparrow(\{s\}, \beta)\tilde{W}_\downarrow(\{s\}, \beta) \tag{2.19}$$

and

$$\tilde{W}_\sigma(\{s\}, \beta) = \det[I + \prod_{i=1}^{L} \Omega_{\sigma i}]. \tag{2.20}$$

Green's functions and the updating procedure are obtained similarly as in the canonical-ensemble approach[13,14]. We note that W and $\tilde{W}$ can usually be real quantities.

To obtain ground-state properties, the canonical-ensemble algorithm is more efficient because the convergence to the ground state is obtained at smaller β if one takes a better trial state $|\Phi\rangle$. In fact, if one chooses a trial state, β is no longer the inverse temperature. Physical quantities obtained at β from the canonical ensemble with a trial wave function can show faster convergence than the thermodynamic quantities at finite temperature $T = 1/\beta$, if the trial state is appropriate. The rate of convergence to the ground state has been examined in detail for a variety of initial trial states by Furukawa and Imada[16]. The optimized initial trial state obtained from the unrestricted Hartree-Fock solution provides an example of fast convergence. In §4, we discuss useful results obtained by this choice of the trial state. Another choice of the trial state has been discussed by Sorella[17], who chose the quantum number so as to prohibit low-lying excitations.

3 Sign Problem

In fermion simulations at low temperatures, we are frequently faced with the minus-sign problem. At sufficiently low temperatures, the partition function Z should be proportional to $\exp[-\beta E_g]$ where E_g, is the ground-state energy. However, it has been shown in many cases that a finite portion of W(or $\tilde{W}$) in samples of simulations has amplitudes proportional to $\exp[-\beta E_g^*]$, where $E_g^* < E_g$. In fact, in many cases, $\sum_{\{s\}} |W(\{s\}; \beta; \varphi)|$ and $\sum_{\{s\}} |\tilde{W}(\{s\}; \beta)|$ grow in proportion to $\exp[-\beta E_g^*]$ with $E_g^* < E_g$. This is only possible when the samples showing the amplitude proportional to $\exp[-\beta E_g^*]$ have in some cases positive but in the other cases negative signs, with cancellations in the summation leading to the final result $Z \propto e^{-\beta E_g}$. The statistical sampling can be performed only if the sampling weight is a positive definite probability function. In ordinary cases, this probability function is chosen as the absolute value of the weight. In the auxiliary-field algorithm, it is $|W(\{s\}; \beta; \varphi)|$ (or $|\tilde{W}(\{s\}; \beta)|$). Hereafter we only refer to the canonical-ensemble case, though the grand-canonical case may be discussed by replacing W with $\tilde{W}$. According to the importance sampling algorithm, the sample with the field configuration $\{s\}$ is employed with the probability $|\mathcal{W}|$, while a spin-dependent physical quantity A_σ is calculated as

$$\langle A_\sigma \rangle = \frac{\frac{1}{L}\sum_l \langle \text{sign}\mathcal{W}\det[\langle L_\sigma(l)|A_\sigma|R_\sigma(l)\rangle]/W_\sigma\rangle}{S}, \qquad (3.1)$$

where the average sign $S \equiv \langle \text{sign}\mathcal{W}\rangle = \sum_{\{s\}} \mathcal{W}(\{s\};\beta;\varphi)/\sum_{\{s\}} |\mathcal{W}(\{s\};\beta;\varphi)|$ should follow the form

$$S \propto \exp[-\beta\Delta] \qquad (3.2)$$

with $\Delta = E_g - E_g^*$. Here $\langle\rangle$ denotes the sample average. Because the average sign S decreases exponentially with the increase of β, the evaluation of (3.1) becomes increasingly difficult for large β due to statistical errors of both the denominator S and the numerator.

To reduce this sign problem, the optimization of the initial trial wave function φ turns out to be very important to reach the convergence to the ground state. In fact, as we discuss in §4, ground state properties of the two-dimensional Hubbard model have been obtained at relatively small β using this optimization before the sign difficulty gets serious.

Another prescription for reducing the sign problem is to measure Δ from β-dependence of S. Because we can easily estimate E_g^* by neglecting the sign, the true ground-state energy E_g is obtained[18] from $E_g = E_g^* + \Delta$. Other physical quantities are obtained similarly; this is called the ratio-correction method[19]. A reweighting method has also been proposed to reduce the sign problem, in which the probability function in the sampling is modified so as to reduce the minus-sign ratio[20].

4   Application to the Hubbard Model

Because of the simplicity of the model in addition to its fundamental importance, the Hubbard model is the most celebrated and thoroughly investigated system among various lattice fermion models. Although the Hubbard model was proposed more than thirty years ago as one of the simplest strongly correlated models[21], its ground-state and finite temperature properties are still under vital debate. In particular, the Hubbard model on a square lattice has been extensively studied for several years, because it has turned out to be one of the standard models for high-T_c cuprate superconductors.

In this section, we discuss several important properties of the Hubbard model defined by Eq.(2.1) on a square lattice, clarified recently in numerical simulations[7,15,22-29]. At half filling, the auxiliary-field algorithm provides a way of simulation without the minus-sign problem, if t_{ij} in (2.1) is restricted to the nearest-neighbor pairs. In this case, because of the electron-hole symmetry, $\mathcal{W}$ and $\tilde{\mathcal{W}}$ may be taken positive definite. Extrapolation of finite-size results to the thermodynamic limit indicates the existence of the antiferromagnetic long-range order for nonzero U at half filling[15,22,28]. The Fourier transform of the spin correlation, namely

$$S(\mathbf{q}) = \frac{1}{3N} \sum_{ij} \langle \mathbf{S}_i \cdot \mathbf{S}_j \rangle e^{i q (r_i - r_j)} \qquad (4.1)$$

for an N-site system, shows the scaling of the peak at $\mathbf{q} = \mathbf{Q} \equiv (\pi, \pi)$ as $S(\mathbf{Q}) \propto N$. When the next-nearest-neighbor transfer t' is nonzero, the electron-hole symmetry and the perfect nesting are not satisfied at half filling. Although $\mathcal{W}$ is not positive definite in this case, the simulation to obtain ground-state properties has been successfully done by using the optimized initial trial state. According to the simulation result [22], $S(\mathbf{Q})$ is not sensitively dependent on t'/t. The antiferromagnetic order seems to be present even at $t'/t \sim 0.5$. Because the Fermi surface structure of the noninteracting system drastically changes and the nesting is completely destroyed with the increase of t'/t from 0 to 0.5, it is not easy to understand the existence of the antiferromagnetic order even at $t'/t \sim 0.5$ using the weak-correlation approach such as RPA. However, it is easier to understand it using the strong-coupling limit because the ratio of the exchange coupling constant between the next-nearest-neighbor pair $J' \propto t'^2/U$ and the nearest-neighbor pair $J \propto t^2/U$ is only 0.25 even at $t'/t = 0.5$.

The square lattice Hubbard model with only the nearest-neighbor transfer has been shown to have a charge gap at half filling if the on-site interaction U is nonzero[22,27]. Here, the charge gap is defined as

$$\Delta_c \equiv \frac{N}{2} |E_G(\tfrac{1}{2}N, \tfrac{1}{2}N) - E_G(\tfrac{1}{2}N - 1, \tfrac{1}{2}N - 1)|, \qquad (4.2)$$

where N is the number of lattice sites and $E_G(m_\uparrow, m_\downarrow)$ is the ground-state energy per site of $m_\uparrow + m_\downarrow$ fermions with $m_\uparrow$ up and $m_\downarrow$ down spins. If U is small, the amplitude of the charge gap Δ_c is consistent with the essentially singular behavior

$$\Delta_c \sim t \exp[-2\pi\sqrt{t/U}] \qquad (4.3)$$

expected in the Hartree-Fock approximation[29]. On the contrary, for large U, the gap Δ_c deviates from the Hartree-Fock prediction.

It is clear that the half-filled Hubbard model is in the Mott insulating state. The nature of the metal — Mott-insulator transition (M-MI transition) has been a long-standing controversial issue. The transition from a metal to an insulator is represented by vanishing of the conductivity at $T = 0$. The Drude formula for the conductivity is given by

$$\sigma(\omega) = \frac{n_c e^2 \tau}{m_c^*(\omega^2 + \tau^2)}. \qquad (4.4)$$

Here e, n_c and m_c^* are the carrier charge, carrier concentration and charge effective mass, respectively. The transport relaxation time is described by τ. At zero temperature, the Drude weight defined by

$$D = \int_0^{\omega_c} d\omega \sigma(\omega) = \frac{n_c e^2}{m_c^*} \qquad (4.5)$$

suggests that the transition from a metal to an insulator occurs either by the divergence of m_c^* or by the vanishing of n_c.

The so-called Hubbard III[30] has derived the critical value of $U(= U_c)$ above which the Hubbard model is in the insulating state at half filling. According to this approximation, the transition from a metal to a Mott insulator is essentially the same as that from a metal to a band insulator. If one approaches the insulating state away from half filling, the carrier number n_c decreases to zero. This approximation does not satisfy the Luttinger sum rule. The volume surrounded by the Fermi surface vanishes at the transition.

On the contrary, Brinkman and Rice[31] have used the Gutzwiller approximation to treat the transition at U_c at half filling. In this approximation, the metallic state is always described by the Fermi liquid. However, the transition is caused by the divergence of the electron effective mass. In other words, the insulating state is realized when the electron renormalization factor z vanishes. If the Fermi liquid state with the Luttinger sum rule is realized close to the insulating state, the continuous decrease of the conductivity near the metal-insulator transition is possible only when the effective mass is enhanced.

Mott[32] has speculated that the transition might be of the first order with a discontinuous jump of n_c from a finite value to zero. This is because a small concentration of holes and electrons favors the formation of their bound states due to the lack of screening at low density, which leads to an instability below a threshold concentration. This argument does not apply to the case away from half filling.

Generally speaking, in three-dimensional systems, an antiferromagnetic metal may be realized in an extended region between a paramagnetic metal and a Mott insulator. In this case a sudden change of the Fermi surface structure should happen at the paramagnetic - antiferromagnetic boundary.

In this paper, we mainly discuss the case of the M-MI transition caused by the change in the filling. In particular, in the two-dimensional Hubbard model, as we explain below, the antiferromagnetic order is confined to the Mott insulating state. Therefore, in this case, the M-MI transition caused by the change in the filling is a direct transition from a paramagnetic metal to a Mott insulator.

The problem of a hole in a Mott insulator has been extensively studied by a variety of approximations as well as by numerical calculation. It should be noted, however, that the antiferromagnetic order present in two- and three-dimensional insulators does not disappear in the one-hole case in the thermodynamic limit. Therefore, this problem is completely different from the case of a small but finite concentration of holes in the paramagnetic phase. In the presence of the antiferromagnetic order, the one-hole problem is still an open question, though we do not discuss it in this paper.

Let us consider ground-state properties of the Hubbard model on a square lattice, revealed by the Monte Carlo calculations. At a finite concentration of holes, the global Fermi level structure has been shown to satisfy the Luttinger sum rule[22,33]. In fact, finite-size calculations show a characteristic feature of the shell structure derived only from a "large Fermi surface". It should also be noted that the "Fermi level structure"

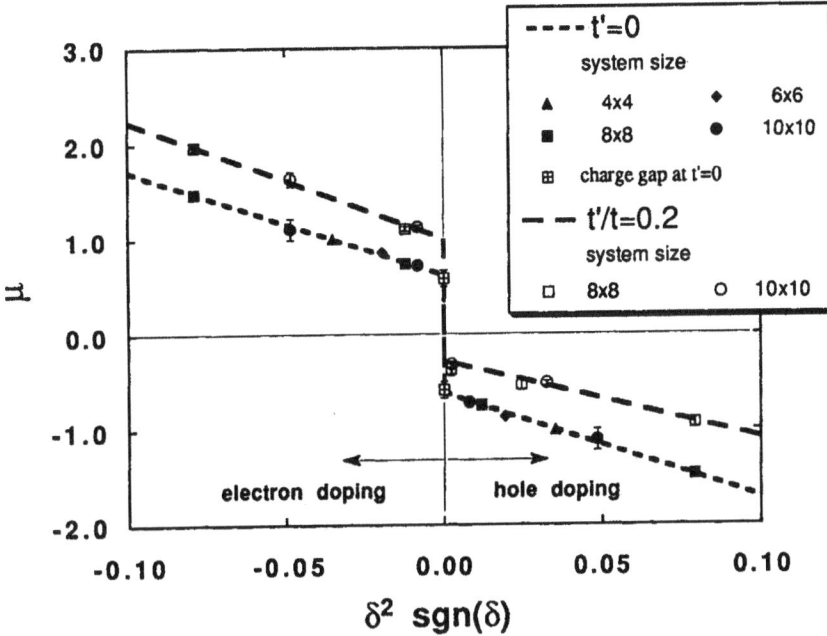

Figure 1: Doping-concentration dependence of the chemical potential for the two-dimensional Hubbard model. The band structure completely changes when the next-nearest-neighbor transfer t' increases. However, μ always changes in proportion to δ^2.

describes only a global feature in a relatively large energy scale. Therefore, the existence of the "large Fermi surface" does not directly imply the existence of the Fermi liquid state, as we discuss below.

The charge susceptibility defined by

$$\chi_c = \frac{\partial n}{\partial \mu}, \tag{4.6}$$

where μ is the chemical potential and n is the filling, probes the charge effective mass. It is clear that $\chi_c = 0$ at half filling because of the presence of a charge gap. In grand-canonical simulations, the behavior of χ_c away from half filling was a controversial issue. Although the filling dependence of χ_c seemed to suggest a continuous decrease in χ_c with the decrease in the doping[33], the temperature dependence at a fixed filling implied the enhancement at low temperatures in the low-doping region[34]. The ground-state simulation in the canonical ensemble has provided very clear evidence of the enhancement and settled this controversy[22]. In fact, in the low-doping region, χ_c at $T = 0$ has been shown to behave as

$$\chi_c \propto \delta^{-1}, \tag{4.7}$$

where $\delta \equiv 1 - n$ is the doping concentration from half filling. Figure 1 illustrates that the chemical potential follows the form

$$\mu = -(\text{sign}\delta)(\Delta_c + a\delta^2) \qquad (4.8)$$

for $t'/t = 0$ and $U/t = 4$. The form (4.7) is derived from Eq.(4.8).

The singular divergence of χ_c close to half filling is an indication of the charge mass divergence. The Monte Carlo simulation result has revealed that the M-MI transition is in fact caused by the divergence of m_c^*. Further, it has been shown that Eq.(4.7) is satisfied even when t'/t is nonzero, as seen in Fig 4.1[27]. This indicates that the singular divergence has nothing to do with the electron-hole symmetry or the van Hove singularity.

The spin susceptibility χ_s is another fundamental quantity measured in the simulation. At finite temperatures, at half filling, χ_s has a broad peak and substantially decreases below the peak with the decrease of temperature, very similarly to the case of the Heisenberg antiferromagnet[15]. Away from half filling, at $T = 0$, χ_s is always a smooth function of doping and is quantitatively well reproduced by a renormalized RPA[22]. This is in sharp contrast with the singular behavior of χ_c. Because χ_c and χ_s are both simply proportional to the effective mass in the Fermi liquid state, a serious question arises about the validity of the Fermi liquid state. In fact, these susceptibilities are similar as in the case of interacting one-dimensional systems[35] where the Tomonaga-Luttinger liquid is realized. The singular divergence of χ_c in the Monte Carlo result has also raised a serious question about the M-MI transition in the cuprate superconductors. In cuprates, various experimental indications show a fundamental difference from the M-MI transition in the two-dimensional Hubbard model. We have no indication for the enhancement of the charge mass in cuprate superconductors[36,37]. This problem is now one of the central issues of the normal-state properties in the cuprates[38].

Near half filling, antiferromagnetic correlation develops even in the paramagnetic phase. The spin structure factor $S(\mathbf{q})$ defined by Eq.(4.1) has degenerate peaks at four incommensurate wave numbers for $t'/t = 0$, while it has a broad peak at (π, π) for large t'/t. If one of the wave numbers at the peaks is denoted by $\mathbf{Q}$, $S(\mathbf{Q})$ has been shown to follow

$$S(\mathbf{Q}) \propto \delta^{-1}, \qquad (4.9)$$

as we see in Fig.2. This scaling is derived from the existence of a characteristic length scale $\xi = \delta^{-1/2}$ below which the spin correlation follows essentially the same form as the Heisenberg model. This means that the antiferromagnetic correlation $\langle \mathbf{S}(0) \cdot \mathbf{S}(\mathbf{r}) \rangle$ is given by

$$\langle \mathbf{S}(0) \cdot \mathbf{S}(\mathbf{r}) \rangle \propto \begin{cases} \text{constant} & |\mathbf{r}| \ll \xi, \\ 1/\mathbf{r}^\gamma & |\mathbf{r}| \gg \xi \end{cases} \qquad (4.10)$$

where $\gamma > 2$. It is not clear whether γ is equal to the Fermi liquid value ($\gamma = 3$). An important point is that the antiferromagnetic order appears not as a result of the decrease in γ but due to the increase in ξ. The asymptotic behavior at long distance

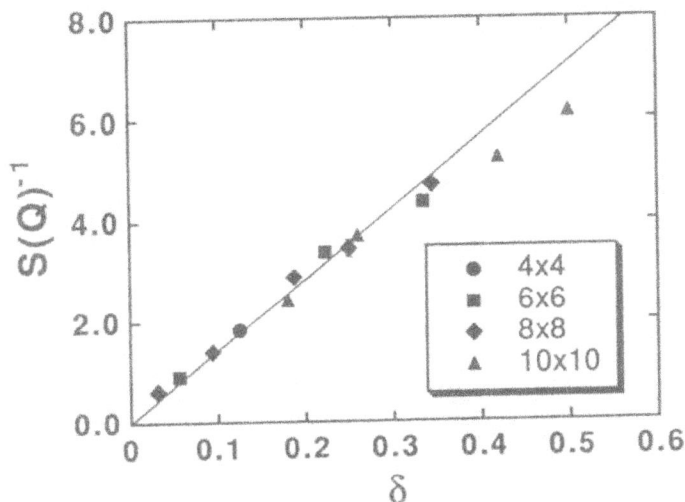

Figure 2: Doping-concentration dependence of the inverse of $S(Q)$ for the two-dimensional Hubbard model. It follows the scaling form Eq.(4.9). Although $S(Q)^{-1}$ slightly deviates from the scaling Eq.(4.9) close to half filing, this it is due to finite-size effects.

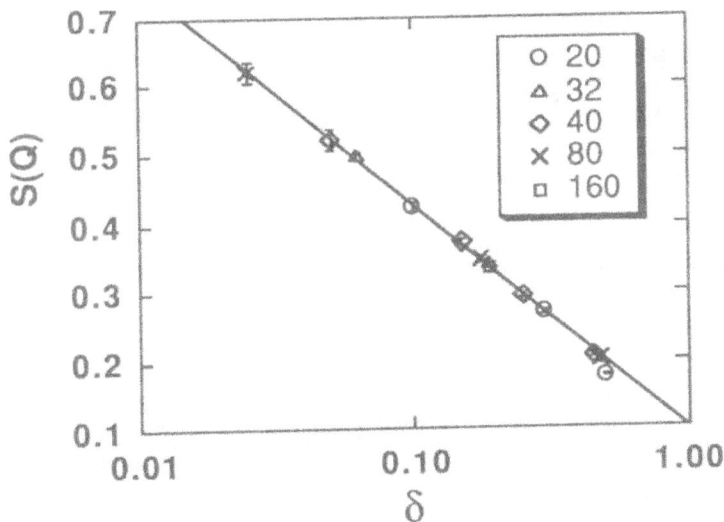

Figure 3: Monte Carlo results for doping-concentration dependence of $S(Q)$ in the one-dimensional Hubbard model. It follows the scaling form Eq.(4.11).

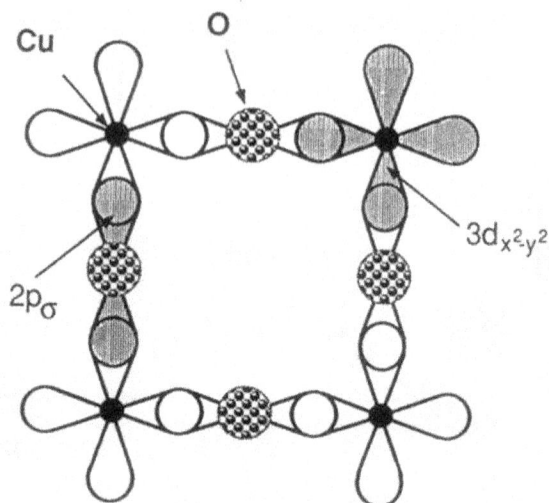

Figure 4: Electronic orbitals in the CuO$_2$ plane of cuprate superconductors.

does not determine the appearance of the antiferromagnetic order. This leads to the existence of crossovers also in energy and temperature. These crossovers are not accounted for by the weak-correlation approaches such as RPA. Essentially the same behavior with the existence of $\xi = \delta^{-1}$ has also been shown in the one-dimensional Hubbard model[26]. This is derived from the fact that the scaling form of $S(\mathbf{Q})$ is given by

$$S(\mathbf{Q}) \propto - \log \delta, \qquad (4.11)$$

as we see in Fig.3.

5   Applications to Other Strongly Correlated Systems

Since the discovery of high-T_c cuprates, much effort has been concentrated on numerical simulations of multi-band systems to gain insights into the nature of the CuO$_2$ plane in the cuprates. Cuprate superconductors are known to have an antiferromagnetic insulating state in the mother materials. In contrast with the single-band Hubbard model, the charge excitation gap of this insulating state is determined not from U but from the charge transfer energy Δ of a hole from the Cu $3d_{x^2-y^2}$ orbit to the in-plane O $2p_\sigma$ orbit, because Δ is smaller than U. Thus, this insulating state belongs to a different region than that in the single-band Hubbard model[39]. The possible importance of charge-transfer-type excitation spectra, which can be examined only in multi-band systems, has been suggested.

The model of the CuO$_2$ plane as a network of Cu $3d_{x^2-y^2}$ orbits and O $2p_\sigma$ orbits, illustrated in Fig.4, has been investigated extensively. This model is called the d-p model. In a realistic range of parameters, Monte Carlo results suggest that the d-p model has the antiferromagnetic order at "half filling"[40]. Here, "half filling" is defined by the filling of one electron per Cu atom. Therefore, the Mott insulating state at "half filling" in cuprates may have more or less similar properties to the Mott-Hubbard case. Away from half filling, Dopf, Muramatsu and Hanke[40] have suggested that the Fermi level structure satisfies the Luttinger sum rule. The structure in the k-space is in fact in overall agreement with the angle-resolved photoemission data in the cuprate superconductors.

Various types of pairing correlations in the d-p model have been examined[40-42] to clarify the difference between the d-p model and the Hubbard model, because the absence of indications for the superconducting long-range order has been widely observed in various simulations of the Hubbard model. Imada[41] and Dopf *et al.*[40] have shown that several types of pairing correlations have a small but finite enhancement in contrast to the Hubbard model. So far, however, the simulation of the d-p model has been done only for relatively small sizes. It makes it very difficult to observe system-size dependences because they seem to be rather small. Another problem is that the physics and the origin of the small enhancement seen in the simulation is not clear enough to devise a more efficient way of improved simulations at lower-energy scale.

By taking account of the effect of $3d_{3z^2-r^2}$ or $2p_\pi$ orbits, which are not included in the d-p model, different types of multi-band systems have also been proposed by Asai[43] and by Aoki and Kuroki[44] for studying pairing mechanisms. Although the d-p model is one of the multi-band models, it has a rather strong hybridization between $3d_{x^2-y^2}$ and $2p_\sigma$ orbits. In the above cases, however, Cu $3d_{x^2-y^2}$ and O $2p_\pi$ or Cu $3d_{3z^2-r^2}$ form rather well-separated bands without hybridization. Therefore the pairing mechanism investigated in these multi-band simulations has close relation to the excitonic mechanism.

Heavy fermion systems provide another challenging and interesting field for studying strongly correlated electrons. Various problems such as the multi-channel Kondo problem, the two-impurity problem, a competition between a crystal field and the Kondo effect, and the Kondo effect in systems with a gap have been recently investigated by numerical methods combined with analytic approaches. Although the impurity Anderson model is known to have the exact solution by Bethe Ansatz, its correlation functions and dynamical properties such as the spectral weight are not tractable by the exact solutions. The finite-temperature crossover to a Kondo singlet state has been investigated[45,46] by the so-called impurity Anderson algorithm developed by Hirsch and Fye[45]. The two-impurity Anderson model has also been the subject of intesive studies because it provides the simplest way of examining the competition between an intersite antiferromagnetic correlation and a local singlet formation. Although a discontinuous change at a boundary has been suggested to occur between the antiferromagnetic region and the local singlet region in the parameter space[47], detailed examinations of two-impurity models have clarified that the singularity only occurs under a special parity symmetry. Monte Carlo simulations[48,49]

combined with recent numerical renormalization-group analysis[50] have given us useful information. Small clusters of the periodic Anderson model and the Kondo lattice
have also been investigated with the aim of observing the competition between an
antiferromagnetic order and the Kondo singlet[51,52]. The competition between the
Kondo effect and a gap, namely the band gap or the superconducting gap, also leads
to an interesting crossover[53].

Novel pairing mechanisms have been the subject of recent intensive studies, motivated by a variety of anomalous properties of high-T_c cuprates. In particular, the
spin-gap phenomena[54-56] observed at substantially higher temperatures than the superconducting transition temperature T_c have inspired a variety of approaches clarifying this fluctuation. Although several one-dimensional Mott insulators are known
to have a spin gap in the ground state, the formation of a spin gap in the metallic
state is a nontrivial and interesting issue. Imada[57] has shown that the dimerized t-J
model defined by the Hamiltonian

$$
\begin{aligned}
\mathcal{H} = & -t \sum_{\langle ij \rangle} P_d (c_{i\sigma}^\dagger c_{j\sigma} + c_{j\sigma}^\dagger c_{i\sigma}) P_d \\
& + J_1 \sum_{\langle ij \rangle \in A} \mathbf{S}_i \cdot \mathbf{S}_j + J_2 \sum_{\langle ij \rangle \in B} \mathbf{S}_i \cdot \mathbf{S}_j
\end{aligned}
\tag{5.1}
$$

shows a gap in the spin excitation even in the metallic phase; here A and B denote
subsets of nearest-neighbor bonds which form a dimerized lattice. In one dimension,
A and B bonds form an alternating structure. Here P_d is the projection operator prohibiting the double occupation of fermions represented by the creation (annihilation)
operator $c_{i\sigma}^\dagger (c_{i\sigma})$ at the i-th site with the spin σ. In fact, in one dimension, a detailed
study of (5.1) has shown a strong enhancement of pairing correlations in comparison
with critical exponents of other correlations.

Ogata et al.[58] have studied the t-J model with the next-nearest-neighbor exchange
coupling (t-J-J' model) to observe the spin-gap formation in the metallic phase. The
two-chain t-J model and other variations of the models with a spin gap have also
been the subject of intensive studies[59].

In one dimensional systems, the local constraint prohibiting the double occupancy
of fermions can easily be implemented in the world-line algorithm of quantum Monte
Carlo simulation, as clarified by Imada and Hatsugai[7]. In fact, they have applied such
algorithm to observe temperature dependences of various quantities such as magnetic
and pairing correlations. In two or more dimensions, the world-line algorithm is practically useless in metallic phases because of the minus-sign problem. An application to
the study of the spin-gap phenomena is so far restricted to the Mott insulating state.
Katoh and Imada[60] have investigated the transition between an antiferromagnetic
state and a spin-gap state by the Monte Carlo simulation of the two-dimensional
dimerized Heisenberg model. A way of implementing the local constraint in the
auxiliary-field algorithm has been derived from the coherent-state representation, as
proposed by Zhang, Abrahams and Kotliar[61]. Because of a rather serious minus-sign

problem and poor statistical accuracy, its application at low temperatures remains restricted; thus some improvements allowing its practical applications are particularly desirable.

6   Summary and Future Problems

Simulations of correlated fermions are one of the most challenging problems in recent condensed-matter theory. Recent developments of algorithms have clarified that it is in fact possible to obtain low-temperature properties of correlated fermions. In particular, the Monte Carlo results have revealed a variety of interesting properties of the Hubbard model in relation to the metal-insulator transition and the antiferromagnetic transition. An extensive application of simulations for searching superconducting mechanisms has also turned out to be fruitful. More detailed studies of more complicated systems such as the periodic Anderson model and the degenerate Hubbard model are desirable. To make wider and wider applications possible, it is important to make efforts to reduce further the sign problem as well as to reduce the computation time, proportional to N^3 in the case of the present auxiliary-field algorithm.

In this paper, we only described the simulations of lattice fermions. Applications to electrons in the continuum space will provide very challenging problems in the future. To overcome various difficulties in applications of electronic-structure calculation, using the so-called local-density functional, to correlated fermions in real materials, it is desirable to develop an efficient way of simulation by the path-integral approach.

Dynamical properties have special importance for clarifying the nature of various metallic states. Recently, the maximum-entropy method has been employed yielding some useful information on the spectral weight[62]. However, more and more accurate and stable algorithms of extracting dynamical properties are needed to discuss low-energy excitations. We need more efforts to make quantitative comparison of numerical results with the transport coefficients such as the frequency-dependent conductivity and the Hall coefficient as well as the inelastic neutron scattering data, the NMR data, etc. These are extremely important future problems.

Acknowledgements
The author would like to thank N. Furukawa for fruitful discussions. He would like to thank Dr. D. Lipowska for careful reading of the manuscript.

References

1. *Quantum Simulations of Condensed Matter Phenomena*, ed. J. E. Gubernatis

(World Scientific, Singapore, 1990).

2. *Computational Approaches in Condensed-Matter Physics*, ed. S. Miyashita, M. Imada and H. Takayama (Springer, Berlin, 1992) and references therein.

3. K. E. Schmidt and M. H. Kalos, *Applications of the Monte Carlo Method in Statistical Physics*, ed. K. Binder (Springer, Berlin, 1984) Chap. 4.

4. M. Takahashi and M. Imada, J. Phys. Soc. Jpn. **53**(1984) 963, 3765.

5. M. Barma and B. S. Shastry, Phys. Rev. B**18**(1978) 3351.

6. J. E. Hirsch *et al.*, Phys. Rev. B**26**(1982) 5033.

7. M. Imada and Y. Hatsugai, J. Phys. Soc. Jpn. **58**(1989) 3752.

8. M. Imada, J. Phys. Soc. Jpn. **59**(1990) 4121.

9. H. F. Trotter, Proc. Am. Math. Soc. **10**(1959) 545.

10. M. Suzuki, Prog. Theor. Phys. **56**(1976) 1454.

11. J. E. Hirsch, Phys. Rev. B**28**(1983) 4059.

12. S. Sorella, R. Car, S. Baroni and M. Parrinello, Europhys. Lett.**8**(1989) 663.

13. R. Blankenbecler, D. J. Scalapino and R. L. Sugar, Phys. Rev. D**24**(1981) 2278.

14. J. E. Hirsch, Phys. Rev. B**31**(1985) 4403.

15. S. R. White, D. J. Scalapino, R. L. Sugar, E. Y. Loh,Jr., J. E. Gubernatis and R. T. Scalettar, Phys. Rev. B**40**(1989) 506.

16. N. Furukawa and M. Imada, J. Phys. Soc. Jpn. **60**(1991) 3669.

17. S. Sorella, Phys. Rev. B**46**(1992) 11670; Ref.2, p.55.

18. E. Y. Loh,Jr., J.E. Gubernatis, R. T. Scalettar, S. R. White, D. J. Scalapino and R. L. Sugar, Phys. Rev. B**41**(1990) 9301.

19. N. Furukawa and M. Imada, J. Phys. Soc. Jpn. **60**(1991) 810.

20. T. Nakamura, N. Hatano and H. Nishimori, J. Phys. Soc. Jpn. **61**(1992) 3494.

21. J. Kanamori, Prog. Theor. Phys. **30**(1963) 275; J. Hubbard, Proc. Roy. Soc. A**276**(1963)238.

22. N. Furukawa and M. Imada, J. Phys. Soc. Jpn. **61**(1992) 3331.

23. N. Furukawa and M. Imada, J. Phys. Soc. Jpn. **60**(1991) 3669.

24. M. Imada and N. Furukawa, Ref.2, p.63.

25. N. Furukawa and M. Imada, Physica B**186-188** (1993).

26. M. Imada, N. Furukawa and T. M. Rice, J. Phys. Soc. Jpn. **61**(1992) 3861.

27. N. Furukawa and M. Imada, J. Phys. Soc. Jpn. **62** (1993) 2557, and unpublished.

28. J. E. Hirsch, Phys. Rev. B**38**(1988) 12023.

29. N. Furukawa and M. Imada, unpublished.

30. J. Hubbard, Proc. Roy. Soc. London A**281**(1964) 4302.

31. W. F. Brinkman and T. M. Rice, Phys. Rev. B**2**(1970) 4302.

32. N. F. Mott, *Metal-Insulator Transition* (Taylor & Francis, London, 1974).

33. A. Moreo, D. J. Scalapino, R. L. Sugar, S. R. White and N. E. Bickers, Phys. Rev. B**41**(1990) 2313.

34. H. Otsuka, J. Phys. Soc. Jpn. **59**(1990) 2916.

35. N. Kawakami and A. Okiji, Phys. Rev. B**40**(1989) 7066.

36. T. Nishikawa, J. Takeda and M. Sato, Physica C**209**(1993) 553.

37. H. Takagi *et al.*, Phys. Rev. B**40**(1989) 2254; N. P. Ong *et al.*, Phys. Rev.

B35(1987) 8807.

38. M. Imada, J. Phys. Soc. Jpn. **62**(1993) 1105; to be published.
39. J. Zaanen, G. A. Sawatzky and J. W. Allen, Phys. Rev. Lett. **55**(1985) 418.
40. G. Dopf, A. Muramatsu and W. Hanke, Phys. Rev. B41 (1990) 9264; Phys. Rev. Lett. **68**(1992) 353.
41. M. Imada, J. Phys. Soc. Jpn. **57**(1988) 3128.
42. R. T. Scalettar, D. J. Scalapino, R. L. Sugar and S. R. White, Phys. Rev. B44(1991) 770.
43. Y. Asai, Physica C**185-189**(1991) 1633; Ref.2, p.124.
44. K. Kuroki and H. Aoki, Phys. Rev. Lett. **69**(1992) 3820.
45. J. E. Hirsch and R. M. Fye, Phys. Rev. Lett. **56**(1986) 2521.
46. J. E. Gubernatis, Phys. Rev. B36(1987) 394; M. Jarrell, J. E. Gubernatis and R. N. Silver, Phys. Rev. B44(1991) 5347, and references therein.
47. B. A. Jones, C. M. Varma and J. W. Wilkins, Phys. Rev. Lett. **61**(1988) 125.
48. R. M. Fye, Phys. Rev. B41(1990) 2490.
49. T. Saso, Phys. Rev. B44(1991) 450.
50. O. Sakai and Y. Shimizu, Ref.2, p.92.
51. T. Saso and Y. Seino, J. Phys. Soc. Jpn. **55**(1986) 3729.
52. R. M. Fye and D. J. Scalapino, Phys. Rev. B44(1991) 7486.
53. T. Saso and J. Ogura, Physica B to appear; K. Satori, H. Shiba, O. Sakai and Y. Shimizu, J. Phys. Soc. Jpn. **61**(1992) 3239.
54. H. Yasuoka, T. Imai and T. Shimizu, *Strong Correlation and Superconductivity* ed. H. Fukuyama *et al.*, (Springer, 1989) p.254.
55. M. Takigawa *et al.*, Physica C**162-164**(1989) 853.
56. J. Rossat-Mignod *et al.*, Physica C**185-189**(1991) 86.
57. M. Imada, Phys. Rev. B48 (1993) 550.
58. M. Ogata, M. U. Luchini and T. M. Rice, Phys. Rev. B44 (1991) 12083.
59. E. Dogotto, Ref.2, p.84.
60. N. Katoh and M. Imada, J. Phys. Soc. Jpn. **62** (1993) No.10.
61. X. Y. Zhang, E. Abrahams and G. Kotliar, Phys. Rev. Lett. **66**(1991) 1236.
62. J. E. Gubernatis, M. Jarrell, R. N. Silver and D. S. Sivia, Phys. Rev. B44(1991) 6011; Ref.2, p.105.

Note added in proof: After the completion of this review article the author received two review articles on numerical calculation of correlated systems : W. von der Linden, Phys. Rep. 220 p.53 (1992), E. Dagotto, submitted to Rev. Mod. Phys.

DIRTY BOSONS IN 2D: PHASES AND PHASE TRANSITIONS

Nandini Trivedi[1] and Miloje Makivić[2]

[1]Materials Science Division, Bldg. 223
Argonne National Laboratory
9700 South Cass Avenue, Argonne, IL 60439, USA
[2]Department of Physics, Ohio State University
Columbus, OH 43210, USA
e-mail: trivedi@hexi.msd.anl.gov

Abstract

The phase transition in an interacting boson system in 2D from a superfluid to an insulator with increasing amounts of disorder is emerging as a paradigm for a large class of problems. Among them are ^{4}He adsorbed in porous media, superconductor-insulator transition in homogeneous and granular films, Josephson junction arrays, vortices in bulk type II superconductors, the quantum Hall transition from a Hall liquid to a Hall insulator, and random magnets. We review our work on the the application of quantum Monte Carlo methods (checker board and path integral algorithms) to the dirty boson problem. We present first qualitative results for a model of bosons on a 2D square lattice with a random potential of strength V and on-site repulsion U. By calculating the superfluid density and the excitation spectrum we show the existence of three phases at $T = 0-$ (i) a superfluid phase, (ii) a localized or 'Bose glass' phase with gapless excitations, and (iii) a Mott phase with a gap to excitations (found only at commensurate densities). We discuss unusual effects arising from the interplay between interaction and disorder effects. We next investigate the critical properties of the superfluid to Bose glass transition for a hard core bose system as a function of disorder. Using finite size scaling on 64×64 lattices we find the dynamical exponent $z = 0.5 \pm 0.1$, the correlation length exponent $\nu = 2.2 \pm 0.2$, and a non-zero compressibility κ at the transition. Our conclusions differ from the existing scaling theory as well as from simulations on simplified models argued to be in the same universality class. Our results are suggestive of new low lying collective excitations in the disordered system that are modified from usual phonons.

1 Dirty Boson Problem

The dirty boson problem is the localization problem for bosons. This problem is interesting in its own right since it combines disorder and interaction effects in a simple Hamiltonian, whose solution is nevertheless non-trivial. The fermion counterpart of this problem has been studied actively over the past decade to understand the nature of metal-insulator transition in disordered systems, and has so far proved to be rather intractable[1]. The fermion problem has one definite advantage over the boson problem, in that there exists a well-defined non-interacting limit (known as Anderson localization) around which it is possible to study interaction effects perturbatively. In the boson problem, on the other hand, the non-interacting limit is singular and therefore analytical treatments have proved difficult. The singular nature of the non-interacting limit arises because the lowest lying state in the disordered system is localized and can accommodate a macroscopic number of bosons when they are non-interacting. Thus the introduction of even the smallest of repulsion between the particles renders the system unstable.

Numerically, however, bosons have a tremendous advantage over fermions, in that Monte Carlo methods which have been most successful, do not suffer from the so-called sign problem for bosons and yield results which, in principle, are 'exact' with only statistical errors. In this article, we focus on Monte Carlo techniques applied to bosons and the results obtained with regard to the nature of the phases and the excitation spectrum as a function of the strength of the disorder potential and the repulsive interactions between the particles. We also explore the phase transitions between the different phases and obtain the exponents characterizing them. We hope to convey to the reader that the dirty boson problem is extremely rich with possible relation to a large number of experiments discussed below. While Monte Carlo simulations have only just begun to yield results, we believe that with the advances in parallel computers (which allow larger simulations and also speed up the averaging over a large number of realizations of the random potential) the numerical solutions to the problem will go a long way toward directly confronting experiments. We expect numerical simulations to play a major role in the area of dirty bosons, just as they did in the study of spin glasses[2].

What makes the dirty boson problem even more interesting and important is that it is also emerging as a paradigm for a large number of systems shown in Fig. 1 in which phase transitions are seen at $T = 0$ as a function of a tuning parameter like disorder, interaction, magnetic field or density. Similar to the finite temperature phase transitions that arise due to a competition between the disordering effects of temperature and the ordering effects of, say, the exchange coupling between the spins in a magnet, the quantum phase transitions arise from a competition between disorder and quantum tunneling (which is controlled by varying the density).

It is not our intent to convey through Fig. 1 that the physics of all the systems listed is equivalent or identical. We however wish to make a serious suggestion, worthy of future investigation, that as far as certain questions are concerned they *may* be answerable within the framework of a boson Hamiltonian. This is very much in the spirit of classical critical phenomena where we know that very diverse systems, e.g., liquids, binary alloys and magnets, as far as their critical properties are concerned, can be described by a simple Hamiltonian. After much work based on renormalization group techniques, series expansions, Monte Carlo simulations and comparisons with

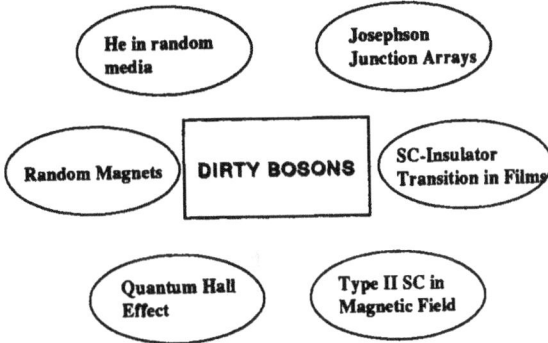

Figure 1: The dirty boson problem is emerging as a paradigm for quantum phase transitions in the systems indicated here.

experiments the idea of universality classes was firmly established. For a classical system, it was established that the criteria determining the universality class[3, 4] are (i) spatial dimensionality, (ii) symmetry of the order parameter, (iii) range (short range or long range) of the interparticle interactions, and for dynamical behavior, (iv) the conservation laws and the Poisson-bracket relations among the order parameter and the conserved densities. What determines universality classes in a quantum system? Compared to classical critical phenomena, quantum critical phenomena is still in its infancy. Theoretical and computational efforts are also just beginning along this direction. The situation is complicated by the fact that it is much harder to get reliable data to probe the zero temperature transitions.

Let us go back to Fig. 1 and briefly discuss how bosons might be relevant in the different systems, besides the obvious case of ⁴He. At low coverages, ⁴He on a random substrate (like vycor, mylar or graphite with adsorbed elements)[5] is localized. When the density of helium is increased beyond a critical coverage the system becomes superfluid. This transition can be viewed equivalently in terms of reducing the disorder for a fixed density of ⁴He (at least for cases where the topology of the substrate is not relevant), which drives the localized system into a superfluid state.

In Josephson junction arrays[6] or granular films[7], superconductivity arises when the phases of the individual (superconducting) grains get locked by Josephson coupling. The individual grains can be viewed as bosons, provided single particle tunneling is negligible. Disorder disrupts the phase coherence driving the system into an insulating state. In homogeneous films[8, 9, 10], the situation is more delicate since there isn't an apparent separation between pair formation and coherence. Nevertheless, as a first step, it can be argued that at the superconductor-insulator transition, on the scale of the diverging correlation length which is much larger than the Cooper pair size, the Cooper pairs can be approximated by bosons[11]. This is still somewhat controversial but only further work will determine how important the Fermi degrees of freedom are. There is certainly some indication from the data on In-InOx films[9] that even in fairly homogeneous films the temperature at which pairs form T_p and then condense T_c are separated because of large fluctuations in 2D[12]. With increasing disorder, both T_p and T_c are driven to zero. The basic question is whether there exists a window in disorder, where T_c has vanished but T_p is non-zero. Recent tunneling measurements[10] indicate that if such a window exists it is very small indeed. However, in the magnetic field driven superconductor-insulator transition[13, 9], from measurements of the longitudinal and Hall resistivity, there are clear indications that the superconductor-insulator transition takes place into an insulator of Cooper pairs (or bosons) at lower fields and at higher fields there is a crossover to an insulator of fermions.

In a bulk type II superconductor in a magnetic field vortices are preferentially nucleated at or near point and extended defects above the lower critical field. At high temperatures, vortices get depinned generating flux flow resistance. An open question is whether there is a finite temperature T_g at which the vortices 'freeze' so that the linear resistivity ρ_L is zero for $T < T_g$? The possible existence of a normal to superconductor transition in a magnetic field has been a topic of tremendous activity recently in the context of high temperature superconductors[14]. The above problem of a finite temperature phase transition in a 3D vortex system can be 'mapped' onto a 2D quantum problem of boson world-lines, where the role of temperature is played by tunneling[15].

There is an exact mapping of the hard core boson Hamiltonian (which allows a site to be occupied by up to one particle) onto a S=1/2 spin problem[16]. For larger S, the equivalent boson problem has a restriction of maximum site occupancy of 2S. Disorder in the hopping of bosons appear as random exchange couplings in the magnet and on-site disorder for bosons is equivalent to a random field.

In the quantum Hall effect of electrons in a magnetic field, it is possible to represent each electron by a boson with an odd number of flux quanta attached to it[17]. At special filling factors, where the quantum Hall effect is manifested, these flux quanta cancel the external magnetic field and we are left with the problem of bosons in a random potential. With increasing disorder the quantum Hall liquid is driven into a Hall insulator.

The dirty boson problem saw a revival through the work of M. P. A. Fisher, Weichman, Grinstein, and D. S Fisher[18]. They proposed the existence of different phases as a function of disorder and interaction and obtained relations between the exponents characterizing the phase transitions based on scaling ideas. They built on earlier work by Ma et. al. [19] and on the 1D renormalization group calculations by Giamarchi and Schulz[20]. For the superfluid-Mott transition in the clean problem, Sheshadri et. al. have done a mean field theory[21]. Singh and Rokhsar[22] have analyzed the disordered boson problem within the Bogoliubov approximation by solving self-consistently for the non-uniform condensate. Exact diagonalization studies[23] on 4×4 systems have been performed for hard core bosons. Zhang[24] has calculated the modification of the speed of sound and the lifetime of Bloch phonons for a weakly disordered bose system to first order in a 1/S expansion (where 2S is the maximum number of bosons at a site). Real space renormalization group techniques have been used in the hard core or nearly hard core limit[25] to obtain the phase diagram and exponents. Besides these, Monte Carlo simulations have been performed in 1D[26] and 2D for soft core[27, 28, 29] and hard core[30] bosons as well as for a related quantum rotor model[31].

The organization of the rest of the article is as follows: We give the model in section 2. In section 3 we give a brief description of the path integral quantum Monte Carlo technique which has been reviewed elsewhere[32]. This is followed by a more detailed discussion of the checker board algorithm in section 3 and its implementation on a parallel computer in section 4. Some qualitative results obtained by path integral MC are described in section 5. Results for the critical phenomena at the superfluid-Bose glass transition obtained by the checker board algorithm are given in section 6 and their implications are discussed in section 7. This is followed by concluding remarks and future work in section 8.

2 Boson Model

We consider a simple model of bosons on a 2D square lattice $N = L \times L$, described by the Hamiltonian H given by

$$H = -t \sum_{<i,j>} (a_i^\dagger a_j + h.c.) + \sum_i (V_i - \mu)n_i + \frac{U}{2} \sum_i n_i(n_i - 1) , \qquad (1)$$

where a_i $(a_i^\dagger)$ is a boson annihilation (creation) operator at a site i and $n_i = a_i^\dagger a_i$. The first term describes the hopping of bosons between nearest neighbor sites with strength t. The second term is the interaction of the bosons with a random potential, with site energies V_i chosen from a a uniform distribution $[-V, V]$, μ is the chemical potential, and the last term describes the on-site repulsive interaction between bosons with strength U. A special case of the interaction term in Eq. 1 is the hard core limit for which $U \rightarrow \infty$ and is implemented as a no double occupancy constraint with $n_i = 0$ or 1.

This model can easily be augmented, depending on the physical situation, by including short range interaction (nearest and next-nearest neighbor) between bosons, and/or long range interaction, different types of disorder, e.g., unbounded disorder or uncorrelated disorder in the time direction (important for pinning studies in the vortex problem), or disorder in the hopping term. Long range interactions are pertinent for modeling the vortex-vortex interaction and may also be important for quantum melting.

3   Quantum Monte Carlo: Algorithm

To begin, let us write the partition function at inverse temperature $\beta = 1/T$, given by $Z = \text{Tr}\exp(-\beta H)$. In quantum statistical mechanics β is viewed as 'imaginary time' and Z describes the amplitude for the system starting in a given state to return to the same state after 'imaginary time' $\tau = \beta$. It is possible to make judicious approximations regarding the short time dynamics. The first step in the algorithm, therefore, is to discretize the total time interval $[0, \beta]$ into M segments or Trotter slices of width $\delta\tau$ such that $M\delta\tau = \beta$ and insert complete sets of states. Thus the problem is reduced to summing over a large number of intermediate states, which are efficiently sampled by Monte Carlo techniques. The location of a particle along the different time slices represents a Feynman path or a world line. The sum over all the intermediate states amounts to summing over all possible paths of the particles. Thus in the end, we have converted a d-dimensional quantum problem into a d+1 dimensional classical problem of polymers or Feynman paths. At least three algorithms have recently been used to simulate bosons depending on the nature of the states used to perform the trace– (i) Path integral QMC, (ii) Checker board QMC, and (iii) Hubbard-Stratonovich QMC.

In the path integral Monte Carlo complete sets of states in the position basis are used, i.e., the trace is taken over all configurations $|r_1(\tau), r_2(\tau), \ldots r_{N_b}(\tau) >$ of the N_b-particle system, where $r_i(\tau)$ is the location of the ith particle on time slice τ. In order to incorporate the indistinguishability of the particles, it is also necessary to sum over permutations of the particles. Details about the choice of the elemental density matrices (over a time interval $\delta\tau$) and the Monte Carlo technique to perform the sum over configurations at the intermediate time slices as well as the sum over permutations is given in Ref.[31]. For bosons, the absence of the sign problem is seen from the fact that all paths contribute with a positive sign as opposed to the fermion case where they contribute with a sign $(-1)^P$ that depends upon the permutation P. In the latter case, there are large cancellations from adjacent paths, especially at low temperatures, producing results with very poor statistics.

In the checker board QMC, complete sets of states of the occupation number basis are used $|n_1(\tau), n_2(\tau), \ldots n_N(\tau) >$, where $n_i(\tau)$ is the occupancy of the ith site on time slice τ. Further, the Hamiltonian is partitioned into a checker board pattern that allows an easy evaluation of the elemental matrix elements over the interval $\delta\tau$. This is often called the 'world-line' algorithm in the literature. We, however, do not use this terminology since world lines arise in the path integral Monte Carlo method described above as well. The distinguishing feature is, of course, the states used to perform the trace. We will discuss this algorithm in detail below.

In the Hubbard-Stratonovich algorithm[32] coherent states are used to perform the trace and it is therefore limited to the grand canonical ensemble. The basic idea is to decouple the interaction term involving 4-boson operators by the introduction of auxiliary fields $\{x_i(\tau)\}$ at site i on time slice τ. The boson trace is then performed analytically leaving a classical simulation to be performed in the auxiliary fields.

The 'probability' that is sampled is a determinant of the auxiliary fields and can be shown to be positive for bosons, once again guaranteeing no sign problem. This method has the advantage that it naturally gives both the time-independent and time-dependent Green functions which are difficult to evaluate by the other methods. It does, however, suffer from the draw back that the action in terms of the auxiliary fields (after integration over the bosons) is non-local, therefore the computation time scales as $V^{(d+1)}$ as opposed to linearly with V in the other two approaches.

3.1 Checker Board Monte Carlo

This is a well-established method[34] for treating quantum systems and has been applied in many cases. We review here only the essential aspects that pertain to simulations of dirty bosons in 2D. Upon evaluating that trace in the occupation number basis, the partition function is given by

$$Z = Tre^{-\beta H} = \sum_{\{n_i\}} < n_1, n_2, \ldots, n_N | e^{-\beta H} | n_1, n_2, \ldots n_N > , \qquad (2)$$

where the sum is over all combinations of the occupation number n_i at site i of an $N = L \times L$ lattice. We impose periodic boundary conditions in the spatial and temporal directions. We next break the Hamiltonian into four pieces[35] as shown in Fig. 2, H_{ox} containing odd bonds along the x-direction, H_{oy} containing odd bonds along the y-direction, H_{ex} containing even bonds along the x-direction, and H_{ey} containing even bonds along the y-direction.

Then

$$e^{-\beta H} = \left(e^{-\delta\tau H} \right)^M = \left[e^{-\delta\tau(H_{ox}+H_{oy}+H_{ex}+H_{ey})} \right]^M . \qquad (3)$$

In general the different Hamiltonians do not commute, however if we neglect corrections of order $(\delta\tau)^2$ in the trace (known as the Suzuki-Trotter approximation), we obtain,

$$Z \approx \sum_{\{C\}} < C_1 | e^{-\delta\tau H_{ox}} | C_2 >< C_2 | e^{-\delta\tau H_{oy}} | C_3 >< C_3 | e^{-\delta\tau H_{ex}} | C_4 >< C_4 | e^{-\delta\tau H_{ey}} | C_5 >$$
$$\cdots < C_{4M} | e^{-\delta\tau H_{ey}} | C_1 > , \qquad (4)$$

where $|C_k> = |n_1(\tau_k), n_2(\tau_k), \ldots n_N(\tau_k) >$ is the configuration on the kth time slice. If we look at a particular term in the trace above we see that the problem factorizes into the evolution of a two-site problem from τ to $\tau + 1$, of the form, $P(n_i(\tau), n_j(\tau); n_i(\tau + 1), n_j(\tau + 1)) = < n_i(\tau)n_j(\tau)| \exp -\delta\tau H_{ij} | n_i(\tau + 1)n_j(\tau + 1) >$, where i, j correspond to sites on a particular bond and H_{ij} is the part of the Hamiltonian pertaining to the two sites. For hard core bosons, because of the conservation of particles on a bond, the only non-zero elements are

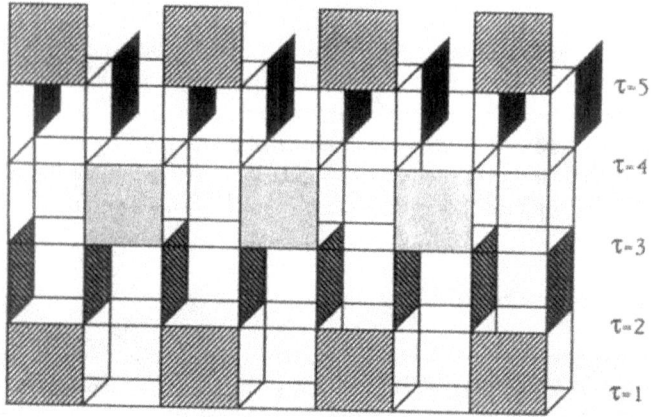

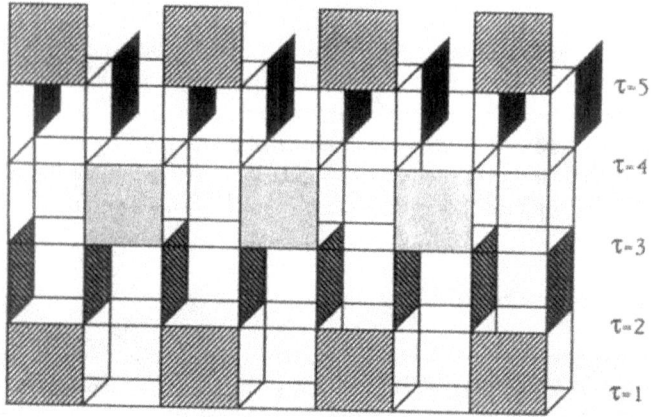

Figure 2: Checker board decomposition in 2D– Between any two time slices τ and $\tau+1$ the particles move only along the shaded plaquettes. Between $\tau = 1$ and $\tau = 2$ only the odd bonds along x-direction, described by H_{ox} are involved. Between $\tau = 2$ and $\tau = 3$, odd bonds along y-direction (H_{oy}) are involved; between $\tau = 3$ and $\tau = 4$, even bonds along x-direction (H_{ex}) evolve and between $\tau = 4$ and $\tau = 5$, even bonds along y-direction (H_{ey}) evolve and the pattern is repeated thereafter.

$$
\begin{aligned}
P(0,0;0,0) &= 1 \\
P(1,1;1,1) &= e^{-\delta\tau(W_i/4+W_j/4)} \\
P(0,1;0,1) &= e^{-\delta\tau W_j/4}\cosh(\delta\tau t) \\
P(1,0;1,0) &= e^{-\delta\tau W_i/4}\cosh(\delta\tau t) \\
P(1,0;0,1) &= \sinh(\delta\tau t) = P(0,1;1,0) ,
\end{aligned}
\tag{5}
$$

where $W_i = V_i n_i + (U/2)n_i(n_i - 1)$. Notice that the matrix elements are all positive implying no sign problem for the bose case.

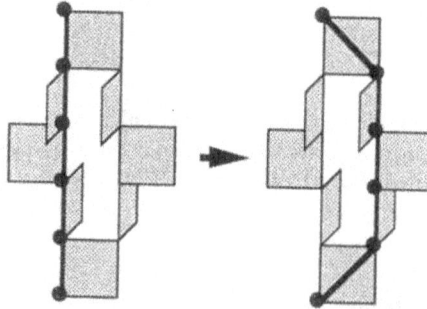

Figure 3: A new configuration is generated by plucking the world line to the right.

Generating world line configurations

If we connect the sites with the bosons along the interacting plaquettes, it is possible to uniquely assign a world line to each particle for the case of hard core bosons, since the hard core constraint prevents two lines from crossing one another. The procedure to generate new configurations from a given configuration can now be viewed in terms of moves of the world lines as described below.

Distorting a world line – A new configuration is generated locally by distorting or plucking a world line as shown in Fig. 3.

World-line exchange – Consider the non-interacting plaquette with alternately occupied sites in Fig. 4. Upon moving the bosons to the adjacent unoccupied positions the two world lines get exchanged or twisted around each other. Due to this permutation the world lines do not necessarily close on themselves, although the configuration is periodic in time.

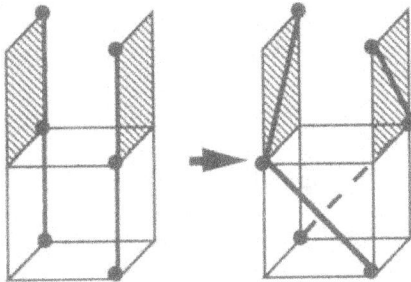

Figure 4: Two world lines are twisted around each other or exchanged. This is achieved by the local hops of two bosons on the unshaded plaquette to neighboring positions. Note that besides the two shaded plaquettes shown, the front and back squares also represent interacting plaquettes but for clarity reasons have not been shaded.

Changing number of world lines – This is a global update that changes the number of particles in the system as seen from Fig. 5. It is essential for generating fluctuations of the particle number that is a measure of the compressibility of the system. We found these moves very useful in achieving equilibrium quickly. One searches for a straight world line and annihilates it or one adds a particle to a site that is unoccupied on all time slices. In general, moves that change the winding number can also be incorporated, but due to the low acceptance rates of these moves, they were not used in the simulations described here.

4   Implementation on a parallel computer

For reliable finite size scaling, it is important to have data on large lattices that are well separated in size. In order to study large systems we use a word packing

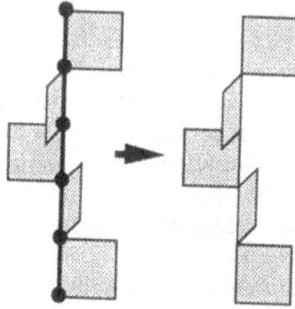

Figure 5: A boson is removed and its associated world line is destroyed. This is a global move that changes the number of particles in the system and is necessary for the calculation of the compressibility.

algorithm[35] to investigate the special case of hard core bosons by checker-board Monte Carlo. Since an integer word has 32-bits, it is possible to store the occupancy of a site (which for hard core bosons can be only 0 or 1) on 32 time slices in one integer word. Thus the information about a site on $M = 8$ Trotter slices or $4M$ fine time slices (from the break-up of the Hamiltonian into 4 pieces) is packed in one word. All the moves are implemented through bit-wise logical operations on words.

Further, the world-line algorithm is parallelized by distributing regions of the lattice among the nodes of a concurrent processor. Both the word packing and the parallelization of the code allowed us to simulate up to $L^2 = 64 \times 64$ lattice with $M = 48$ Trotter slices, significantly larger than previous simulations on the CRAY XMP (limited to about 100 sites and 32 time slices). This was possible because we partitioned a 64×64 lattice into sections of size 64×4 (for example) and each section was stored on a single processor. Thus 16 processors were required for the full lattice.

For a given disorder configuration, 3×10^5 sweeps of the $L^2 \times M$ lattice are performed and at each value of the disorder, we average over many realizations of the random potential. For the 64^2 lattice 8 realizations of disorder are used. In addition, the stiffness on smaller lattices is averaged over all the sublattices of that size contained within the 64^2 lattice This procedure yields 32 disorder realizations for the 32^2 lattice, going up to 2048 realizations for the smallest 4^2 lattice. As a test of equilibration, we verified that the results are independent of the initial boson configurations. We tested our code against exact diagonalization results for small systems. We also have a rather stringent test in that the code was checked against the predictions of Kosterlitz-Thouless theory[36, 26].

5   Interplay of disorder and interactions

In order to distinguish between the superfluid and localized phases, we calculate the stiffness Υ, which is a measure of the rigidity of the many-body wave function to a twist in the boundary conditions and is defined as $\Upsilon = < W_x^2 + W_y^2 >$. The winding number is given by[37]

$$W_\mu = \frac{1}{L} \sum_{\ell=1}^{N_b} [r_{\ell\mu}(0) - r_{\ell\mu}(\beta)] \tag{6}$$

along $\mu = x$ or y, N_b is the average number of bosons and $r_{\ell\mu}(\tau)$ is the position of the ℓth boson at time τ. We define a reduced stiffness by $\overline{\Upsilon} = \Upsilon/2\rho\beta t$, where $\rho = N_b/L^2$.

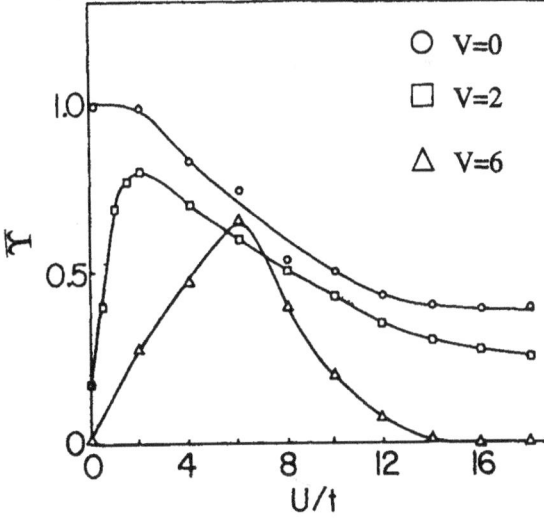

Figure 6: Interplay of disorder and interaction for incommensurate densities. The superfluid density $\overline{\Upsilon}$ vs. interaction strength U/t is shown in a 6×6 system of density $\rho = 0.75$ and $\beta t = 4$. The values of the disorder parameter are $V/t = 0$ (triangles), 2 (squares) and 6 (circles). In the clean system, the effect of U is always to decrease $\overline{\Upsilon}$. In the presence of disorder, U initially enhances $\overline{\Upsilon}$ until $U \approx V$ and then causes $\overline{\Upsilon}$ to decrease beyond that.

We present first qualitative results[27] obtained for a small system using path integral Monte Carlo techniques. These simulations were performed primarily on Silicon Graphics workstations. The behavior of the superfluid stiffness $\overline{\Upsilon}$ as a function of U, the repulsion between the particles, at a density $\rho = 0.75$ is shown in Fig. 6. The surprising feature is that while the stiffness decreases monotonically with increasing U in the clean system, it shows a very non-monotonic behavior in the disordered system. Increasing the repulsion between the particles for small U leads to an *enhancement* of the superfluid density. This can be understood because the first few monolayers of bosons in localized states screen the disorder potential so that the subsequently added bosons feel a smoother potential. The effective random potential V_{eff} is given by the sum of the initial random potential plus a Hartree repulsion from the localized particles,

$$V_{eff}(i) = V(i) + UN|\phi_0(i)|^2 , \tag{7}$$

where ϕ_0 is the non-uniform condensate[21]. At large U, if the disorder is sufficiently strong, the system gets localized into a Bose-glass state with gapless excitations.

Another interesting aspect of the interplay is seen in Fig. 7. At a commensurate density of one particle per site, the pure system shows a transition from a superfluid to a Mott insulator with a gap at a $U \approx 9t$. What is surprising in the presence of disorder is that in the plot of Υ vs. U, the curve with disorder crosses the curve without disorder. This implies that at a $U \approx 10t$ (for example), the clean system is a Mott insulator but the disordered system is driven into a superfluid state. A possible explanation is the large density fluctuations generated by the random potential in the otherwise inert clean system at large U with one particle at every site.

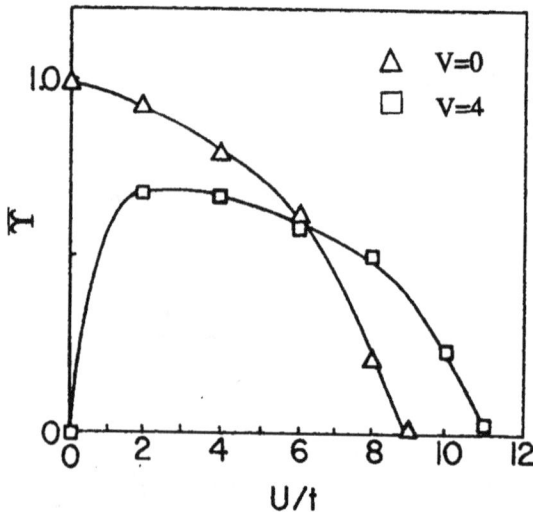

Figure 7: Competition between Bose glass and Mott phases. Superfluid density Υ *vs.* interaction strength U/t in a 10×10 system of density $\rho = 1$ and $\beta t = 4$. The values of the disorder parameter are $V/t = 0$ (triangles) and 2 (squares). Note that for $8.5 < U/t < 11$, the clean system is in a Mott phase, but in the disordered system becomes superfluid.

6   Quantum critical phenomena

Let us now turn to an investigation of the critical phenomena at the superfluid-Bose glass transition at a density $N_b/L^2 = 1/2$ for hard core bosons[29] by studying the size-dependence of the superfluid stiffness. From a single simulation in the zero winding number sector on a 64×64 lattice, we obtain the stiffness on lattice sizes ranging from 4×4 to 32×32 by measuring the winding numbers on various sublattices (which are not constrained to $W = 0$). As the superfluid-insulator transition, occurring at a critical disorder $V = V_c$ and $T = 0$, is approached, the spatial correlation length ξ diverges as $\xi \sim \delta^{-\nu}$ where $\delta = | V - V_c | /V_c$. Here ξ defined by $\lim_{\ell \to \infty} < a_\ell^\dagger(\tau)a_0(\tau) > - < a_\ell^\dagger(\tau) >< a_0(\tau) > \sim \exp(-\ell/\xi)$ is the distance over which the order parameter correlations decay on a fixed time slice. Furthermore, since we are looking at quantum critical phenomena, the statics and dynamics are linked by the Hamiltonian. In other words, the Hamiltonian directly dictates the dynamics, in contrast to the situation in the classical case. The correlation time also diverges as $\xi_\tau \sim \xi^z \sim \delta^{-z\nu}$. ξ_τ is defined by the decay time of the temporal correlations by $\lim_{\tau \to \infty} < a_0^\dagger(\tau)a_0(0) > - < a_0^\dagger(\tau) >< a_0(0) > \sim \exp(-\tau/\xi_\tau)$. Our aim is to determine the exponents z and ν that characterize the transition.

6.1 Finite Size Scaling

The finiteness of the simulation is sensed when the correlation lengths are cut off by the $L^2 \times L_\tau$ system, i.e., in the space direction $\xi \approx L$ or the correlation time $\xi_\tau \approx L_\tau$; $(L_\tau = \beta)$. Near the transition, we make the following finite size scaling *Ansatz*[38] for the stiffness

$$\Upsilon(L, L_\tau, \delta) = (L^{2-d}/L_\tau)\mathcal{F}[L/\xi, L_\tau/\xi_\tau] \quad (8)$$

We want to obtain the appropriate form of the function $\mathcal{F}(x, y)$ in 2D. First let us consider the limit where L, ξ, ξ_τ are fixed and finite and let $L_\tau \to \infty$. In this limit, Υ cannot depend on L_τ which implies $\lim_{y \to \infty} \mathcal{F}(x, y) = yg(x)$. Now let $L \to \infty$. Since Υ cannot depend on L in this limit, we must have $g(x \to \infty) = $ const. This implies that

$$\Upsilon(L \to \infty, L_\tau \to \infty, \delta) \sim \frac{1}{\xi_\tau} \sim \delta^{-z\nu} \quad (9)$$

which is the Josephson scaling relation in 2D.

For the finite size scaling regime, we are interested in the opposite limit. By redefining variables, it can be shown that

$$\Upsilon(L, L_\tau, \delta) = \frac{1}{L_\tau}\mathcal{G}(\delta L_\tau^{1/z\nu}, L/L_\tau^{1/z}) \quad (10)$$

To begin, let us assume that $z = 2$ as suggested by the scaling theory of Fisher et. al.[18]. In this case for the size of the simulations we have studied, $L^z > L_\tau$ and the divergence of the correlation length is limited by L_τ rather than by L. In the spatial direction this implies that the correlations cannot grow beyond $L_\tau^{1/z}$. In Fig. 8 we plot ΥL_τ vs. V for two values of L_τ and correspondingly for two different L such that the second argument $L/\sqrt{L_\tau}$ of $\mathcal{G}$ is fixed. A signature of a superfluid to insulator transition with $z = 2$ should show up as an intersection of the two curves indicating the size independence at the critical point. The curves fail to show any sign of an intersection. The possibility of an intersection of the curves improves considerably if z is reduced below unity and provides the first hint that $z < 1$.

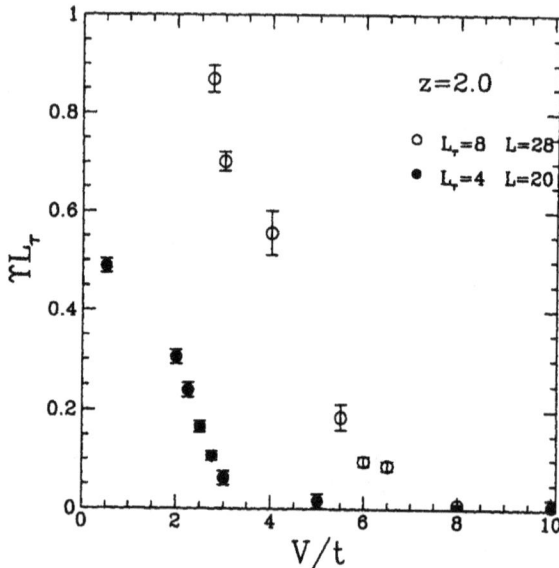

Figure 8: $\overline{\Upsilon}L_\tau$ vs. disorder V for two temperatures such that $L_{\tau 1} = 8/t$ and $L_{\tau 2} = 4/t$ and linear lengths $L_1 = 28$ and $L_2 = 20$ chosen such that the second argument of the scaling function in Eq. 2, $L/L_\tau^{1/z}$ with $z = 2$ is fixed. The value of $z = 2$, suggested by the scaling theory, does not yield an intersection of the curves, indicating the absence of a superfluid-insulator transition for this value of z.

In order to extract the value of z we perform an equivalent finite size scaling analysis in the regime $z < 1$ by noting that now L provides the cut-off for the diverging correlation length and $L_\tau > L^z$. Let us examine the limit $L/\xi \to 0$ and $L_\tau/\xi_\tau \to \infty$. We have from the analysis given above, $\overline{\Upsilon} \sim (1/\xi_\tau)g(L/\xi)$. As $\xi \to \infty$, at the critical point $\overline{\Upsilon} \sim L^{-\lambda}$, where λ is an exponent that we determine below. We want $\overline{\Upsilon}L^\lambda \sim (L^\lambda/\xi_\tau)g(L/\xi)$ to be independent of L which is possible only if $g(x \to 0) \sim x^{-\lambda}$. This implies that for $\overline{\Upsilon}L^\lambda \sim \xi^\lambda/\xi_\tau$ to be a constant, the condition $\lambda = z$ must be satisfied. Thus the scaling function $\mathcal{F}$ can be written as $\mathcal{F}(x,y) = yx^{-z}f(x, x^z/y)$, where $x = L/\xi$, $y = L_\tau/\xi_\tau$ and $f(x_1, x_2)$ is *analytic* at $(0,0)$. We approach the finite size scaling regime, given by the limit $x \to 0$, $y \to 0$, by keeping the argument $x^z/y \ll 1$ (or equivalently $L_\tau \gg L^z$). Keeping only the zeroth order term in the Taylor expansion of $f(x_1, x_2)$ around $x_2 = 0$, the dependence on L_τ drops out, and we obtain

$$\overline{\Upsilon}(L,\delta)L^z = g(\delta L^{1/\nu}) \ . \tag{11}$$

In Fig. 9 we plot $\overline{\Upsilon}(L)L^z$ vs. V for various lattice sizes at a fixed $T = 0.25t$. We have tried various values of z and find that we get a very distinct intersection for $z = 0.5 \pm 0.1$ (we use $z = 0.5$ in Fig. 9). We caution that the scaling analysis is difficult because we have to determine two exponents. Nevertheless, based on the analysis in Figs. 8 and 9 we rule out $z = 2$, that is indicated by the scaling theory[18].

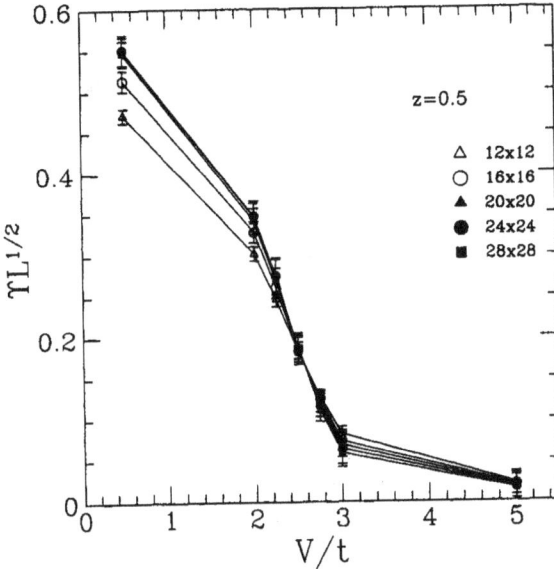

Figure 9: We obtain the dynamical exponent z by plotting $\overline{\Upsilon}(L)L^z$ vs. the disorder V for $L = \{12, 16, 20, 24, 28\}$ at $T = 0.25t$. The value of z used is 0.5. The curves cross at $V = 2.53t$ which identifies the critical disorder because at this point $\overline{\Upsilon}(L)L^z \equiv g(0)$ becomes independent of the lattice size.

Linearizing Eq. 11 around the critical point $V_c = 2.5t$, we get $\overline{\Upsilon}L^z = g(0) + g'(0)L^{1/\nu}\delta$. This equation allows us to extract $\nu = 2.2 \pm 0.2$ (see Fig. 10).

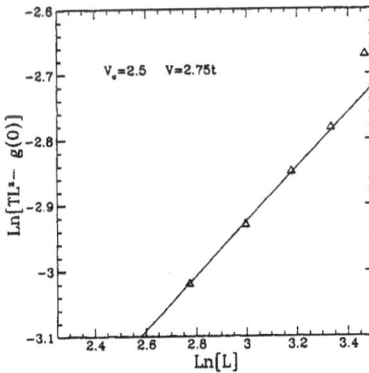

Figure 10: By linearizing around the critical point found from Fig. 9, we extract $\nu = 2.2 \pm 0.02$.

The value of ν obtained satisfies the bound of $\nu \geq 2/d = 1$ in $d = 2$.[39] The exponent for the stiffness, $\Upsilon \sim \delta^\zeta$, where $\zeta = (d - 2 + z)\nu \approx 1$. It is interesting to note that our value of $\nu \approx 7/3$ is also seen in the quantum hall experiments[40].

6.2   Critical region

In order to ascertain that the temperature for the data in Fig. 9 is low enough to probe the quantum fluctuations, we have repeated the simulations at $T = 0.125t$ and find that though V_c changes to $5.2t$, the exponents at both temperatures agree within error and the data from both temperatures can be collapsed onto a single scaling function $g(x)$ as shown in Fig. 11. We therefore believe that the transition is governed by the T=0 fixed point. The change in the critical disorder with temperature is given by the relation $T^{1/z\nu} \sim |V_c(T) - V_c(0)|$, where the crossover exponent $z\nu \sim 1.1$. Using a linear extrapolation we deduce that $V_c(0) \sim 8$.

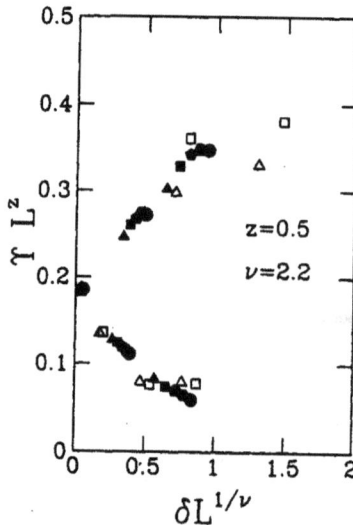

Figure 11: Scaled plot of $\rho_s(L)L^z$ vs. $L^{1/\nu}\delta$ for different lattice sizes and for two temperatures $T = 0.25t$ and $T = 0.125t$. Here $z = 0.5$ and $\nu = 2.2$ for both temperatures; $V_c = 2.5t$ at $T = 0.25t$ and $V_c = 5.2t$ at $T = 0.125t$. The collapse of the data on a single scaling curve shows that we are in the critical regime for quantum fluctuations.

7   Implications of Results

It is interesting to compare our results with recent simulations of the quantum rotor model[30]. In the pure system, the boson Hamiltonian and the rotor model are in the same universality class and reproduce the Kosterlitz-Thouless transition. The quantum rotor model is obtained by quantizing the 2D XY model $H_{XY} = -K\sum_{<i,j>} \cos(\theta_i - \theta_j)$. For short-range repulsive interactions, the dynamical exponent $z_{QR} = 2$ is assumed in Ref.[30] and a transition is found characterized by

$\nu_{QR} = 1.0\pm0.1$. Thus when compared with our results ($z = 0.5\pm0.1$ and $\nu = 2.2\pm0.2$) it can be immediately seen that in the presence of disorder the boson model and the quantum rotor model are in *different* universality classes.

What is the origin of this difference? The quantum rotor model can be approximately obtained[41] from the boson Hamiltonian Eq. 1 by replacing $a_j^\dagger \rightarrow \sqrt{n_{oj}} \exp(i\theta_j)$ and assuming that $t\sqrt{n_{oi}n_{oj}} = K = $ const. Here n_{oj} is the condensate density at site j which is highly non-uniform in the disordered system, as discussed in section 5. By assuming that K is constant in the quantum rotor model one is neglecting the specific *coupling* between amplitude and phase degrees of freedom.

Recently, Sachdev et. al.[42] have studied two models and arrived at similar conclusions. They analyzed the disordered quantum Heisenberg magnet (which exactly maps onto a hard core boson model with nearest neighbor interactions) and the disordered quantum rotor model with infinite range exchange couplings chosen from a Gaussian distribution. The models were treated by large N and large S techniques. They found surprisingly, that these models which are in the same universality class for the pure case, are in different universality classes for the disordered case. The excitation spectra also in both cases was also found to be very different– gapless for the Heisenberg magnet and gapped for the rotors.

Let us reexamine the scaling argument of Fisher et. al.[18] which gives a value for the dynamical exponent $z = 2$ in two dimensions. Their argument is based on a low energy and long wavelength effective action for a disordered superfluid of the following form,

$$S_{eff} = \frac{1}{2} \int d^d x \int_0^\beta d\tau [2\rho i\dot\phi + \kappa\dot\phi^2 + (\rho_s/m)(\nabla\phi)^2] \,, \qquad (12)$$

where ϕ is the phase of the order parameter, The basic assumption behind Eq. 12 is that the only low energy excitations in the disordered superfluid are phonons and the effect of disorder is to renormalize the compressibility and the superfluid density. Using simple scaling one can show from Eq. 12 that

$$\kappa \sim \delta^{\nu(d-z)} \,. \qquad (13)$$

Now κ is finite in both the superfluid and Bose-glass phases because of the presence of low energy excitations. If we further assume that κ is finite at the transition we obtain the relation $z = d$ for the dynamical exponent from Eq. 13.

If we use our value of $z \approx 1/2$ in Eq. 13, we conclude that the compressibility should vanish at the transition. We, however, directly compute the compressibility from the density fluctuations, $\kappa = <\rho^2> - <\rho>^2$ and find that it is *finite* at the transition (see Fig. 12), contrary to the scaling conjecture[18]. Our simulation results of a finite κ at the transition in conjunction with a dynamical exponent $z < 2$, suggests that the the effective action in Eq. 12 is missing important degrees of freedom and the low lying excitations are modified from usual phonons in the presence of disorder. Such a conjecture is further supported by the low temperature specific heat obtained deep within the superfluid phase, which is found to deviate from the expected $C_v \sim T^2$

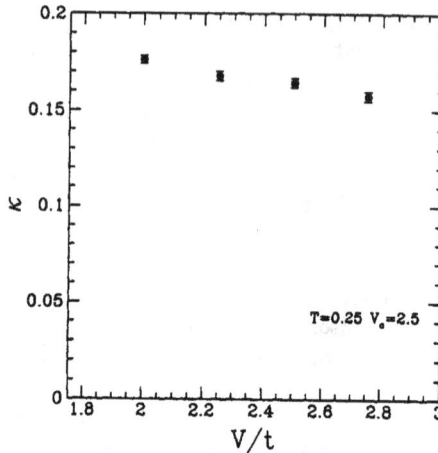

Figure 12: Compressibility κ vs. disorder V for $T = 0.25t$ Our result showing that κ is non-zero at the transition is in contradiction with the scaling theory which gives $\kappa \sim \delta^{\nu(d-z)}$ and would imply that for our $z \approx 1/2$, κ should vanish at the transition.

behavior for phonons in 2D[43]. The anomalous specific heat behavior could arise from diffusive density fluctuations modes in the disordered system[44]. Another possible source is remnant localized states in the superfluid[45].

7.1   Implication for vortex glass with columnar defects

Our results have bearing for the pinning of vortices (equivalent to the boson world lines) by extended-twin boundary or columnar defects (equivalent to the time-independent disorder potential) in disordered type II superconductors in a magnetic field. One of the open questions is whether or not there is a true superconducting phase in a magnetic field at a temperature $T < T_g \neq 0$ in which the linear resistivity ρ_L is indeed zero. While the answer for arbitrary pinning potential is not known, our simulations have shown that for columnar pins there is a Bose glass phase for $V > V_c \neq 0$, which, by the analogy suggested in Ref.[15] implies that there is indeed a true superconducting phase at finite temperatures. The exponents at this transition and a comparison with experiments is discussed in Ref[29].

8   Future Work

While the simulations above give the first hint that the dynamical exponent may deviate from the expected value of 2, more careful and systematic simulations are required to ascertain that one is indeed in the critical regime for the $T = 0$ transition. In this regard it is necessary to systematically track the phase transitions in the disorder-temperature plane. Starting with the evolution of the Kosterlitz-Thouless transition at small disorder, its crossover to the quantum limit can be followed. We are currently investigating this question. It is also necessary to simulate $L \times L \times L^z$ lattices with $z = 1/2$ in order to perform the finite size scaling with less ambiguity.

There remain several interesting open questions regarding the nature of the low lying excitations in the superfluid phase (alluded to above) as well as in the glass phase. Also the nature of the glass state is not clear at the moment. Another point that we have not addressed here is the suggestion[11] that at the superconductor-insulator transition the system is metallic with a universal conductivity on the order of the quantum of conductance $\sigma_c \approx (2e)^2/h$. Experiments on uniform films are roughly in agreement with this prediction. The situation with the simulations on the boson and the rotor models needs to be analyzed further.

9 Acknowledgments

This work was supported in part by the U.S. Department of Energy under grant W-31-109-ENG-38 (N.T.) and NSF grants ASC-9015000 and DMR-8857341 (M.M). We also acknowledge computer time on the parallel processors Intel Gamma and the Intel Touchstone Delta System operated by Caltech on behalf of the Concurrent Supercomputing Consortium. Access to this facility was provided by Argonne National Laboratory.

References

1. P. A. Lee and T. V. Ramakrishnan, *Rev. Mod. Phys.* **57**, 287 (1985).
2. K. Binder and A. P. Young, *Rev. Mod. Phys.* **58**, 801 (1986).
3. M. E. Fisher, "Scaling, Universality and Renormalization Group Theory", Lecture Notes in Physics, **186** 'Critical Phenomena', edited by F. J. W. Hahne, Springer-Verlag 1983.
4. P. C. Hohenberg and B. I. Halperin, *Rev. Mod. Phys.* **49**, 435 (1977).
5. B. C. Crooker, B. Hebral, E. N. Smith, Y. Takano, and J. D. Reppy, *Phys. Rev. Lett.* **51**, 666 (1983); M. H. W. Chan, K. I. Blum, S. Q. Murphy, G. K. S. Wong, and J. D. Reppy, *Phys. Rev. Lett.* **61**, 1950 (1988).
6. J. E. Mooij, B. J. van Wees, L. J. Geerligs, M. Peters, R. Fazio, and G. Schön, *Phys. Rev. B* **65**, 645 (1990); A. van Otterlo, K.-H. Wagenblast, R. Fazio, and G. Schön, *Phys. Rev. B* **48**, 3316 (1993).
7. H. M. Jaeger, D. B. Haviland, B. G. Orr and A. M. Goldman, *Phys. Rev. B.* **40**, 182 (1989); J. M. Valles and D. C. Dynes, *Mat. Res. Soc. Symp. Proc.* **195**, 375 (1990).
8. D. B. Haviland, Y. Liu and, A. M. Goldman, *Phys. Rev. Lett.* **62**, 2180 (1989); Y. Liu, D. B. Haviland, B. Nease, and A. M. Goldman, *Phys. Rev. B* **47**, 5931 (1993).
9. A. F. Hebard and M. A. Paalanen, *Phys. Rev. Lett.* **65**, 927 (1990); *Phys. Rev. Lett.* 69, 1604 (1992).
10. J. M. Valles, D. C. Dynes, and J. P. Garno, *Phys. Rev. Lett.* **69**, 3567 (1992).
11. M. P. A. Fisher, G. Grinstein, and S. M. Girvin, *Phys. Rev. Lett.* **64**, 587 (1990).
12. M. Randeria, N. Trivedi, A. Moreo, and R. T. Scalettar, *Phys. Rev. Lett.* **69**, 2001 (1992).
13. M. P. A. Fisher, *Phys. Rev. Lett.* **65**, 923 (1990).
14. D. S. Fisher, M. P. A. Fisher, and D. A. Huse, *Phys. Rev. B* **43**, 130 (1991).
15. D. R. Nelson and H. S. Seung, *Phys. Rev. B* **39**, 9153 (1989); M. P. A. Fisher and D. H. Lee, *Phys. Rev. B* **39**, 2756 (1989).
16. T. Matsubara and H. Matsuda, *Prog. Theor. Phys.* **16**, 569 (1956).
17. S. C. Zhang, *Int. J. Mod. Phys.* **B6**, 25 (1992).
18. D. S. Fisher and M. P. A. Fisher, *Phys. Rev. Lett.* **61**, 1847 (1988); M. P. A. Fisher, P. B. Weichman, G. Grinstein, and D. S. Fisher, *Phys. Rev. B* **40**, 546 (1989).
19. M. Ma, B. I. Halperin, and P. A. Lee, *Phys. Rev. B* **34**, 3136 (1986).
20. T. Giamarchi and H. J. Schulz, *Europhys. Lett.* **3**, 1287 (1987); see also L. Zhang and M. Ma, *Phys. Rev. A* **37**, 960, 1988.
21. K. Sheshadri, H. R. Krishnamurthy, R. Pandit, and T. V. Ramakrishnan, *Europhys. Lett.* to appear.
22. K. G. Singh and D. S. Rokhsar, (preprint 1993).
23. K. J. Runge, *Phys. Rev. B.* **45**, 13136 (1992).
24. L. Zhang, *Phys. Rev. B.* **47**, 14364 (1993).

25. L. Zhang and M. Ma, *Phys. Rev. B* **45**, 4855, 1992; K. G. Singh and D. S. Rokhsar, *Phys. Rev. B* **46**, 3002 (1992); L. Zhang and X-Q. Wang, (preprint, 1993).

26. G. G. Batrouni, R. T. Scalettar, and G. T. Zimanyi, *Phys. Rev. Lett.* **65**, 1765 (1990); R. T. Scalettar, G. G. Batrouni, and G. T. Zimanyi, *Phys. Rev. Lett.* **66**, 3144 (1991).

27. W. Krauth and N. Trivedi, *Europhys. Lett.* **14**, 627 (1991).

28. W. Krauth, N. Trivedi, and D. M. Ceperley, *Phys. Rev. Lett.* **67**, 2307 (1991).

29. N. Trivedi and S. Ullah, *J. Low Temp. Phys.* **89**, 67 (1992).

30. M. Makivić, N. Trivedi, and S. Ullah, *Phys. Rev. Lett.* **71**, 2307 (1993).

31. E. S. Sørensen, M. Wallin, S. M. Girvin, and A. P. Young, *Phys. Rev. Lett.* **69**, 828 (1992).

32. N. Trivedi, in *Computer Simulation Studies in Condensed Matter Physics V*, eds. D. P. Landau, K. K. Mon, and H. B. Schüttler, Springer Proceedings in Physics, Springer Verlag Berlin, Heidelberg, New York 1993.

33. J. W. Negele and H. Orland, "Quantum Many-Particle Systems", Addison-Wesley, 1988.

34. M. Barma and B. S. Shastry, *Phys. Rev. B* **18**, 3351 (1978); J. E. Hirsch, R. L. Sugar, D. J. Scalapino, and R. Blankenbecler, *Phys. Rev. B* **26**, 5033 (1982); M. Suzuki, *Phys. Lett.* **113A**, 299 (1985).

35. For an application of this algorithm in 1D, see J. Tobochnik, G. G. Batrouni, and H. Gould, *Computers in Physics* **6**, 673 (1992); G. G. Batrouni and R. T. Scalettar, *Phys. Rev. B* **46**, 9051 (1992).

36. M. Makivić and H. Q. Ding, *Phys. Rev. B.* **43**, 3562 (1991).

37. H. Ding and M. Makivić, *Phys. Rev. B.* **42**, 6827 (1990); M. Makivić, *Phys. Rev. B.* **46**, 3167 (1992).

38. E. L. Pollock and D. M. Ceperley, *Phys. Rev. B* **36**, 8343 (1987).

39. K. Binder, in "Finite Size Scaling and Numerical Simulation of Statistical Systems", edited by V. Privman (World Scientific, 1990).

40. J. T. Chayes, L. Chayes, D. S. Fisher, and T. Spencer, *Phys. Rev. Lett.* **57**, 2999 (1986).

41. S. Koch, R. J. Haug, K. V. Klitzing, and K. Ploog, *Phys. Rev. B* **46**, 1596 (1992).

42. M. C. Cha, M. P. A. Fisher, S. M. Girvin, M. Wallin, and A. P. Young, *Phys. Rev. B* **44**, 6883 (1991); S. M. Girvin, M. Wallin, E. S. Sørensen, and A. P. Young, *Physica Scripta* **T42**, 96 (1992).

43. S. Sachdev and J. Ye, *Phys. Rev. Lett.* **70**, 3339 (1993); J. Ye, S. Sachdev, and N. Read, *Phys. Rev. Lett.* **70**, 4011 (1993).

44. N. Trivedi and M. Makivić, unpublished.

45. A. J. Leggett, *Ann. Phys.* **72**, 80 (1972).

46. For related ideas on local moment formation in disordered Fermi systems see M. Milovanović, S. Sachdev and R. N. Bhatt, *Phys. Rev. Lett.* **63**, 82 (1989); R. N. Bhatt and D. S. Fisher, *Phys. Rev. Lett.* **68**, 3072 (1992).

PATH-INTEGRAL QUANTUM MONTE CARLO STUDIES OF THE STATIC AND TIME-DEPENDENT THERMODYNAMICS OF THE VIBRATIONAL PROPERTIES OF CRYSTALS

Arthur R. McGurn
Department of Physics
Western Michigan University
Kalamazoo, Michigan 49008 U.S.A.

Abstract

A review is given of recent efforts at using the path-integral formulation of the quantum Monte Carlo to study the static and dynamic properties of crystals. The computation of the static properties of energy, specific heat and lattice constant as functions of temperature for systems of atoms interacting by Lennard-Jones (12-6) potentials is first considered. Results for fcc systems appropriate to neon and argon are presented and these are compared with analytical results from the effective potential and improved self-consistent theories and with results from the harmonic approximation. In addition, some earlier investigations on the static properties of a one-dimensional chain of atoms are also considered and compared with the exact solutions for these properties in the corresponding classical system. A discussion of the time-dependent thermodynamic properties of a one-dimensional quantum chain of atoms is made in terms of the response functions which describe the inelastic neutron scattering from the chain. A continued fraction representation, based on the work of Mori [1], is presented for these response functions and the coefficients in the continued fraction are computed using the static path-integral quantum Monte Carlo method. Improvements on Mori's method applied to the quantum chain are made by using molecular dynamics results for the corresponding classical chain of atoms to guide in terminating the continued fraction representation of the quantum response functions.

1) H. Mori, Prog Theor Phys **33** (1965) 423 and **34** (1965) 399.

1 Introduction

One recent application of path-integral Quantum Monte Carlo (PIQMC) techniques has been to studies of the thermodynamic properties associated with the motion of atoms in solids. In comparison with the long history in twentieth century physics of the study of atomic motion or vibrational properties of solids [1-4], the techniques of PIQMC are a relatively recent development [5,6]. Nonetheless, the PIQMC has shown itself effective in adding to the understanding of these vibrational properties [7-13]. Unlike all previous techniques used in studies of vibrational properties, the PIQMC correctly treats both the quantum nature of atomic motion and the nonlinearity in the inter-atomic interactions. Both of these aspects of atomic systems become important at low temperatures and also in solids formed from elements of low atomic mass such as solid Ne, which we consider below [10]. The fundamental limitations of the PIQMC are the limitations of computer memory and CPU time whereas past analytical and numerical treatments have always been limited by some fundamental assumption which is of a theoretical nature (i.e., restrictions to a perturbation expansion or to treating the system in a classical or harmonic approximation).

The vibrational models of solids treated by the PIQMC to date have been simple fcc systems in which the atomic interactions are described by a Lennard-Jones (12-6) pair potential [7-13]. In addition, to reduce the CPU time involved, all studies have been limited to models in which only nearest neighbor atoms have been considered to interact with one another. These models then roughly approximate the properties of inert gas solids such as Ar and Ne. As the vibrational properties of inert gas solids have been extensively studied using a variety of theoretical and experimental techniques, there exists a large amount of material to compare with the results from the PIQMC computations on the fcc Lennard-Jones model [14]. In addition, the PIQMC has been used in a series of related studies of one-dimensional chains of atoms interacting by nearest neighbor Lennard-Jones (12-6) pair potentials [11-13,15,16]. This system is not really a crystalline system but is a one-dimensional gas which was looked at in studies preliminary to the PIQMC work on the fcc Lennard-Jones solids. We shall discuss below PIQMC work on both the fcc and one-dimensional models.

Work using the PIQMC has concentrated on models related to the inert gas solids because these solids are the simplest types of vibrational systems. The interactions between atoms in these systems are short ranged, which limits the requirements on memory and CPU time, and the form of the potential energy is fairly well known and expressible in simple analytical forms. As we shall see below the PIQMC is very effective in computing the static properties of energy, lattice constant at zero pressure and to a lesser extent the specific heat as a function of temperature for Ar and Ne solids.

In addition to the static thermodynamic properties, some preliminary efforts on the computation of the time-dependent response functions of Lennard-Jones systems at non-zero temperature have been made [11-13], and these properties will be discussed in the second part of this review. Specifically, the response functions related to the neutron scattering from one-dimensional chains have been considered. The method of computation used is based on Mori's [17] expansion of the time-dependent response functions in terms of the frequency moments of their spectral representations. The frequency moments of the spectral representation can be expressed in terms of static thermodynamic averages of functions of the atomic position and momenta variables and these static averages can be easily handled in a PIQMC formulation. We shall see below that the continued fraction representation, when used in conjunction with classical molecular dynamics results for the classical limit of the quantum system of interest, can yield effective representations of the quantum mechanical response functions [11-13]. This is found to be the case particularly if the spectral representations at a given momentum transfer are dominated by a small number of frequency peaks.

2 PIQMC of the Static Properties

2.1 Theory

The PIQMC is based on the path-integral formulation of the partition function for a many atom system. In this formulation the Trotter identity [18],

$$\exp\left[\sum_{i=1}^{n} A_i\right] = \lim_{M \to \infty} \left(e^{A_1/M} e^{A_2/M} \cdots e^{A_n/M}\right)^M \tag{1}$$

for $\{A_1, A_2, \ldots, A_M\}$ a set of non-commuting operators, is used to express the quantum mechanical partition function

$$Z = tr \; e^{-\beta(H_0 + H_1)} \tag{2}$$

for a system of N atoms with

$$H_0 = \sum_{i=1}^{N} \frac{P_i^2}{2m} \tag{3a}$$

$$H_1 = \frac{1}{2}\sum_{i \neq j} v\left(|\vec{r}_i - \vec{r}_j|\right) = V\left(\vec{r}_1, \vec{r}_2, \ldots \vec{r}_N\right) \tag{3b}$$

$$[H_0, H_1] = -\frac{\hbar^2}{2m}\sum_{i=1}^{N}\sum_{j \neq i}\left\{2\nabla_i v\left(|\vec{r}_i - \vec{r}_j|\right)\cdot\nabla_i + \nabla_i^2 v\left(|\vec{r}_i - \vec{r}_j|\right)\right\} \tag{3c}$$

and $\beta = 1/k_\beta T$, in terms of a path-integral involving classical (commuting) variables. When this is done to Z in Eq. 2 for a system of atoms condensed into a solid, Z is then expressed as the path-integral [10]

$$Z = \int\prod_{i=1}^{N} D\vec{r}_i(\tau)e^{-S[\vec{r}(\tau)]/\hbar} \tag{4}$$

where $\vec{r}(\tau) = \vec{r}_1(\tau), \vec{r}_2(\tau), \ldots, \vec{r}_N(\tau)$ are the atomic coordinates, and the action

$$S[\vec{r}(\tau)] = \int_0^{\beta\hbar}\left[\sum_{i=1}^{N}\frac{1}{2}m\dot{\vec{r}}_i^2(\tau) + V(\vec{r}(\tau))\right]d\tau \tag{5}$$

where $V(\vec{r})$ is the total inter-atomic potential. The path-integral in Eq. 4 represents an integral over all possible closed paths in the space of $\vec{r}(\tau)$ and the formulation in Eq. 4

is made for a system of distinguishable atoms. In treating the atomic motions in most solids corrections for the indistinguishability of the atoms are negligible. Exceptions to this are solid H or He systems and these systems will not be treated in this review.

An important point to note about the path-integral formulation for the partition function in Eqs. 4 and 5 is that the $\vec{r}(\tau)$ are commuting variables so that Z in Eq. 4 can be treated as a problem in classical statistical mechanics. All of the methods applicable to the study of classical mechanical partition functions are now available for the evaluation of Z.

The discretized form of Eq. 4, which is used as a starting point for all quantum Monte Carlo simulations, is

$$Z = \lim_{M \to \infty} Z_M \qquad (6a)$$

$$Z_M = \left[\frac{m}{2\pi\hbar\Delta\tau}\right]^{3NM/2} \int \prod_{j=1}^{M} \prod_{i=1}^{N} [d\vec{r}_i(\tau_j)] \exp\left\{-\sum_{j=1}^{M}\left[\sum_{i=1}^{N} \frac{1}{2}m|\vec{r}_i(\tau_{j+1})\right.\right.$$
$$\left.\left. -\vec{r}_i(\tau_j)|^2/\Delta\tau^2 + V(\vec{r}(\tau_j))\right]\frac{\Delta\tau}{\hbar}\right\} \qquad (6b)$$

where $\Delta\tau = \beta\hbar/M$, $\tau_j = j\Delta\tau$ and $\vec{r}(\tau_{N+1}) = \vec{r}(\tau_1)$. M in Eq. 6 is known as the Trotter number and is essentially the same M as in Eq. 1 for the case in which $A_1 = H_0$, $A_2 = H_1$, $A_i = 0$ otherwise. Our discussion below will center on the Monte Carlo evaluation of the properties of Z_M for atomic solids and the extraction of their $M \to \infty$ limits.

The thermodynamic variables of energy and specific heat are obtained from Eq. 6 as the $M \to \infty$ limit of the temperature derivatives of $\ln(Z_M)$ and the pressure, P, as a function of the temperature is obtained as the $M \to \infty$ limit of

$$P = \frac{1}{\beta}\frac{\partial \ln Z_M}{\partial V} \qquad (7)$$

In most PIQMC computer programs, the lattice constant, a, along with the temperature is specified at the beginning of a PIQMC run. Choosing a by trial and error, one can approximately, for a given temperature, set the value of the average pressure as computed by the PIQMC.

Unfortunately, except for systems of non-interacting atoms or atoms interacting by harmonic forces, an exact evaluation of the $M \to \infty$ limit of the above expressions is not possible. For finite M, however, a numerical evaluation of Eq. 7 and of the temperature derivatives of the logarithm of Eq. 6 is possible. Due to the classical nature of the variables in Eq. 4, this numerical evaluation can be handled by standard classical Monte Carlo methods. Then by studying the numerical evaluations for finite M with increasing

M, the M → ∞ limit can be extracted. The rate of convergence with increasing M is, hence, of fundamental importance in any successful Monte Carlo evaluation of the properties of Z_M.

We shall see below that the numerical results for vibrational systems can be made to converge very rapidly with increasing M < ∞ to the M → ∞ limit if some minor changes are made in the argument of the exponential in Eq. 6. For the remainder of Section 2.1 we shall be concerned with the methodology for increasing the convergence of the properties of the M < ∞ system to the M → ∞ limit.

In all treatments of the Monte Carlo evaluation of Eq. 7 and the energy and specific heat from Eq. 6, the rate of convergence with increasing M to the M → ∞ limit is very important for determining the M → ∞ thermodynamics. A number of changes to the form of Z_M in Eq. 6b for M < ∞ have been suggested so as to increase this rate of convergence while yielding the same M → ∞ limit [19,20]. The origin of the discrepancy between the thermodynamics of the M → ∞ and the M < ∞ systems has to do with the difference between the Trotter identity of Eq. 1, which is used to obtain $\lim_{M\to\infty} Z_M$, and the approximation

$$\exp\left(\sum_{i=1}^{n} A_i\right) = \left(e^{A_1/M} e^{A_2/M} \dots e^{A_n/M}\right)^M \tag{8}$$

for finite M, which is used to obtain Z_M for M < ∞. The left and right hand sides of Eq. 8 differ by terms of order $(1/M)^2$ and this difference gives rise to a non-zero difference in $\left(\lim_{M\to\infty} Z_M\right) - Z$ which is of order $\beta^3 M^{-2}$. However, by replacing $A_1, A_2, \dots, A_n$ on the right hand side of Eq. 8 by correctly chosen $B_1, B_2, \dots, B_n$, we can write

$$\exp\left(\sum_{i=1}^{n} A_i\right) = \left(e^{B_1/M} e^{B_2/M} \dots e^{B_n/M}\right)^M \tag{9}$$

correct to $(1/M)^n$ for n > 2. By using Eq. 9 in the formulation of Z_M from Eq. 2, then a more rapid convergence with increasing M to the M → ∞ limit is obtained. Using these ideas for improving M convergence, Takahashi and Imada [19] have shown that Z_M can be reformulated so that

$$Z_M = \left[\frac{m}{2\pi\hbar\Delta\tau}\right]^{3NM/2} \int\prod_{j=1}^{M}\prod_{i=1}^{N}\left[d\vec{r}_i(\tau_j)\right]\exp\left\{-\sum_{j=1}^{M}\left[\sum_{i=1}^{N}\frac{1}{2}m\,|\,\vec{r}_i(\tau_{j+1})\right.\right.$$
$$\left.\left. -r_i(\tau_j)|^2/\Delta\tau^2 + V^{(1)}(\vec{r}(\tau_j))\right]\frac{\Delta\tau}{\hbar}\right\} \tag{10a}$$

where

$$V^{(1)}(\vec{r}) = V(\vec{r}) + \frac{1}{24}\frac{\hbar^2}{m}\left(\frac{\beta}{M}\right)^2 \sum_{i=1}^{N} |\nabla_i V|^2 \qquad (10b)$$

In this formulation for Z_M the difference $\left(\lim_{M\to\infty} Z_M\right) - Z$ is of order $\beta^5 M^{-4}$ and the thermodynamic averages, computed from Eqs. 7 and 10 above, are correct to order $\beta^5 M^{-4}$. Both formulations of Z_M, that in Eq. 6 and that in Eq. 10, are found to give the same thermodynamics in the $M \to \infty$ limit. In the work presented below the formulation for Z_M in Eq. 10 has been used because of its increased convergence in M.

2.2 Results and Discussion

The first use of the static formulation in Eq. 10 for computing the quantum statistical mechanics of a three dimensional vibrational system was given by Maradudin, et al. for a model approximating the properties of fcc solid Ar [7,8]. Specifically, the atoms in this model were crystallized in an fcc lattice and interacted with one another through nearest-neighbor only interactions given by the Lennard-Jones (12-6) pair potential, $v(r) = 4\epsilon[(\sigma/r)^{12} - (\sigma/r)^6]$. The parameters of the Lennard-Jones potential, ϵ and σ, and the atomic mass were chosen such that $\alpha = \hbar/\sqrt{m\epsilon}\,\sigma = 0.03$ which is close to the value $\alpha = 0.0295$ of these parameters used to model Ar. The parameter α arises naturally in the theory of the quantum statistical mechanics of vibrational Lennard-Jones systems and is a measure of the quantum mechanical fluctuations in the system. For large α quantum corrections to the vibrational thermodynamics of the system are important. For $\alpha = 0$ the results of classical statistical mechanics are obtained. In Lennard-Jones models of solid Ar, α is small and quantum effects are not as important as they are in the Lennard-Jones models for solid Ne which we discuss below.

Using an fcc array of 32 atoms subject to periodic boundary conditions Maradudin, et al. [7,8] computed at zero pressure the low temperature energy, specific heat and lattice constant as functions of temperature. In Figure 1 results are presented for the energy versus temperature at zero pressure for $\alpha = 0.03$. The dots represent the PIQMC data where the error in the data is smaller than the size of the dots and the solid curve represents a least squares fit of the form $E/N\epsilon = \epsilon_0 + \epsilon_1 T^4 + \epsilon_2 T^6 + \epsilon_3 T^8$ to the data. For $0.20 > k\beta T/\epsilon > 0.05$ the lattice constant at zero pressure versus temperature was found to obey

$$a(T)/\sigma = 0.046097(k_b T/\epsilon) + 1.1347201 \qquad (11)$$

with an error of less than one percent. The data in Figure 1, obtained taking $M = 10$ in Eq. 10, are expected, from comparison with results for $M = 15$, to be accurate to within a few percent of the $M \to \infty$ limit.

Rather than using Monte Carlo methods to evaluate the second temperature derivative of $\ln Z_M$, the low temperature specific heat in Maradudin et al. was computed by taking the temperature derivative of the fitted energy form shown by the solid line in Figure 1.

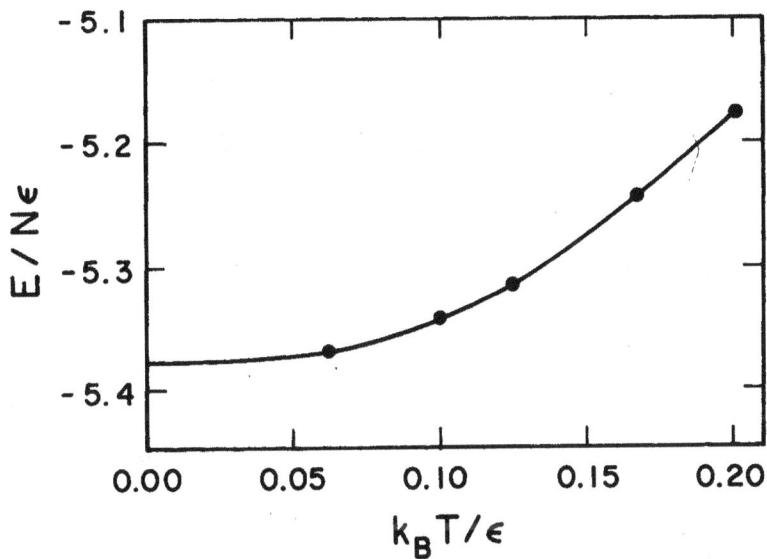

Figure 1: PIQMC results (solid curves) for the internal energy, E/Nε, versus temperature, T, of a 32 atom fcc Lennard-Jones crystal, plotted for the system at zero pressure.

The standard method of evaluation based on the Monte Carlo computation of the second temperature derivative of ln Z_M involves the computation of the variance of the energy in the statistical distribution of states. This computation requires many more Monte Carlo configuration than the computation of the average energy alone and could not be reasonably done. A comparison of the specific heat from the fitted form was made by Maradudin et al. with the harmonic approximation to indicate the small anharmonic corrections in the system. The reader is referred to Ref. [7,8] for a discussion of these results.

The static thermodynamic properties of the fcc nearest-neighbor Lennard-Jones system have also been studied by Cuccoli, et al. [9] using PIQMC methods. In their work a system of 108 atoms subject to periodic boundary conditions was considered for $\alpha = 0.0294$ appropriate to solid Ar, and PIQMC results for this system were generated as function of 1/M in order to extract the $M \to \infty$ limit. Corrections for contributions from atomic pairs at greater separations than nearest-neighbor pairs were approximated using static lattice results, and these corrections in large part account for the differences between the results of Cuccoli, et al. and those of Maradudin, et al., for these Lennard-Jones systems. The energy and the equation of state were investigated by Cuccoli, et al., and found to converge rapidly with increasing M to the $M \to \infty$ limit and to be in surprisingly good agreement with experimental values for solid Ar. The specific heat was computed both by standard methods and as the numerical temperature derivative of the energy data. In Table 1 we present some of the results of Cuccoli et al. [9].

TABLE 1. PIQMC Results of Cocculi et al. for Ar

Temperature (ϵ/k)	Pressure (ϵ/σ^2)	Energy (ϵ)	Specific Heat (Constant V) (k)
0.08347	0.077	-7.775	0.54
0.1669	0.089	-7.699	1.54
0.339	0.111	-7.343	2.36
0.5008	0.111	-6.853	2.61

One object of the work of Cuccoli, et al. [9], which shall be of importance later in discussing time-dependent properties, was to test a new and very promising approach, the effective potential-Monte Carlo (EPMC), at approximating the thermodynamics of Z_M in Eq. 10. This was done by comparing results for the thermodynamics of the Lennard-Jones model for solid Ar generated using the EPMC with corresponding data obtained using the PIQMC.

The EPMC is an improved version of a variational approach originally developed by Feynman for approximating path-integrals of the form shown in Eq. 4 above. In this approach the action $S[\vec{r}(\tau)]$ is replaced by a trial action $S_0[\vec{r}(\tau)]$ containing a number of variational parameters. The variational parameters in $S_0[\vec{r}(\tau)]$ are determined by minimizing the right hand side of the Feynman-Jensen inequality [10,21]:

$$F \le F_0 + \frac{1}{\beta} \left\langle S - S_0 \right\rangle_0 \qquad (12)$$

where in Eq. 12 $Z = e^{-\beta F}$ for Z defined in Eq. 4, $Z_0 = e^{\beta F_0}$ for Z_0 defined by,

$$Z_0 = \int \prod_{i=1}^{N} D\vec{r}_i(\tau) e^{-S_0[\vec{r}(\tau)]/\hbar} \qquad (13)$$

and $\langle \quad \rangle_0$ represents a thermodynamic average computed using the trail action $S_0[\vec{r}(\tau)]$. Once $S_0[\vec{r}(\tau)]$ has been determined in this way, an approximation for the free energy of the system is obtained by taking $F = F_0 + \left\langle S - S_0 \right\rangle_0/\beta$.

In Feynman's original studies, he took $S_0[\vec{r}(\tau)]$ to be that for a system of free-particles [21]. In the EPMC, $S_0[\vec{r}(\tau)]$ is taken to be that of a system of atoms interacting by a generalized quadratic potential energy in the atomic displacements from their equilibrium positions and in this approximation it is found that $\left\langle S - S_0 \right\rangle_0 = 0$ [10]. The EPMC was shown by Cuccoli, et al. [9] to be effective in yielding accurate and computationally rapid approximations of the Ar system. Below, we shall see that this method has also proven useful in studies of solid Ne and in the study of time-dependent properties.

In another series of studies, Lui, et al. [10] treated the nearest-neighbor Lennard-Jones model of Maradudin, et al. considering parameters appropriate to Ne[22] at zero pressure ($\alpha = 0.0759$). Both the PIQMC for a system of 108 atoms subject to periodic boundary conditions and EPMC were computed by Lui, et al. as well as results for the improved self-consistent (ISC) theory which can be described in terms of a diagrammatic perturbation expansion in the anharmonic interaction. The PIQMC results were found to converge rapidly with increasing M and are believed for M = 20 to be within less than 1% of the M → ∞ limit. The results for the PIQMC, shown in Table 2, were used by Lui, et al. as standards to test the validity of the EPMC and ISC approximations. In general the ISC energy and lattice constants as functions of temperature were found to agree well with those of the PIQMC while the EPMC energy and lattice constants agreed less well with the PIQMC results. Approximations for the specific heat are possible in both the ISC and EPMC but the PIQMC does not yield reliable specific heat data and will not be given here. The importance of the PIQMC, here as in the work of Cuccoli, et al. [9], is that it yields reference values for well defined theoretical models and these reference values can be used to test less computationally demanding methods based on approximation schemes.

TABLE 2.   Summary of PIQMC Results (Lui, et al.)

Temperature (ϵ/k)	Spacing σ	Number of atoms	Trotter number	Millions of configs.	Pressure (ϵ/σ^3)	Energy (ϵ)
0.134	1.16816	108	30	13	-0.08±0.01	-4.426±0.006
0.249	1.17075	108	20	26	-0.02±0.01	-4.329±0.005
0.421	1.1834	108	10	14	0.01±0.02	-3.984±0.006

The conclusions, then, on the present state of the PIQMC applied to the nearest-neighbor Lennard-Jones solid are that:

1) For M as small as 10 or 20 an accurate (<1% error) computation using Eq. 10 of the energy and lattice constants as functions of temperature and pressure can be obtained at temperatures for which both quantum mechanical and non-linear effects are important.

2) The PIQMC yields accurate enough values for the energy and lattice constants of the Lennard-Jones model so as to be useful in checking results of analytical theories on these solids.

3) Surprisingly good agreement with the experimentally determined properties of solid Ar can be obtained by using correctly interpreted PIQMC results, i.e., results including estimates of further than nearest-neighbor pair interactions.

4) The PIQMC is found to be least effective in computing specific heats. This is due to the fact that the specific heat is related to the variance of the energy and this variance is a higher moment of the statistical distribution of states than is just the average energy of the system.

To conclude this section on static properties we mention in passing some earlier work on nearest neighbor Lennard-Jones atomic chains [15,16]. This work was done as preliminary to studies of the fcc system considered above, and the atomic chain model was treated both by the PIQMC and EPMC methods. For simplicity, the atoms were taken only to move parallel to the chains and the chains were subject to periodic boundary conditions. While this system is not crystalline, at low temperatures it has a significant amount of short range order so that in these regions of temperature results for the energy and specific heat in the harmonic approximation are found to be in fair agreement with PIQMC data. In addition, in the classical limit the thermodynamics of this model is exactly solvable [22]. In Figure 2 we present the specific heat as a function of temperature for fixed lattice constant $a/\sigma = 2^{1/6}$. The linear chain model is of interest as it was the first many-atom model of atomic motions to be treated in the PIQMC and indicated that the PIQMC could be effectively employed in the study of such systems. We shall now turn to a consideration of this chain model in the new developing field of PIQMC modeling of time-dependent properties of atomic systems.

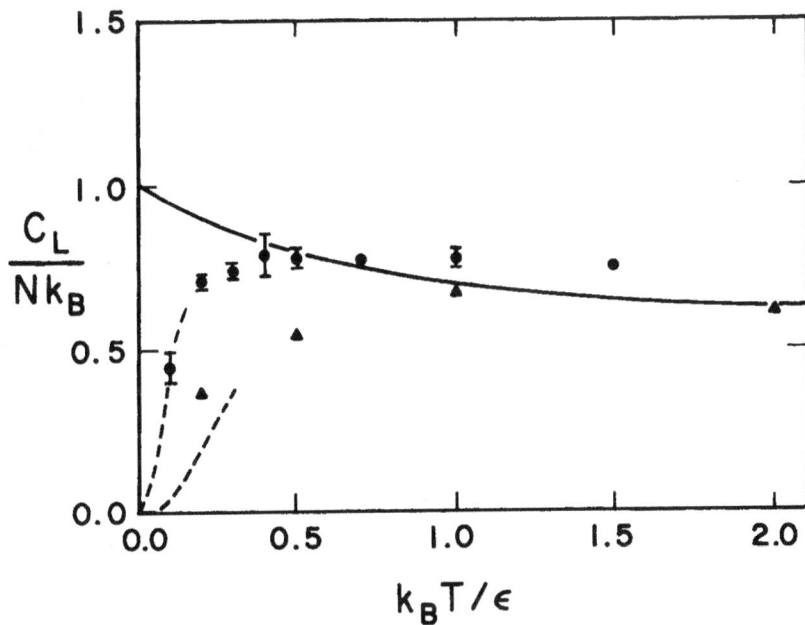

Figure 2:   The specific heat at constant length, C_L/Nk_B, versus absolute temperature, T, for Lennard-Jones chains.   Quantum Monte Carlo results for $\alpha = 0.3$ (circles) and $\alpha = 0.1$ (triangles) are shown. The low temperature harmonic approximation results for $\alpha = 0.03$ and 0.1 are indicated by dashed curves.   The classical result is given by the solid curve.

3 PIQMC of the Time-Dependent Properties

Recent efforts of PIQMC computations of the time-dependent properties have concentrated on the properties of the one-dimensional chain system discussed at the end of Section 2 [8,11,13,14]. The spectral density related to neutron scattering from this system is given for a chain of N atoms subject to periodic boundary conditions by

$$C(k,\omega) = \frac{1}{N} \sum_{i,j} e^{-ika(i-j)} \int dt\, e^{i\omega t} \langle (x_i(t) - x_j(0))^2 \rangle \qquad (14)$$

where $< >$ indicates a thermodynamic average, x_i is the coordinate of the ith atom along the chain and $C(k,\omega)$ is an even function of k and ω. At low temperatures we expect $C(k,\omega)$, for a fixed k, to be peaked about two k dependent frequencies, $\pm\omega_k$, in the system. These peaks should be broadened by the nonlinearity of the system as well as by the thermal fluctuations. As the temperature increases the peaks should eventually be washed out as $C(k,\omega)$ becomes a complicated function of ω.

In the low temperature limit, the two peaked ($\pm\omega_k$) behavior of $C(k,\omega)$ allows for the possibility of approximating $C(k,\omega)$ by a continued fraction representation such as that proposed by Mori [17]. In Mori's formulation [8,11,13,14,17]

$$C(k,\omega) = \frac{\mu_0(k)}{\pi}\, Re\big[\psi_0(k,i\omega)\big] \qquad (15a)$$

$$\psi_n(Z) = \big[Z + \delta_{n+1}\psi_{n+1}(Z)\big]^{-1} \qquad (15b)$$

where δ_n are related to the frequency moments $\mu_{2n}(k) = \int_{-\infty}^{\infty} d\omega\, \omega^{2n}\, C(k,\omega)$ such that $\delta_1 = \mu_2/\mu_0$, $\delta_2 = \mu_4/\mu_2 - \mu_2/\mu_0$, $\delta_3 = \big[\mu_6/\mu_2 - (\mu_4/\mu_2)^2\big]$, etc. The continued fraction representation in Eq. 15 is found to be exact for an atomic chain when the interatomic forces are harmonic, and in the harmonic case the δ_n vanish for $n > 1$. For weakly anharmonic systems, such as the Lennard-Jones chain, it then is anticipated that the knowledge of a limited number of δ_n will suffice in the description of $C(k,\omega)$ using Eq. 15 provided that some effective termination of the continued fraction in Eq. 15 is used.

The frequency moments μ_{2n} which generate the δ_n in Eq. 15 can be computed using static PIQMC or EPMC methods. This follows from the fact that μ_{2n} can be expressed in terms of the time derivatives of the $\{x_i\}$. The time derivatives of the $\{x_i\}$ are expressible from the quantum commutation relations in terms of functions involving the atomic positions, the interatomic potentials and the spatial derivatives of the interatomic potentials. The computation of the moments then represent a static problem in the PIQMC.

Using both PIQMC and EPMC methods, Cuccoli, et al. [11,13] have computed the first three δ_n and the first four μ_{2n} as functions of the temperature. Good agreement between the PIQMC and the EPMC moments were obtained so that ultimately the EPMC computed moments, which are easiest to determine, were used in a continued fraction representation of $C(k,\omega)$. As only a finite number of moments can be computed by either PIQMC or EPMC methods, some termination scheme must be used on the continued fraction in Eq. 15. The two commonly known termination schemes are the Gaussian approximation [23] and the n pole approximation [24]. In the Gaussian approximation, at the n order it is assumed that $\delta_{n+1+m} = (m + 1)\delta_{n+1}$ for m > 0 and in the n pole approximation it is assumed that $\delta_n\psi_n(z) \cong \delta_n\psi_n(0) = 1/\tau$ where τ does not depend on z. Both of these approximate termination schemes were studied as terminations of the continued fraction in Eq. 15 for the computation of $C(k,\omega)$ of the quantum chain system and were found not to yield significantly different results.

Results using these two termination schemes for the first six frequency moments can be found in Ref. [11]. In Ref. [11] continued fraction representations for $C(k,\omega)$ are computed for both the quantum chain of atoms and the corresponding classical system. These two systems are found to yield surprisingly different behaviors. The quantum system is found to generally have narrower peaks in $C(k,\omega)$ at higher frequencies than does the classical system. The results in Ref. [11] can, however, be improved upon, and before plots are presented for $C(k,\omega)$ let us consider these improvements.

In their most recent study of $C(k,\omega)$, Cuccoli, et al. [13] used an improved method of terminating Eq. 15. In this study Cuccoli, et al. first considered a continued fraction representation for $C(k,\omega)$ for the classical mechanical chain of atoms. The parameters in the Gaussian or n pole terminations of the continued fraction representation were then adjusted by Cuccoli, et al. to fit data from molecular dynamics computations of $C(k,\omega)$ for the classical mechanical system. The Gaussian or n pole terminations computed in this way for the classical system were then used to terminate the continued fraction representations of the corresponding quantum mechanical $C(k,\omega)$. It is felt that this yields a good representation of the quantum mechanical $C(k,\omega)$ as higher order δ_n are thought to be affected less by the quantum fluctuations in the system.

In Figure 3 we present results for $F(k,\omega) = C(k,\omega)/\mu_0$ at ka = 0.5π, for $k_BT/\epsilon = 0.1$, 0.3 and 0.8 computed as per the discussion in the previous paragraph [13]. The continuous and dashed lines are the classical and quantum 4-pole approximations, respectively, obtained by using the values of the pole approximation termination deduced by fitting classical molecular dynamics data. Again we see here that the quantum and classical systems exhibit different behaviors. The quantum system, particularly at lower temperatures, displays narrower peaks at higher frequencies in $C(k,\omega)$ that does $C(k,\omega)$ computed for the classical system. At the Brillouin Zone edge, the results in Ref. [13] for the quantum system differ very little from those for the quantum system in Ref. [11]. Significant differences in the results from Ref. [11] and [13] occur only for k near the center of the Brillouin Zone.

The results obtained by Cuccoli, et al. for the quantum chain are quite encouraging that continued fraction representations of simple spectral densities for many atom quantum systems can be computed using PIQMC or EPMC determined frequency moments of these spectral densities. This should be true for spectral densities which are dominated by two or less frequency peaks. Currently, efforts are underway at making such a representation of the response functions related to the low temperature neutron scattering from fcc Lennard-Jones solids.

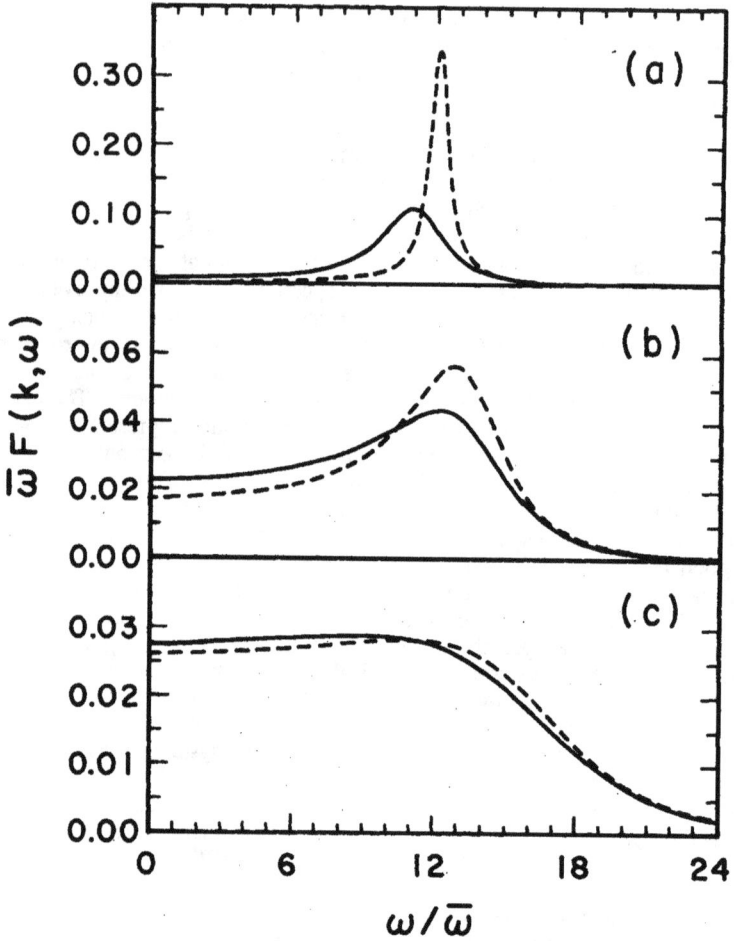

Figure 3: Classical and quantum relaxation function $F(k,\omega)$ at $k_BT/\epsilon = 0.1$ (a), $k_BT/\epsilon = 0.3$ (b), and $kT/\epsilon = 0.8$ (c). In these plots $\bar{\omega} = (\epsilon/m\sigma^2)^{1/2}$.

Acknowledgement:

Manuscript preparation supported in part by NSF Grant No. DMR 92-13793. Thanks to Department of Physics, University of California, Irvine, for hospitality during manuscript preparation.

References:

1. M. Born and K. Huang, Dynamical Theory of Crystal Lattices (Oxford University Press, New York, 1954).
2. Lattice Dynamics, Ed. A. A. Maradudin (Benjamin, New York, 1969).
3. Physics of Phonons, Ed. T. Paskiewcz (Springer-Verlag, Berlin, 1987).
4. Lattice Dynamics and Semiconductor Physics, Ed. J. Xia, Z. Gun et al. (World Scientific, Singapore, 1990).
5. M. Suzuki, Prog Theor Phys **56**, (1976) 1454.
6. M. Suzuki, S. Miyashita and A. Kuroda, Prog Theor Phys **58** (1977) 1377.
7. A. A. Maradudin, A. R. McGurn, R. F. Wallis, M. S. Daw, A. J. C. Ladd, in Lattice Dynamics and Semiconductor Physics, Ed. J. Xia, et al. (World Scientific, Singapore, 1990).
8. A. R. McGurn, A. A. Maradudin and R. F. Wallis in Microscopic Aspects of Non-linearity in Condensed Matter, Ed. A. J. Bishop and V. Tognetti (Plenum, 1992).
9. A. Cuccoli, A. Macchi, M. Neumann, V. Tognetti and R. Vaia, Phys Rev **B45** (1992) 2088.
10. S. Lui, G. K. Horton, R. E. Cowley, A. R. McGurn, A. A. Maradudin and R. F. Wallis, Phys Rev **B45** (1992) 9716.
11. A. Cuccoli, V. Tognetti, A. A. Maradudin, A. R. McGurn and R. Vaia, Phys Rev **B46** (1992) 8839.
12. A. Cuccoli, A. Macchi, V. Tognetti, A. A. Maradudin, A. R. McGurn and R. Vaia, invited talk at Symposium in honour of Hiroomi Umeyawa, Perugia, May 28-31, 1992.
13. A. Cuccoli, V. Tognetti, A. A. Maradudin, A. R. McGurn and R. Vaia, accepted for publication in Phys Rev B.
14. M. L. Klein and J. A. Venables, Rare Gas Solids (Academic Press, 1976).
15. A. R. McGurn, P. Ryan, A. A. Maradudin and R. F. Wallis, Phys Rev **B40** (1989) 2407.
16. A. Cuccoli, V. Tognetti and R. Vaia, Phys Rev **B41** (1990) 9588.
17. H. Mori, Prog Theor Phys **33** (1965) 423 and **34** (1965) 399.
18. H. F. Trotter, Proc Am Math Soc **10** (1959) 545.
19. M. Takahashi and M. Imada, J Phys Soc Jpn **53** (1984) 963 and **53** (1984) 3765.
20. H. DeRaedt and A. Lagendyk, Phys Rpts **127** (1985) 234.
21. R. P. Feynman, Statistical Mechanics (Benjamin, Reading, MA, 1972).
22. F. Gursey, Proc Cambridge Philos Soc **46** (1950) 182.
23. K. Tomita and H. Tomita, Prog Theor Phys **45** (1971) 1407; H. Tomita and H. Moshiyama, Prog Thor Phys **48** (1972) 1133.
24. S. W. Lovesey and R. A. Mescroe, J. Phys **C6** (1972) 79.

RELAXATION OF QUANTUM SYSTEMS IN FLUCTUATING MEDIA

Masako Takasu

Department of Physics, Kanazawa University,

Kakuma, Kanazawa 920-11 Japan

e-mail takasu@icews1.ipc.kanazawa-u.ac.jp

Abstract

We discuss numerical methods for calculating time-correlation functions for quantum systems. We review several methods including analytical continuation methods, stationary phase methods and exact enumeration. We present some results for the relaxation of a quantum spin in a fluctuating medium obtained by exact enumeration.

1   Introduction

Quantum dynamics is an important and interesting topic, but it is not yet fully explored due to the difficulties of numerical calculations. One method for investigating quantum dynamics is quantum Monte Carlo simulation[1], which has been found powerful for equilibrium calculations for quantum systems. Using Suzuki-Trotter decomposition formula[2], we can transform a d-dimensional quantum system into a $(d+1)$-dimensional classical system and calculate expectation values of physical quantities. However, if we want to obtain information on time-dependent phenomena from Monte Carlo simulation, we encounter the sign problem[3, 4, 5, 6, 7, 8, 9, 10] because the complex phases cancel each other and give bad statistics. This problem is similar to the negative sign problem for equilibrium calculations for frustrated spin models[11, 12, 13, 14, 15] or fermion systems[16, 17, 18, 19, 20, 21, 22]; both problems originate from the decomposition[2] of a Boltzmann operator into local operators, and a bigger number of partitions or larger time (inverse temperature in equilibrium case) gives worse statistics. Various methods such as analytical continuation methods[3, 23, 24, 25] and stationary phase methods[4, 8, 26, 27] have been used. These methods work well for certain parameter regions but does not work very well for regions where the cancellation due to phases becomes very large.

Another method is exact enumeration, based also on the decomposition[2] of operators. This method works well for any amount of quantum coherence, but is limited to the studies of short-time behaviors, because the CPU time required for exact enumeration increases exponentially as the number of decompositions increases. A rough idea of the applicability of Monte Carlo method and exact enumeration is shown in

Fig. 1. One should decide which method to use depending on the parameters and the total time.

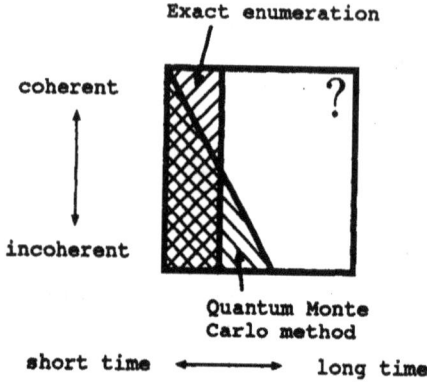

Fig. 1 The applicability of the quantum Monte Carlo method and exact enumeration. The horizontal axis is time t, and the vertical axis is a parameter that determines quantum coherence. The question mark denotes the region of large t for a coherent case, where it is most difficult to obtain good statistics.

In this paper, we review these methods for calculating time-dependent quantities for quantum systems. In particular we use an example of spin-boson model[28, 8, 9, 29, 30, 31], which is a model of a spin relaxing in an environment of an infinite number of bosons. For this model, we explain the basic procedures for calculating quantum time-correlation functions and relaxation functions. We describe mainly the computational methods and briefly mention some of our results for this model obtained by exact enumeration. The details of the results are in other papers[30, 31, 32].

2   Time Correlation Functions

A quantity of interest for time-dependent phenomena is the following time correlation function of a physical quantity A, whose behavior is governed by a Hamiltonian H:

$$\langle A(0)A(t)\rangle = \frac{1}{Z}Tr\{\exp(-\beta H)A\exp(iHt/\hbar)A\exp(-iHt/\hbar)\} \qquad (1)$$

where $1/\beta = k_B T$, $Z = Tr\exp(-\beta H)$ and the trace runs over all degrees of freedom.

Another interesting quantity is the following non-equilibrium quantity that describes the relaxation of A, with the initial condition that A is in state 1 at $t \leq 0$:

$$\bar{A}(t) = \frac{1}{Z_1}Tr\exp(-\beta H_1)\exp(iHt/\hbar)A\exp(-iHt/\hbar). \qquad (2)$$

Here, the trace is the same as the trace in Eq. 1, H_1 is the Hamiltonian with A fixed to the state 1, and $Z_1 = Tr\exp(-\beta H_1)$. The difference between Eq. 1 and Eq. 2 should be small for high temperatures, but not for low temperatures.

Let us assume that the matrix H is a very large one and is not numerically diagonalizable. If H is decomposed into

$$H = H_a + H_b, \tag{3}$$

where H_a and H_b can be diagonalized separately (in general, $[H_a, H_b] \neq 0$), then we can make use of the Suzuki-Trotter formula[2],

$$\exp(Q_a + Q_b) \sim \{\exp(Q_a/N)\exp(Q_b/N)\}^N, \tag{4}$$

where Q_a and Q_b are operators, and N is often called the Trotter number. We apply Eq. 4 to the three operators in Eq. 1, $\exp(-\beta H)$, $\exp(iHt/\hbar)$ and $\exp(-iHt/\hbar)$. We have complex weights because of the last two operators.

In the case of the spin-boson model[28, 8], we have both the spin degree of freedom s and the boson degrees of freedom $\{x_j\}$:

$$H = H_0(\mathbf{s}) + H_B(\{x_j\}) + H_{int}(\mathbf{s}, \{x_j\}), \tag{5}$$

The general three-level systems are considered in Refs. [30, 31, 32, 33]. In this paper, we consider only the case of two-level system[8, 31, 32]: $H_0 = \begin{pmatrix} E_1 & K \\ K & 0 \end{pmatrix}$ is the bare Hamiltonian of the electronic states, and $H_B = \sum_j (\frac{m_j x_j^2}{2} + \frac{p_j^2}{2m_j})$ is the bath Hamiltonian. H_{int} is the interaction between the electron and the bath $\mathcal{E}$: $H_{int} = s^z\mathcal{E}$. Here, $s^z = \begin{pmatrix} 1 & 0 \\ 0 & -1 \end{pmatrix}$ and $\mathcal{E} = \sum_j c_j x_j$, where c_j is a coupling constant.

We assume here that a physical quantity A is a diagonal matrix in s and independent of $\{x_j\}$. By the decomposition formula of Eq. 4, the orginal spin-boson model is transformed to N classical spins $\{s_i\}$ and N sets of classical harmonic oscillators, $\{x_{ji}\}$ ($i = 1, ..., N$). Since the density matrix of harmonic oscillators is Gaussian in $\{x_{ji}\}$, and the interaction between the spin and the harmonic oscillators is linear, we can integrate out $\{x_{ji}\}$[34]. Finally, we obtain

$$\langle A(0)A(t)\rangle = \frac{\sum A(s_{p+1})A(s_{p+q+1})w(\{s_i\})}{\sum w(\{s_i\})}. \tag{6}$$

Here the summation runs over the configurations of classical spins $\{s_i\}$, where $s_i \in \{1, -1\}$. p is the Trotter number for $\exp(-\beta H)$, and q is the Trotter number for $\exp(\pm iHt/\hbar)$. The classical spins $\{s_i\}$ are numbered as in Fig. 2. The weight $w(\{s_i\})$ is written as

$$w(\{s_i\}) = \exp(\varphi(\{s_i\})). \tag{7}$$

Here, the action $\varphi(\{s_i\})$ is

$$\varphi(\{s_i\}) = \sum_{ij} J_{ij}s_i s_j, \tag{8}$$

where J_{ij} is a function of β and t. In general, J_{ij} takes a complex value and is non-zero for most pairs of (ij). The explicit form of J_{ij} for general three-level model can be found in Ref. [32]. The non-local nature of J_{ij} comes from the fact that we have integrated out the boson degrees of freedom. In Fig. 2, we show a view of the classical spins $\{s_i\}$ for $\langle A(0)A(t)\rangle$ and $\bar{A}(t)$. For the calculation of $\langle A(0)A(t)\rangle$, we need a total of $(p+2q)$ spins: p spins on the temperature axis, and q spins on both the positive time and negative time axis. For the calculation of $\bar{A}(t)$, we need $(2q-1)$ spins, because the $(p+1)$ spins on the temperature axis are fixed to the up-state. Although the spins are numbered one-dimensionally, there are interactions between any pair of spins. The exact solution for this model in the general case has not been found. Also, a usual procedure such as the transfer matrix method does not apply here.

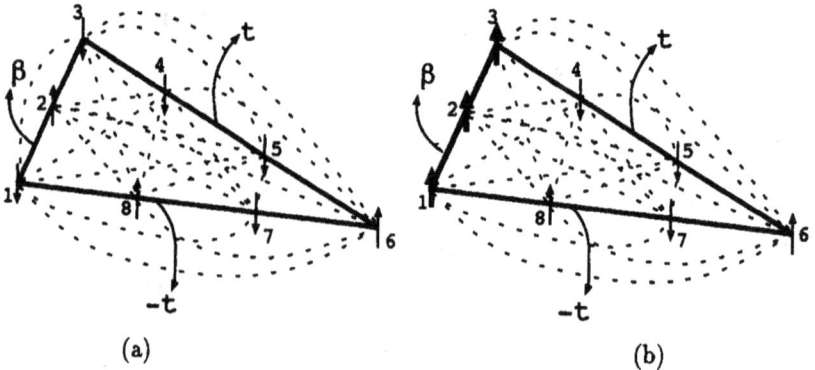

(a)   (b)

Fig. 2   A view of the classical representation of $\langle A(0)A(t)\rangle$ (a) and $\bar{A}(t)$ (b) for $p=2$ and $q=3$. β denotes the temperature axis, and t and $-t$ denote the positive and negative time axes, respectively. The solid lines are nearest-neighbor interactions and the dotted lines are other interactions. Note that in (b), all the spins on the temperature axis (s_1, s_2 and s_3) are fixed to the up-state.

3   Methods of Calculations

As we have seen in the previous section, a spin-boson model can be transformed to a set of classical spins $\{s_i\}$ with interactions J_{ij}'s, which take complex values and are non-zero for most pairs of (ij). It is difficult to perform Monte Carlo simulations for such models, as we will explain in Section 3.1. Various methods to overcome this difficulty are reviewed in Section 3.2 through 3.5.

3.1 Straightforward Quantum Monte Carlo Simulation

A straightforward way to perform Monte Carlo simulations for a system with complex weights is to take the modulus of a complex weight and use $p^{(k)} = C|w^{(k)}|$ as the probability of generating the state k. (The constant C is unimportant because we use only the relative probability $p^{(k+1)}/p^{(k)}$ in the Metropolis algorithm[35].) Since $\sum_k w^{(k)} \propto \sum_k p^{(k)} \exp(i\phi^{(k)})$, where $\exp(i\phi^{(k)}) = w^{(k)}/|w^{(k)}|$, we have to accumulate $\exp(i\phi^{(k)})$ as we do the Monte Carlo simulation. To calculate the Monte Carlo average of a quantity A, we use the formula

$$\langle A \rangle = \frac{\sum_k p^{(k)} \exp(i\phi^{(k)}) A^{(k)}}{\sum_k p^{(k)} \exp(i\phi^{(k)})}, \tag{9}$$

where $A^{(k)}$ is the value of A at state k. The sign-problem occurs because of the factor $\exp(i\phi^{(k)})$. The nature of the sign problem is similar to that observed in frustrated spin systems[12]; the cancellation increases exponentially as the system size (the physical size or the Trotter number) increases.

Since the straightforward Monte Carlo simulation of time-correlation functions causes much trouble, various methods have been applied. Here we review several of them: analytical continuation methods[3, 23, 24, 25], stationary phase methods[4, 8, 26, 27], filtering method[9] and exact enumeration[36, 31, 32].

3.2 Analytical Continuation

Thirumalai and Berne[3] developed the analytical continuation method to calculate the real-time correlation function from the imaginary-time correlation function. For example, for the time-correlation function $F(t) = \langle A(t)A(0) \rangle$, the Fourier transform is given by

$$F(t) = \int_{-\infty}^{\infty} \frac{d\omega}{2\pi} \exp(-i\omega t) f(\omega). \tag{10}$$

The imaginary part of the susceptibility $\chi''(\omega)$ is defined by

$$\chi''(\omega) = i\{1 - \exp(-\beta\hbar\omega)\} f(\omega). \tag{11}$$

Since $F(0) = F(-i\beta\hbar)$, $F(-i\lambda)$ $(0 \leq \lambda \leq \beta\hbar)$ can be expanded into the Fourier series:

$$F(t) = \frac{1}{-i\beta\hbar} \sum_n \exp(-iZ_n t) \tilde{f}(Z_n), \tag{12}$$

where $t = -i\lambda$, $0 \leq \lambda \leq \beta\hbar$, and $Z_n = 2\pi n/(-i\beta\hbar)$. After analytically continuing $\tilde{f}(Z_n)$ to all non-real values of Z, Thirumalai and Berne[3] obtained

$$\chi''(\omega) = i\{\tilde{f}(\omega + i\epsilon) - \tilde{f}(\omega - i\epsilon)\}. \tag{13}$$

Thus, if we calculate the imaginary-time correlation, $F(-i\lambda)$ $(0 \le \lambda \le \beta\hbar)$, which does not involve complex weights, we can obtain $\tilde{f}(Z_n)$ from the inversion of Eq. 12. Then, by analytical continuation to all values of Z in $\tilde{f}(Z)$, we can obtain $\chi''(\omega)$ from Eq. 13 and $f(\omega)$ using Eq. 11. Finally, we can obtain real-time correlation function $F(t)$ (t: real) from Eq. 10.

Knowing the values $\tilde{f}(Z_n)$, one can numerically perform the analytical continuation using the Padé approximation[3]. To obtain $\tilde{f}(Z_n)$ by this method, one needs accurate estimates of the imaginary-time correlation function $F(-i\lambda)$ for $0 \le \lambda \le \beta\hbar$. Schüttler and Scalapino[23] used least-squares-fit method to obtain real-frequency correlation functions from imaginary-time Monte Carlo data. White et al.[24] proposed a method based on the least-squares-fitting procedure, imposing smoothness on real-frequency quantities. Silver et al.[10] suggested applying the maximum entropy method to extract real-time properties from imaginary-time data for several simulations. Lang et al.[25] applied the same method to nuclear shell model.

Despite these developments, analytical continuation methods are numerically unstable, and several modifications for the straightforward Monte Carlo simulation have been proposed. Thirumalai and Berne[3] suggested calculating the complex-time correlation, $\langle A(0)A(t + i\beta\hbar/2)\rangle$ instead of the real-time correlation function $\langle A(0)A(t)\rangle$. Behrman and Wolynes[5] used this to perform Monte Carlo simulations and compared the results with several approximations. Another modification is the stationary phase method, which is explained in the next subsection.

3.3 *Stationary Phase Methods and Related Techniques*

Doll et al.[4] proposed using the stationary phase method for continuous models. Suppose one wants to calculate the following integral by a Monte Carlo method:

$$I(t) = \int dx \exp(itf(x)), \tag{14}$$

where $\int dx$ is in general a summation over all the possible paths. For the sake of simplicity, we will consider $\int dx$ as a one-dimensional integral over variable x in this subsection.

Equation 14 is equivalent to

$$I(t) = \int dx D(x, t) \exp(itf(x)), \tag{15}$$

where

$$D(x, t) = \int dy P(y) \exp\{it(f(x + y) - f(x))\}. \tag{16}$$

Here, $P(y)$ is an arbitrary function normalized as $\int dy P(y) = 1$. For example, Doll et al.[4] chose $P(y) = C \exp(-y^2/2\epsilon^2)$, where C is the normalizing constant. Now, if ϵ is small, $f(x+y)$ in Eq. 16 can be approximated by a second-order expansion. Performing the Gaussian integration, Doll et al. obtained[4]

$$D_\epsilon(x,t) = (1 - it\epsilon^2 f'')^{-1/2} \exp[-(\epsilon t f')^2/2(1 - it\epsilon^2 f'')]. \tag{17}$$

As we can see from Eq. 17, $D_\epsilon(x,t)$ is large when f' is small, i.e., when x is near a stationary point x_{st}, where $f'(x_{st}) = 0$. When f changes rapidly, $D_\epsilon(x,t)$ is small, and this part does not contribute to the calculation of $I(t)$ in Eq. 15. Thus, for finite ϵ, one can obtain a relatively accurate value of the integral,

$$I_\epsilon(t) = \int dx D_\epsilon(x,t) \exp(it f(x)). \tag{18}$$

By calculating $I_\epsilon(t)$ for several values of $\epsilon > 0$, one can extrapolate the results to $\epsilon = 0$ and obtain $I(t)$.

There are also other choices of $D_\epsilon(x,t)$ obtained, for example, by Filinov[26] and by Makri and Miller[27]. Beck et al.[37] suggested using simulated annealing procedures for locating stationary points. The Monte Carlo evaluation of I_ϵ is not difficult if there is only one stationary point. If there are many stationary points, the evaluation of I_ϵ becomes difficult.

Mak and Chandler[8] used the Hubbard-Stratonovich transformation to convert the discrete spin-boson model into a continuous spin-model. They distorted the integration contours so that the contours pass through the stationary points. Then, by applying the stationary phase method, they obtained better statistics for the phases than those from straightforward Monte Carlo simulations.

Distortion of integration contour has also been used by Chang and Miller[38]. For a Hamiltonian consisting of kinetic energy part and potential energy part, they distorted the integration contour in such a way that the kinetic part becomes real. The seriousness of the sign problem in their method depends on the potential part, and for certain potentials they obtained good results.

Doll, Coalson and Freeman[39] also used distortion of integration contour for Monte Carlo simulation, together with partial averaging[40], in which they approximate the average of an exponential by the exponential of the average. They used cumulant expansions to obtain higher-order approximations.

3.4 Other Quantum Monte Carlo Methods for Dynamical Quantities

In this subsection we mention a few other Monte Carlo methods for calculating dynamical quantities for quantum systems. Cline and Wolynes[6] used a quasi-classical Lanvevin equation to generate a set of paths, and then performed Monte Carlo simulation to obtain quantum fluctuations around these paths. Makri[7] used truncated basis set of plane wave functions to obtain propagators which are less oscillatory than those obtained by a straightforward method.

Mak[9] proposed a filtering method, adding some fluctuations along the mean path during his Monte Carlo simulation. In Eqs. 15 and 16, one can regard x as a mean path $\bar{x}$, and y as the fluctuation around $\bar{x}$. Then one obtains

$$I(t) = \int d\bar{x} D(\bar{x}, t) \exp(it f(\bar{x})), \tag{19}$$

where

$$D(\bar{x}, t) = \int dy P(y) \exp(it\{f(\bar{x} + y) - f(\bar{x})\}). \tag{20}$$

Mak's method[9] is to do the path summation of $\bar{x}$ in Eq. 19 by Monte Carlo simulation and numerically add the fluctuations y in Eq. 20. If one adds all the fluctuations y around $\bar{x}$, the expression becomes exact, but numerically unstable due to phase cancellations. Mak[9] proposed the approximate function:

$$D_\epsilon^{(n)} = A^{-1} \exp(\sum_{j=1}^{n} \epsilon^j \xi^{(j)}). \tag{21}$$

Here ϵ is a small positive number, n is typically 2, $\xi^{(1)} = C^{(1)}$ and $\xi^{(2)} = C^{(2)} - C^{(1)2}/2$, where $C^{(j)}$ is the contribution from the j-th order fluctuations. For example, with a configuration $\bar{x}$, one can calculate $C^{(1)}$ from all configurations obtained by flipping only one spin of $\bar{x}$.

If $\epsilon = 0$, the expression is exact but phase cancellation remains. If ϵ is large, there is less phase cancellation, but the results are not exact even with an infinite number of Monte Carlo steps. Thus one should carefully adjust the parameter ϵ to obtain a good result.

3.5   Exact Enumerations

Exact enumeration is a straightforward way of obtaining numerically exact results for a finite number (N) of classical spins by generating 2^N states. ($N = p + 2q$ for the correlation function of Eq. 1 and $N = 2q - 1$ for the relaxation function of Eq. 2.) Due to the development of computers and in particular vector machines, exact enumeration has become a very useful way of calculating dynamical properties of quantum system in some parameter range. As we have discussed in Sec. 1, whether quantum Monte Carlo simulation works for spin-boson model depends on both the amount of quantum coherence (see the next section for details) and the total time t in $\exp(iHt/\hbar)$. Exact enumeration works for short time with any amount of quantum coherence.

The main part of exact enumeration consists of three procedures. Suppose we want to obtain an expectation value of a physical quantity A, $\langle A \rangle = A_1/Z_1$, where $Z_1 = \sum_k w^{(k)}$, $A_1 = \sum_k w^{(k)} a^{(k)}$, and the superscript k denotes a configuration. For each configuration of classical spins, we have the following procedures:

1. Calculate the action, $\varphi^{(k)} = \sum_{ij} J_{ij} s_i^{(k)} s_j^{(k)}$.

2. Calculate the weight, $w^{(k)} = \exp(\varphi^{(k)})$, and add this to Z_1.

3. Calculate $a^{(k)}$ and add this to A_1.

In the actual implementation, we divided the spins into several groups. The grouping of spins is useful for the efficient calculations of the action, $\varphi = \sum_{ij} J_{ij} s_i s_j$, where we omitted the superscript (k) for simplicity. Suppose we have N classical spins and we divide them into two groups, $S_a = \{s_i : i \in T_a\}, S_b = \{s_i : i \in T_b\}$, where $T_a = \{1, 2, ..., n_a\}, T_b = \{n_a + 1, ..., N\}$. The action φ in Eq. 8 can be divided into, $\varphi = \varphi_a + \varphi_b + \varphi_{ab}$, where φ_a denotes the action depending on only S_a, and φ_{ab} denotes the action depending on both S_a and S_b. The calculation of φ_{ab} can be expressed in two stages, $f_i \leftarrow \sum_{j \in T_b} J_{ij} s_j$, $\varphi_{ab} \leftarrow \sum_{i \in T_a} s_i f_i$.

Procedure P: $\varphi_a \leftarrow \sum_{i,j \in T_a} J_{ij} s_i s_j$ for all configurations of S_a.

$\varphi_b \leftarrow \sum_{i,j \in T_b} J_{ij} s_i s_j$ for all configurations of S_b.

$Z_1 \leftarrow 0, \quad A_1 \leftarrow 0.$

Procedure B: $f_i \leftarrow \sum_{j \in T_b} J_{ij} s_j$ for all $i \in T_a$.

Procedure A1: $\varphi_{ab} \leftarrow \sum_{i \in T_a} s_i f_i,$

$\varphi \leftarrow \varphi_a + \varphi_b + \varphi_{ab}.$

Loop A

Loop B

Procedure A2: $w \leftarrow \exp(\varphi),$

$Z_1 \leftarrow Z_1 + w.$

Procedure A3: Calculate $a,$

$A_1 \leftarrow A_1 + wa.$

Fig. 3 An algorithm of exact enumeration. Loop A corresponds to all the configurations of S_a, and loop B corresponds to all the configurations of S_b.

Procedure Q: $\langle A \rangle \leftarrow A_1/Z_1.$

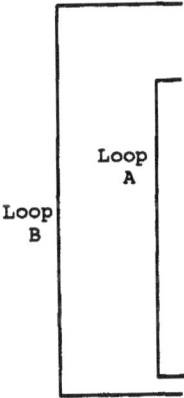

The algorithm for the case of two groups of spins is shown in Fig. 3. The Procedures A1, A2, and A3 correspond to three basic procedures 1, 2, and 3 mentioned above. The configurations of spin group S_a are used for the inner loop A, and S_b for the outer loop B. The procedures A1, A2 and A3 in the inner loop A are referred 2^N times, and Procedure B in the outer loop B is referred 2^{n_b} times, where $n_b = N - n_a$. Since f_i's are calculated outside the loop A, the number of operations necessary in Procedure A1 is of the order of $2^N \times n_a$ compared to $2^N \times N^2$ for calculating the whole action $\varphi = \sum_{ij} J_{ij} s_i s_j$ for each configuration. The CPU time for Procedure B and P are negligible, because the numbers of operations necessary are of the order of $2^{n_b}n_a n_b$ and $2^{n_a}n_a^2 + 2^{n_b}n_b^2$, respectively.

The maximum number of spins in each division is determined by the memory size of computers. For vector machines, one should not make n_a too small, because the vectorization will be inefficient if the vector length in loop A, 2^{n_a}, becomes too small. For Cray computers we put 12 (or less) spins in one division, because $2^{12} = 4096$ states is a reasonable size for the allocation of arrays for Cray computers. We calculated up to 29 spins, and we divided the spins into three groups. The extension of the algorithm in Fig. 3 to the case of three groups is straightforward. The innermost loop A is vectorized for Cray computers.

Unlike the diagonalization method, this method is limited by CPU time and not by memory size. If the matrix J_{ij} in Eq. 8 is pre-calculated, a typical calculation of $\bar{n}_1$ for a two-level model of $N = 27$ (1.3×10^8 states) takes 1.8 minutes on a Cray YMP2E at the Institute for Chemical Research, Kyoto University. Forty-eight percent of the time is spent for the exponential operation of $w = \exp(\varphi)$ (Procedure A2), 36% for calculating φ (Procedure A1), and 15% for calculating $\bar{n}_1(t)$ (Procedure A3) and 1% for other calculations.

4   Properties of Spin-boson Model

4.1   Introduction

The spin-boson model[28, 8] can be applied to electron transfer in liquids or in solids by changing the spectral density. Although the model represents both short-distance and long-distance electron transfer in liquid, we will concentrate on the short-distance type in our studies. Short-distance electron-transfer is characterized by a large coupling constant. In this case, the period of the non-dissipative system becomes small, and the computation requires a smaller number of spins.

4.2   Results and Discussions

In this subsection, we present some of our results for the asymmetric two-level model introduced in Section 2. The properties of the bath and the interaction between the bath and the spin are determined by the spectral density, $J(\omega) = \frac{\pi}{2}\sum_j \frac{c_j^2}{m_j\omega_j}\delta(\omega - \omega_j)$. Here we use an Ohmic type spectral density, namely $J(\omega) \sim \omega$ for small ω since this type is appropriate for electron transfer in liquids[41]. In particular we choose the form, $J(\omega) = \eta\omega\exp(-\omega/\omega_c)$, where η and ω_c are parameters independent of ω.

In the absence of the bath, this spin-boson model shows quantum coherence characterized by oscillatory behavior. If the bath is strong, the quantum coherence is destroyed and the relaxation becomes non-oscillatory. If a dynamical quantity $A(t)$ decreases monotonically with time, we call the state incoherent. Otherwise, we call the state coherent. Examples are shown in Fig. 4. For asymmetric two-level models[31, 32], we obtained phase diagram determined by the behavior of $\bar{n}_1$, where n_1 is the number density of state 1. $\bar{n}_1(t)$ is the relaxation of n_1 with the initial condition of fixing the system to state 1, as has been defined in Eq. 2. In Fig. 5, phase boundaries for two cases, $E_1/K = 2$ and $E_1/K = -2$, are shown. The coherent phase is larger in the case of $E_1/K = -2$ than in the case of $E_1/K = 2$. In the parameter region between these two boundaries, the relaxation from higher energy to lower energy shows incoherent behavior, and the relaxation in the reverse direction shows coherent behavior. This is an indication of strong non-linearity of the system in this parameter region.

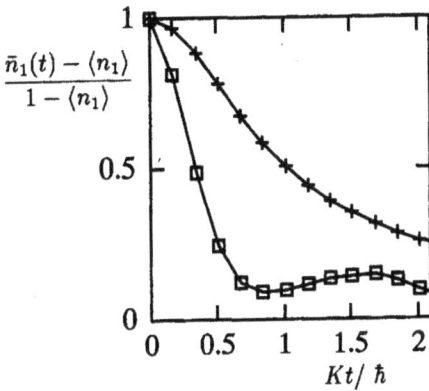

Fig. 4 Examples of the relaxation of asymmetric two-level system for an incoherent case $(+, E_1/K = 2)$ and a coherent case (□, $E_1/K = -2$). $\beta K = 1.0$, $\hbar\omega_c/K = 2.5$, $\eta/\hbar = 1.2566$.

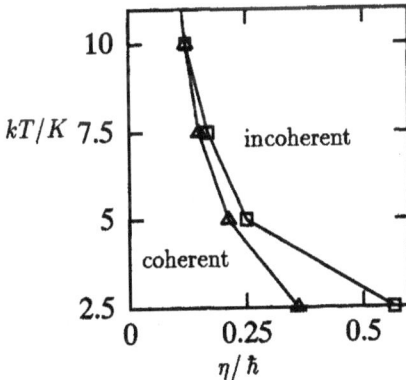

Fig. 5 The phase diagram for the asymmetric two-level system for $\hbar\omega_c/K = 2.5$. $\triangle$ is the phase boundary for $E_1/K = 2$ and □ is for $E_1/K = -2$.

If we look at the dynamics in the incoherent phase, we can observe the effect of the energy gap E_1. In Fig. 6, we plot the quantity

$$z(t) = \ln(Re\ \bar{n}_1(t) - \langle n_1 \rangle). \qquad (22)$$

For large E_1, $z(t)$ lies on a straight line after the initial relaxation, indicating an exponential relaxation. On the other hand, for small E_1, $z(t)$ does not lie on a straight line, indicating a non-exponential relaxation. When a rate constant for a reaction is measured in experiments, an exponential relaxation is usually assumed. One should be careful when the reaction shows a non-exponential behavior. The detailed analysis of these two types of relaxation and the rate constant will be given elsewhere[32].

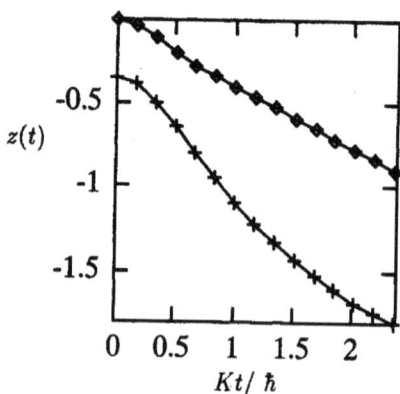

Fig. 6 Relaxation of a two-state model, for $E_1/K = 6$ (◇) and $E_1/K = 1$ (+). $\beta K = 1.0$, $\hbar\omega_c/K = 2.5$, $\eta/\hbar = 1.2566$.

5   Concluding Remarks

We have reviewed several numerical methods for calculating dynamical quantities. Quantum Monte Carlo simulation, with the aid of stationary phase methods or the filtering method, works well for short or intermediate times, with small or intermediate quantum coherence. Exact enumeration works for short times with any amount of quantum coherence. We presented some exact enumeration results for the case of strong coupling with close energy levels. Further development of time-dependent quantum Monte Carlo simulation should enable us to explore the long-time behavior of quantum systems.

Acknowledgements

I would like to thank M. Suzuki, D. Chandler, C. Mak, K. Leung, J. Gehlen, N. Kawashima, and Y. Iba for helpful discussions. I would also like to thank R. Yee, T. Muranaka and K. Uehara for their comments on the manuscript. Most of the numerical calculations were performed on the Cray X-MP at the University of California, Berkeley, the Cray Y-MP8 at the San Diego Supercomputer Center and on the Cray Y-MP2E at the Supercomputer Laboratory, Institute for Chemical Research, Kyoto University. This research is partially supported by the National Science Foundation and Grant-in-Aid from the Japanese Ministry of Education, Science and Culture.

References

1. See, for example, *Quantum Monte Carlo Methods*, ed. by M. Suzuki, Springer Series in Solid-State Physics **74** (1987) and references therein.
2. H. F. Trotter, Proc. Am. Math. Soc. **10**, 545 (1959);
   M. Suzuki, Phys. Lett. **113A** (1985) 299.
3. D. Thirumalai and B. J. Berne, J. Chem. Phys. **81** (1984) 2512;
   D. Thirumalai, B. J. Berne, Comp. Phys. Comm. **63** (1991) 415.
4. J. D. Doll, D. L. Freeman and M. J. Gillan, Chem. Phys. Lett. **143** (1988) 277.
   J. D. Doll, T. L. Beck, D. L. Freeman, J. Chem. Phys. **89** (1988) 5753.
5. E. C. Behrman, P. G. Wolynes, J. Chem. Phys. **83** (1985) 5863.
6. R. E. Cline, P. G. Wolynes, J. Chem. Phys. **88** (1988) 4334.
7. N. Makri, Comp. Phys. Comm. **63** (1991) 389;
   N. Makri, J. Phys. Chem. **97** (1993) 2417.
8. C. H. Mak, D. Chandler, Phys. Rev. A **41** (1990) 5709;
   C. H. Mak, D. Chandler, Phys. Rev. A **44** (1991) 2352.
9. C. H. Mak, Phys. Rev. Lett. **68** (1992) 899.
10. R. N. Silver, D. S. Sivia, J. E. Gubernatis, Phys. Rev. B**41** (1990) 2380;
    J. E. Gubernatis, M. Jarrell, R. N. Silver, D. S. Sivia, Phys. Rev. B**44** (1991) 6011.
11. M. Takasu, S. Miyashita, M. Suzuki, Prog. Theor. Phys. **75** (1986) 1254.
12. M. Takasu, S. Miyashita, M. Suzuki, Springer Series in Solid State Sciences **74** (1987) 114.
13. M. Imada, J. Phys. Soc. Jpn. **56** (1987) 311.
14. T. Nakamura, N. Hatano, H. Nishimori, J. Phys. Soc. Jpn. **61** (1992) 3494.
15. N. Hatano, M. Suzuki, Phys. Lett. A **163** (1992) 246.
16. E. Y. Loh, J. E. Gubernatis, R. T. Scalettar, S. R. White, D. J. Scalapino, R. L. Sugar, Phys. Rev. B **41** (1990) 9301.
17. G. G. Batrouni, R. T. Scalettar, Phys. Rev. **42** (1990) 2282.
    G. G. Batrouni, P. de Forcrand, Phys. Rev. B**48** (1993) 589.
18. F. F. Assaad, D. Würtz, Z. Phys B**80** (1990) 325.
19. N. Furukawa, M. Imada, J. Phy. Soc. Jpn. **60** (1991) 810.

20. M. Frick, H. De Raedt, Z. Phys. B88 (1992) 173.
21. W. H. Newman, A. Kuki, J. Chem. Phys. 96 (1992) 1409.
22. J. H. Samson, Phys. Rev. B47 (1993) 3408.
23. H. -B. Schüttler, D. J. Scalapino, Phys. Rev. Lett. 55 (1985) 1204; Phys. Rev. B34 (1986) 4744.
24. S. R. White, D. J. Scalapino, R. L. Sugar, N. E. Bickers, Phys. Rev. Lett, 63 (1989) 1523.
25. G. H. Lang, C. W. Johnson, S. E. Koonin, W. E. Ormand, preprint (1993). C. W. Johnson, S. E. Koonin, G. H. Lang, W. E. Ormand, Phys. Rev. Lett. 69 (1992) 3157.
26. V. S. Filinov, Nucl. Phys. B271 (1986) 717.
27. N. Makri, W. H. Miller, J. Chem. Phys. 89 (1988) 2170.
28. A. J. Leggett, S. Chakravarty, A.T. Dorsey, M. P. A. Fisher, A. Garg, W. Zwerger, Rev. Mod. Phys. 59 (1987) 1.
29. R. Egger, U. Weiss, Z. Phys. B89 (1992) 97.
30. M. Takasu, D. Chandler, in the Proceedings of the Workshop of Harmonic Oscillators, NASA Conference Publications 3197 (1992) 365.
31. M. Takasu, D. Chandler, in *Computer Aided Innovation of New Materials II*, ed. by M. Doyama et al., Elsevier Science Publishers (1993) 375.
32. M. Takasu, D. Chandler, in preparation.
33. R. Egger, C. H. Mak, preprint.
34. See, for example, R. P. Feynman, *Statistical Mechanics* (Addison-Wesley, 1972), Chapter 2;   D. Chandler, in *Liquids, Freezing and Glass Transition*, Les Houches, Section LI, ed. by J. P. Hansen, D. Levesque and J. Zinn-Justin, (North Holland, 1991).
35. See, for example, *Monte Carlo Methods in Statistical Physics*, ed. by K. Binder, (Springer-Verlag, Berlin, 1979).
36. R. D. Coalson, J. Chem. Phys. 94 (1991) 1108.
37. T. L. Beck, J. D. Doll, D. L. Freeman, J. Chem. Phys. 90 (1989) 3181.
38. J. Chang, W. H. Miller, J. Chem. Phys. 87 (1987) 1648.
39. J. D. Doll, R. D. Coalson, D. L. Freeman, J. Chem. Phys. 87 (1987) 1641.
40. J. D. Doll, R. D. Coalson, D. L. Freeman, Phys. Rev. Lett. 55 (1985) 1. R. D. Coalson, D. L. Freeman, J. D. Doll, J. Chem. Phys. 85 (1986) 4567.
41. J. S. Bader, D. Chandler, Chem. Phys. Lett. 157 (1989) 501. J. S. Bader, R. A. Kuharski, D. Chander, J. Chem. Phys. 93 (1990) 230.

www.ingramcontent.com/pod-product-compliance
Lightning Source LLC
Chambersburg PA
CBHW061619220326
41598CB00026BA/3807